T0297686

Embodiment, Enaction, and Culture

Embodiment, Enaction, and Culture

Investigating the Constitution of the Shared World

edited by Christoph Durt, Thomas Fuchs, and Christian Tewes

The MIT Press
Cambridge, Massachusetts
London, England

This book was set in Stone Sans and Stone Serif by Toppan Best-set Premedia Limited.

Library of Congress Cataloging-in-Publication Data is available.

ISBN: 978-0-262-03555-2 (hardcover)
ISBN: 978-0-262-54925-7 (paperback)

In memory of our friend and TESIS project fellow John J. McGraw (1974–2016), whose generosity of intellect and spirit will be greatly missed

Contents

Introduction: The Interplay of Embodiment, Enaction, and Culture

Christian Tewes, Christoph Durt, and Thomas Fuchs

While traditional theories of cognition tend to conceive of mental capacities as disembodied or merely supervenient on brain states, in recent decades the insight has spread that mental processes cannot be confined to activities inside the skull alone. The paradigm of enactive embodiment endeavors to overcome the limitations of traditional cognitive science by reconceiving the cognizer as an embodied being and cognition as enactive. According to a well-known early definition, cognition depends on "the kinds of experience that come from having a body with various sensorimotor capacities" (Varela, Thompson, and Rosch 1991, 173).

It is important not to overlook the other part of the definition of "embodied" in the enactive sense, namely, that "these individual sensorimotor capacities are themselves embedded in a more encompassing biological, psychological, and cultural context" (ibid.). Since Varela, Thompson, and Rosch's *The Embodied Mind*, a great number of books on biological and psychological aspects of embodiment have been published. The cultural context of enactive embodiment, by contrast, has not yet been explored in an interdisciplinary volume dedicated to this purpose. The present book does exactly this and thereby offers a starting point for more extensive studies of the cultural context of embodiment. It is a multidisciplinary investigation into the role of culture for embodied and enactive accounts of cognition, encompassing fundamental philosophical considerations, as well as the newest developments in the field.

Here we have brought together philosophical, neurophysiological, psychological, psychiatric, sociological, anthropological, and evolutionary studies of the interplay of embodiment, enaction, and culture. The constitution of the shared world is understood in terms of participatory and broader collective sense-making processes manifested in dynamic forms of intercorporeality, collective body memory, artifacts, affordances, scaffolding, use of symbols, and so on. The contributors investigate how preconscious and conscious accomplishments work together in empathy, interaffectivity, identifications of oneself with others through emotions such as shame, we-intentionality, and hermeneutical understanding of the thoughts of others. The shared world is seen as something constituted by intersubjective understanding that discloses things in the shared significance they have for the members of a culture. Special emphasis is put on phenomenological approaches to cognition and culture and their relation to other approaches.

Our introduction explicates the key concepts, relates them to relevant empirical research, raises guiding questions, and explains the structure of the book. Starting with a phenomenological approach to the intertwinement of mind, body, and the cultural world, we continue with an exploration of the concepts of intercorporeality and interaffectivity. The ideas underlying these concepts are put in dialogue with central tenets of enactivism. We then consider further cultural conditions, such as those of cognitive scaffolding, and explain how these cultural conditions in turn depend on the embodied interaction of human beings. Finally, we outline the book's structure and introduce the individual chapters.

1 The Intertwinement of Body, Mind, and Cultural World

The concept of enaction is generally meant to capture the active sensorimotor engagement of the organism with its environment by which the organism makes sense of the environment and potentially changes it. For humans, sense-making is largely a collective activity through which their environment becomes a world of shared significance. Humans collectively *constitute* the world not by creating it in a constructivist sense but by disclosing its intersubjective significance. Cultural forms of constitution include communication as well as collaborative interactions with others; these also shape and change the environment according to the needs of the group or community. Over time, the shared ways of sense-making and interaction are established as rituals, codes, or institutions and as such may be transmitted to subsequent generations. This cultural context impregnates and structures all conscious experience, as Merleau-Ponty explains with his notion of the "intentional arc":

The life of consciousness—cognitive life, the life of desire or perceptual life—is subtended by an "intentional arc" which projects round about us our past, our future, our human setting, our physical, ideological and moral situation, or rather which results in our being situated in all these respects. It is this intentional arc which brings about the unity of the senses, of intelligence, of sensibility and motility. (Merleau-Ponty [1945] 2005, 157)

Anticipating current enactive accounts, Merleau-Ponty regards conscious life as marked by an inherent connection between desire, cognition, perception, and motor agency, which he refers to with the Husserlian expression "I can" (Merleau-Ponty [1945] 2005, 159). These activities of consciousness are united through the intentional arc, which situates the conscious being through the medium of the lived body in the cultural world it inhabits. The arc encompasses one's individual situation, shared habits, and cultural forms of interaction, as well as more reflective cultural accomplishments such as ideology and morals. How deeply these impregnate consciousness can be seen in psychopathological disturbances, such as the disability of the patient Schneider to integrate the different modes of consciousness, which Merleau-Ponty traces back to disturbances of the intentional arc (156–157).

The intentional arc conditions, for instance, the everyday perception of a chair or pencil in connection with possible motor actions, which today are called *affordances* (Gibson 1979; Thompson 2005). The significance of the chair and pencil is disclosed not in a further intentional act separated from the perceptual act but in the perception itself. Husserl (1966) distinguishes here between "passive" and "active" synthesis, which together constitute the intentional objects in the significance in which they are experienced. Conscious perception focuses on certain aspects but is always accompanied by a "horizon" of possibilities that is only "dimly conscious" (Husserl [1913] 1976, §27; see also Moran, this vol.). In the same vein, for Merleau-Ponty, the intentional arc is not just a spotlight on given objects but "makes them exist in a more intimate sense, for us" (Merleau-Ponty [1945] 2005, 157).

Proponents of the "enactive sensorimotor approach" explain the emergence of perceptual patterns as the result of the (en)active exploration and know-how of the agent and specific contingencies of sensory modalities that give rise to sense-specific patterns in perception (cf. O'Regan and Noë 2001). The perceptual pattern or gestalt, however, cannot be separated from its cultural meaning. The intentional arc thus encompasses cultural sense-making processes that are already in play at the prereflective level of motor intentionality, such as learning how to hold a spoon or to climb stairs. This indicates that the "sensorimotor unity of perception and action" needs to be understood in the context of culture.

Cultural sense-making processes build on social interactions that also provide the pragmatic context through which language gains its meaning and significance. The results of sense-making in turn add further layers of significance to the shared world that are expressed in cultural patterns of perceiving, thinking, beliefs, or ideology (cf. Durt, this vol.). Such shared forms of perception and thought are then reflected on in cultural ways, for example, in art and literature. Culture thus permeates sense-making processes from prereflective motor-perceptive levels to the highest forms of significance. The products of culture, such as artifacts, technology, and institutions, in turn become an integral part of sense-making processes.

Culture is constantly changing, and though one can speak of cultural evolution, it is not a unilinear process; it involves contradictory tendencies and competing subcultures. Culture is inherited nongenetically, not merely as "memes," which Richard Dawkins models on genetic inheritance ([1976] 2006, 189–201), but by means of ever new forms of appropriation, which make new sense of the cultural inheritance. Thus the study of the cultural context of cognition needs to go beyond simplistic concepts of culture and instead proceed in differentiated and interdisciplinary ways. It should not surprise us when this approach reveals unexpected connections between culture and embodied cognition.

While culture cannot be reduced to the material products of culture, such as tools or artifacts, these material products are important for sense-making processes themselves. The concept of "material culture" brings to the fore the "meaning-instantiating function" that embeds material things in the culturally shared world (Malafouris 2013, 97). Material things are part of the entire cognitive life because sense-making processes also include relations between human beings and artifacts that are in turn embedded in a wider network of material engagement (Malafouris and Renfrew 2010, 4). The meaning-instantiating character of tools is furthermore asserted by the "manipulation thesis," which claims that tools such as pencils, computers, and sentences allow the manipulation of cognitive accomplishments when rooted in the appropriate action-perception cycles. This allows for cognitive operations that could not be achieved without the help of such "cognitive prostheses" (Menary 2007, 83–85; 2010b, 240).

As an example of how deeply the manipulation processes are entrenched in sociocultural practices, consider the creation and institutionalization of an external storage system such as a library. The use and further development of such a collective memory system requires experts to develop norms on how to store and retrieve information effectively. At least some of these norms need to be passed on to other users and institutions, for instance, via institutionalized storage systems, which in turn have a huge impact on individual memory capacities. Especially those who can read and write are part of a system that considerably enhances the cognitive spatiotemporal boundaries of oral communication practices (Donald 1991, 311).

So far, our outline of the intertwinement of body, mind, and the cultural world has focused on the constitutional function of cultural practices. But how are these realized at the bodily level in the first place; in what way are they embodied? This question leads us to the foundational role of the intercorporeal and interaffective dimensions of human experience and cognitive development, which will be the topic of the next section. Section 3 then explains the contribution of dynamical systems theory and of the concept of participatory sense-making to intercorporeality, and section 4 considers how embodied interactions are constantly shaped by the shared world in which they are embedded.

2 Intercorporeality and Interaffectivity

Merleau-Ponty's reference to the "human setting" in specifying constitutional elements of the intentional arc indicates that the actual encounter with other human beings, between the *I* and the *You*, plays a foundational role for the enculturation of human beings. Dan Zahavi (2014) has argued that every conscious state has the character of mineness, thereby establishing an inseparable "minimal self" at the prereflective level of awareness. This book casts light from different angles upon the pivotal question of how intersubjective relations are related to that basic sense of self-awareness. Is the minimal self the constitutional base for the *We* in joint attention and collective intentionality (cf. Brinck, Reddy, and Zahavi, this vol.)? Or do intersubjective interactions play a foundational role for selfhood at every level of experience (cf. Ratcliffe, this vol.; Ciaunica and Fotopoulou, this vol.)? Or does this claim overlook a "thinner" and "thinnest" concept of the minimal self (cf. Zahavi, this vol.)?

A good starting point for explicating intersubjective relationships is the phenomenological notion of the "extended body," which is formed by means of mutual interactions between expressive and feeling bodies (Froese and Fuchs 2012, 211). The idea is not that other feeling bodies play the role of an extended cognitive device, such as Otto's notebook in the famous thought experiment that Clark and Chalmers (1998) use to illustrate the extended mind hypothesis. The point is rather that a new dynamical whole emerges through bodily interactions and interbodily resonance. Contrary to older strands of emotional theory in the cognitive sciences, affective states and processes are conceived not simply as "cognitive, or mental phenomena" (Colombetti and Thompson 2008, 51) but rather as affective qualities that manifest themselves in the atmosphere of interpersonal encounters (Fuchs 2013, 222–223). This creates an ongoing *interaffective space*, in which the body functions as a medium for emotional and affective states and processes. Emotions such as despair, shame, guilt, hate, and love are intentionally directed toward other persons but are also simultaneously expressed by internal bodily reactions (such as increasing heartbeat rates or specific neural activation patterns) and body postures, gestures, and facial expressions that are directly perceivable in social interactions and lead to interbodily resonance (Froese and Fuchs 2012, 212–213).

Besides emotions that have an intentional direction, there are also existential feelings that prestructure the entire experience of being-in-the-world. An example of an existential feeling is given by Matthew Ratcliffe, who explains:

The world as a whole can sometimes appear unfamiliar, unreal, distant or close. It can be something that one feels apart from or at one with. One can feel in control of one's overall situation or overwhelmed by it. One can feel like a participant in the world or like a detached, estranged observer staring at objects that do not feel quite "there." (Ratcliffe 2008, 37)

Existential feelings are of fundamental importance for the entire constitution of the social realm. They frame the background of our current affective states and also include feelings that concern social settings such as feeling generally respected or unwelcomed, confident or distrustful, and so on (Stephan 2012, 158). They modulate how we perceive and experience social encounters in the first place. The notion of existential feelings further elucidates the inseparability of cognitive processes and emotions and their intra- and interbodily interrelations. Contrary to the classical appraisal theory, cognitive evaluations do not precede or simply trigger the occurrence of emotional episodes (Lewis 2000, 41–42). Rather, cognitive and evaluative processes are always embedded in basic states of mood and episodes of bodily affection.

Some theorists think that empirical findings from studies of fast emotional reactions to not yet fully recognized perceptual patterns imply that emotional reactions and processes are disconnected from cognitive evaluation (Zajonc 1984, 121). They assume a reversed temporal order of emotional and cognitive processes. The term "affective appraisal" refers in this context to organismic reactions to situations that require a quick and automatic evaluation of whether they are a threat or an attack (Robinson [2005] 2007, 42). According to this view, the cognitive evaluation of the perceptual pattern (and early elicited emotional response toward it) occurs only at a later stage of processing. This view seems to be supported by neurophysiological work by LeDoux, who found that different pathways exist for processing, for example, dangerous stimuli. On the one hand, there is the "high road" whereby stimuli reach the amygdala by a route going from the thalamus to the cortex. The functional interpretation of this pathway is that the cortex assesses the afferent information to prevent inappropriate responses (LeDoux [1998] 1999, 163–165). On the other hand, there is the "low road" that stimuli take from the thalamus directly to the amygdala, bypassing the cortex (165).

As Thompson points out, however, the organism must anticipate future states and satisfy its present desires to stay alive and adapt to the environment, thereby *creating* its "own temporal life cycle" (Thompson 2007, 155; cf. Jonas 1966, 86). Affective appraisal is already the beginning of an active encounter with the environment. The adaptive and projective capacities of sense-making processes suggest that emotions and appraisals are strongly interdependent and interconnected. At every temporal stage of pattern formation, self-organizing emotional appraisal processes emerge. These can either acquire a quite stable formation during lifetime, such as in the case of existential feelings, or change very rapidly, such as in the case of atmospheric feelings (Colombetti and Thompson 2008, 58; Stephan 2012, 158).

The emotional space of atmospheric and existential background feelings furthermore prestructures, shapes, and expresses our interactions with others. Empirical research in developmental psychology documents the importance of interaffective attunement or shared emotions between infants and caregivers. From the second month of life, infants and adults share emotions and attune to each other by means of different behavior patterns in "proto-conversations" (Stern 1985, 217; Trevarthen 1989). These are dyadic forms of social interaction, in which caregivers and infants touch, smile, move, or gaze in an affectionate, rhythmic, and turn-taking manner (Trevarthen and Aitken 2001; Tomasello et al. 2005, 681). An example is the rhythmically coupled hand movement of an infant during the speech of an adult (Trevarthen and Aitken 2001, 4).

Such emotions, moods, affections, and feelings do not only concern early forms of development. Intentional understanding, sharing goals, and developing plans together are still embedded in an affective and emotional space. The interaffective exchanges and attunements are essential for the socio-emotional learning processes. Sharing emotions and related evaluations guide children in using cultural artifacts and in acquiring regulative norms, values, and intentions in social life (Trevarthen and Aitken 2001, 16–17).

The relationship between emotional appraisal and evaluation of artifacts has been studied intensively with regard to *social referencing*. Infants use the adult's emotional appraisal of an event or object as a clue to regulate their own behavior toward such objects. For instance, infants are less inclined to play with a toy when mothers show disgust toward it (Hornik, Risenhoover, and Gunnar 1987, 943). Furthermore, emotions become themselves more and more refined due to their embeddedness in the process of enculturation (Carroll 1996, 268): they are associated with specific "paradigm scenarios" (association of objects, emotion types, and normal response reactions) and are further elaborated by storytelling, literature, art, film, and so on (De Sousa 1987, 182).

3 Dynamical Systems Theory and Participatory Sense-Making

The foregoing considerations on the interaffective dimensions of embodiment and the development of socioemotional appraisals challenge the assumption that those processes are appropriately described as linear causal sequences. They are better conceived as reciprocal nonlinear causal loops that can be described by concepts drawn from *dynamical systems theory*, which is consequently a major ingredient of enactivism. The central assumption here is that embodied cognition is a temporal aspect of living and evolving agents (van Gelder 1998; Thompson 2007, 38). As already mentioned, dynamical systems theory no longer conceptualizes cognitive processes as computational linear input–output functions but now understands them as self-organizing wholes, emerging from many interacting nonlinear feedback processes (Kelso 1995). What distinguishes these causal loops, besides their reciprocal structure, is their multilevel relation characterized by top-down (global to local) and bottom-up (local to global) processes across different levels of explanation (Thompson and Varela 2001, 419–420). This allows the explanatory integration of different layers of descriptions such as psychic and neural processes by means of concepts and research tools from dynamics (Lewis 2005, 169).

In the literature on embodied and extended cognition, dynamical systems theory has been applied to intercorporeal relations under the concept of *coupling*. Brains, bodies, and the environment are understood as components of an emergent coupled system, the brain-organism-environment system (Menary 2007, 42; Clark [2008] 2011, 24). An important example of such an emergent whole is the aforementioned extended body, which is created and sustained by means of interbodily resonance dynamics. How to understand the coupling between an organism and its environment is a controversial question, and one may ask if it is a symmetrical or an asymmetrical relation (Menary 2010a, 3–4). Proponents of the extended mind hypothesis have favored the symmetrical interpretation. They point out that a continuous, simultaneous, and reciprocal causal influence exists between two systems (Clark [2008] 2011, 24; Menary 2010b, 233). Enactivists have challenged this view because an organism is a system that not only maintains its own stability within the framework of specific background conditions but also is capable of adjusting its activities according to its needs and the demands of its environment (Campbell 2009, 466).

At this level of complexity, the concepts of *agency* and *autonomous system* come into play. The idea is that, beyond the structural coupling between different systems, the coupling is conceived as a first-order loop that is modulated by means of a second-order loop. Di Paolo specifies the latter as a relational, normative, and asymmetric sense-making process (Di Paolo 2009, 15). A dynamical system that is able to maintain its identity is conceived as an autonomous entity, "a thermodynamically open system with operational closure that actively generates and sustains its identity under precarious conditions" (Thompson and Stapleton 2009, 24).

Enactivism uses the concept of *coordination* to extend the notion of coupling to the social realm. The fruitfulness of applying this concept to intercorporeal interactions is already indicated by an example from Kelso. An adult and a child are walking side by side at a beach without being physically coupled; they are not holding hands and don't continuously touch each other, and they may not be coupled biologically (Kelso 1995, 98). A further level of coordination occurs when they are talking to each other or simply adjust their steps to walk alongside each other, a somewhat fleeting synchronization that nevertheless results in a mutual adjustment of intentional movements with regard to interpersonal dynamics (Dumas, Kelso, and Nadel 2014, 1–2). When connected in coupled actions, the interactors are still engaging in individual sense-making processes. But at the same time, a new autonomous system with new coherent social patterns of significance emerges, for example, in reciprocal speech, dance, or simply joint walking, a process that has been conceptualized as *participatory sense-making* (De Jaegher and Di Paolo 2007; Di Paolo, Rohde, and De Jaegher 2010, 71). This is so because both agents actively modulate but do not entirely control the new emerging system and its global and local features.

Forms of coupled interaction are shaped by culture: language and ways of speaking, styles of dancing, the comfort distance from others, norms of interaction, and even typical gaits differ from culture to culture. But is this the only way in which coupled interactions are dependent on culture? The notion of participatory sense-making singles out the activity of making sense, yet the results of sense-making may in turn have a top-down influence on the sense-making process itself. For instance, dance is a behavior that has significance for the dancers, and the significance it has codetermines the dance they choose and the way they dance it. An example involving explicit reflection is sense-making through reciprocal speaking, in which the significance produced in the conversation guides the further progress of the conversation. Whenever such a top-down influence exists, sense-making behavior and the significance that emerges through it are interdependent. The significance of meaningful behaviors such as dancing and speaking is cultural not only in that it is produced by cultural ways of interaction but also in that it is influenced by patterns of significance that are part of a culture, such as those expressed in worldviews, ideologies, morals, and norms. At least for humans, sense-making needs to be understood in the context of culturally shared significance (cf. Durt, this vol.).

4 Cultural Inheritance and Embodiment

Culture not only guides our interactions and our access to and interpretation of the world we live in but also changes its material composition. The coupling of an organism and its habitat is also one of transformation and production of its environmental setting. Generally speaking, organisms both adapt to their environments and adapt their environment to themselves. They construct their own ecological *niche*, which in turn affects the organism's behavior and development (Sterelny 2010, 470). The web building of spiders, for instance, induces new forms of protective behavior such as camouflage by means of web decorations that conceal the spiders from predators. Bees, ants, wasps, and other insects build nests "that often themselves become the source of selection for many nest heat regulatory, maintenance, and defense behaviour patterns" (Laland, Odling-Smee, and Feldman 2000, 133). In other words, the construction of such complexes frequently results in "downstream consequences": organisms that engage in such activities often reshape the ontogenetic environment and the selection conditions for later generations (Sterelny 2010, 470).

The last aspect has special significance for *cultural* niche construction. Human beings engage in collaborative activities that are frequently mediated, enabled, and structured by artifacts, social institutions, language, and externalized memory systems such as inscribed stone tablets, libraries, or finally electronic devices. This allows the facilitation of skills and knowledge by means of intra- and intergenerational transmission; culture provides humans with "a second nongenetic … inheritance system" (Laland, Odling-Smee, and Feldman 2000, 132). Once new forms of skills and knowledge have developed, they can spread rapidly and are established in a culture. The "cultural ratchet" ensures that jumps in cultural evolution are established by social transmission techniques that enable cumulative cultural evolution (Tomasello 1999, 5–6; Tomasello et al. 2005, 675). The process of internalization and incorporation of cultural skills and knowledge is facilitated by *scaffolding*, which integrates new information with extant knowledge (Williams, Huang, and Bargh 2009, 1257). The transmission of learning strategies enables learners to engage in complex tasks and explorations that would otherwise be beyond their cognitive scope and acquired abilities (Hmelo-Silver, Duncan, and Chinn 2007, 100).

Downstream niche construction is also constituted at the group level; the learning environment of children encompasses norms, traditions, and techniques that are not controlled or transmitted by the individual parents or teachers (Sterelny 2006, 154). This implies that participatory sense-making is nontrivially framed by the entire world shared by a culture. Conversely, social interactions and culture never lose their foundation in ongoing processes of intercorporeality and interaffectivity, from imitative learning in the early ontogenetic process of enculturation up to the complex forms of social behavior and joint enactments of symbolic cultures (Tomasello 1999, 81; Donald 1991, 199–200). Individual forms of habitus, shared intercorporeality, and cultural development are part of an interplay of bottom-up and top-down processes.

The production of tools, larger social institutions, and external memory systems is anchored in and stabilized by the "habitual body" (Casey 1984, 284), which incorporates different levels of constitution through its plasticity, in particular that of the human brain (Clark [2008] 2011, 68). The creation of larger collective memory systems, mediated by new external storage systems, is simultaneously based on the capacities of *habitual body memories*. The key concept of *habitus* can be traced back to Aristotle's use of *hexis*, which is frequently characterized as the socially acquired disposition of thinking, feeling, and acting within a group, including lifestyle, language, taste, and the specific posture of the body. Bourdieu famously describes the bodily basis of the habitus with his concept of "body hexis":

Body hexis speaks directly to the motor function, in the form of a pattern of postures that is both individual and systematic, because linked to a whole system of techniques involving the body and tools, and charged with a host of social meanings and values. (Bourdieu 1977, 87)

Through the acquired skills and enduring dispositions of intercorporeal interactions, cultural evolution is ingrained in the "second nature" of the human body, whose origin is sedimented and thus "forgotten as history" (Bourdieu 1990, 56). A new elaboration of habitual body memories can be found in Thomas Fuchs's concept of "collective body memories" (this vol.). The paradigm of enacted embodiment conceives of the human body not as a mere object but as the "existential ground for culture" (Csordas 1990, 5). The physical body is part of an embodied, living, and experiencing being that, together with other members of a culture, constitutes the shared world it lives in, can become aware of its significance, change it in meaningful ways, and is at the same time shaped by cultural significance.

The foregoing sections have depicted several lines of investigation that are becoming increasingly important for research on embodied and enactive cognition in the culturally constituted world. Each of the book's chapters elaborates further on the depicted lines of investigation. The results and insights the contributors present provide a basis for a better understanding of the interplay of embodiment, enaction, and culture.

5 Contents of the Book

Our anthology brings some of the most renowned scholars in the interdisciplinary study of embodied intersubjectivity together with the latest findings of up-and-coming researchers. Most of them—Ezequiel Di Paolo, Christoph Durt, John Elias, Shaun Gallagher, Vittorio Gallese, Thomas Fuchs, Katrin Heimann, Peter Henningsen, Dan Hutto, Hanne De Jaegher, Alba Montes Sánchez, Vasudevi Reddy, Zuzanna Rucińska, Glenda Satne, and Heribert Sattel—have collaborated for years in the interdisciplinary European research network Towards an Embodied Science of InterSubjectivity (TESIS). This book presents the final outcome of their cooperation in the TESIS network. Other authors contribute their expertise in key areas: Mark Bickhard, Ingar Brinck, Anna Ciaunica, Joerg Fingerhut, Aikaterini Fotopoulou, Duilio Garofoli, Laurence Kirmayer, Dermot Moran, Maxwell Ramstead, and Nicolas de Warren. Though the authors share phenomenological commitments, the underlying explanatory approach is interdisciplinary, bringing together fields such as philosophy, neuroscience, anthropology, psychology, and psychopathology.

The book is divided into four parts. Part 1, "Phenomenological and Enactive Accounts of the Constitution of Culture," explores the philosophical and conceptual foundations of the constitution of culture. The first three chapters explain groundbreaking work by

Husserl, Merleau-Ponty, and Sartre, as well as philosophers from other traditions, such as Wittgenstein and Ryle. Building on these thinkers' ideas, the first two chapters explore the interconnections between embodiment, enaction, intercorporeality, meaning, significance, consciousness, and culture. The third chapter then relates this work to recent debates such as the one around participatory sense making, which in the fourth chapter is formulated anew by two of its best-known proponents. The last chapter of part 1 responds to several recent critiques of radical enactivism and deals with a key topic in embodied research today concerning the foundation of language and culture: how can we explain and understand the phylogenetic and ontogenetic origins of content with its specific properties of truth, reference, and inferential infrastructure within an enactive framework?

In the first chapter, Dermot Moran discusses the key connections in the work of Husserl and Merleau-Ponty between the phenomenological concepts of embodiment and intercorporeality and their role in the constitution of intersubjective sociality and, more generally, culture. From its origins at the outset of the twentieth century to the present day, phenomenology has led the way in exploring not only the first-person experience of lived embodiment (*Leiblichkeit*) but also the first-person and second-person plural experience of "intertwining" (*Verflechtung, l'interlacs*) and "intercorporeality" (*intercorporéité*), the latter being a concept that is found in a few scattered places in Merleau-Ponty's later writings but was first elaborated—although not by name—by Husserl and later Sartre. Moran examines how embodiment and intercorporeal intertwining are necessary steps in the constitution of culture.

In the chapter that follows, Nicolas de Warren explores the meaning and significance of Sartre's concept of "the third" within the social ontology of the *Critique of Dialectical Reason*. Through an examination of three different types of group formation (serial collectives, statutory groups, and sports teams), de Warren provides an analysis of central Sartrean insights into how individual action and collective agency are co-constituted. He also draws attention to the role ascribed to material objects, as well as ideological views and beliefs in the formation of social agency.

Christoph Durt offers a new view on the relation between consciousness and culture by investigating their intertwinement with significance. Against the widespread restriction of consciousness to phenomenal aspects and that of culture to "thick description," Durt argues that consciousness discloses aspects of significance, whereas culture encompasses shared significance, as well as the forms of behavior that enact significance. Significance is intersubjective and constantly reinstantiated in new contexts of relevance rather than belonging to single individuals (cf. Gallagher, this vol.), as well as embedded in the shared world to which we relate by cultural forms of thinking and sense-making. Bringing together insights on the role of consciousness for the constitution of the world from Husserlian phenomenology with those on cultural forms of behavior by Wittgenstein and Ryle, Durt distinguishes different levels of significance accomplished by embodied consciousness and interaction. He contends that the real issue underlying "hybrid" concepts of the mind consists not in embodied versus disembodied systems of production (cf. Di Paolo and De Jaegher, this vol.) but in different levels of significance accomplished by consciousness and culture. Consciousness is embodied on every level, and it integrates different levels of significance.

Ezequiel Di Paolo and Hanne De Jaegher summarize some of the main proposals of the enactive approach to social understanding and discuss some common misreadings of the notion of participatory sense-making. The emphasis on the role played by social interaction in the enactive perspective is, in their view, sometimes misinterpreted as entailing an interactionist stance, whereby individual processes are less relevant. They argue that this is not the case, and proceed to explain the central role played by individual agency, subpersonal processes, and subjective personal experience in the framework of participatory sense-making. Social interaction is defined as involving the co-arising of autonomous relational patterns, not under the full control of any participant, but without loss of individual autonomy for those engaged in the social encounter. Di Paolo and De Jaegher discuss how interactive patterns can sustain a deep entanglement between brain, body, and interactive dynamics during social engagement, as well as the functional role played in some cases by collective dynamics. They argue that hybrid approaches perpetuate dualistic distinctions between mind and body and contend that participatory sense-making, instead, offers precisely the dialectical tools for the self-deployment of the tensions that give rise to the individualist and interactionist frameworks.

Dan Hutto and Glenda Satne contend that radically enactive cognition (REC) does not imply that all forms of cognition are content involving and, especially, not root forms. According to radical enactivists, only minds that have mastered special kinds of sociocultural practice are capable of content-involving forms of cognition. The chapter addresses criticisms that have been leveled at REC's vision of how content-involving cognition may have come on the scene. In the first section, Hutto and Satne respond to the charge that REC faces a fatal dilemma when it comes to accounting for the origins of content in naturalistic terms— a dilemma that arises from REC's own acknowledgment of the existence of the Hard Problem of Content. In subsequent sections, they address the charge that REC entails continuity skepticism, reviewing this charge in its scientific and philosophical formulations. Hutto and Satne conclude that REC is not at odds with evolutionary continuity, when both REC and evolutionary continuity are properly understood. Furthermore, although REC cannot completely close the imaginative gap that is required to answer the philosophical continuity skeptic, it is, in this respect, in no worse a position than its representationalist rivals and their naturalistic proposals about the origins of content.

Part 2, "Intersubjectivity, Selfhood, and Persons," focuses on the conceptual and empirical relationships of the self and encultured full-fledged persons. If the most basic forms of the self are already embedded in intersubjectivity, one may think that there is no part of consciousness that is not impregnated with culture. But is there not something in the self that precedes all intersubjectivity? Dan Zahavi (2014) maintains that there is a "minimal self"; every consciousness experience has a character of mineness. Together with Brinck and Reddy (this vol.), he defends the claim that individual experience is not preceded by we-experience. In contrast, Matthew Ratcliffe as well as Anna Ciaunica and Aikaterini Fotopoulou argue in their chapters in different ways that even the "minimal self" needs to be conceived in intersubjective terms, a challenge taken up again by Zahavi in his subsequent response.

In the first chapter of part 2, Ingar Brinck, Vasudevi Reddy, and Dan Zahavi consider some arguments that could be adopted for the primacy of the *we*, and examine their conceptual and empirical implications. The question of the relation between the collective and the individual has had a long but patchy history within both philosophy and psychology. They argue that the *we* needs to be seen as a developing and dynamic identity, not as something that exists fully fledged from the start. The concept of *we* thus needs more nuanced and differentiated treatment than currently exists, distinguishing it from the idea of a "common ground" and discerning multiple senses of "*we*-ness." At an empirical level, beginning from the shared history of human evolution and prenatal existence, a simple sense of prereflective *we*-ness, the authors argue, emerges from second-person *I-you* engagement in earliest infancy. Developmentally, experientially, and conceptually, engagement remains fundamental to the *we* throughout its many forms, characterized by reciprocal interaction and conditioned by the normative aspects of mutual addressing.

Matthew Ratcliffe addresses the view that schizophrenia involves disturbance of the minimal self, and that this distinguishes it from other psychiatric conditions. He challenges the distinction between a minimal and an interpersonally constituted sense of self by considering the relationship between psychosis and interpersonally induced trauma. First, he suggests that even minimal self-experience must include a prereflective sense of what *kind* of intentional state one is in. Then he addresses the extent to which human experience and thought are interpersonally regulated. He proposes that traumatic events in childhood or in adulthood can erode a primitive form of "trust" in other people on which the integrity of intentionality depends, thus disrupting the phenomenological boundaries between intentional state types. Ratcliffe concludes that a distinction between minimal and interpersonal self is untenable, and schizophrenia should be thought of in relational terms rather than simply as a disorder of the individual.

This intersubjective constitution of the self is explored further in the next chapter, by Anna Ciaunica and Aikaterini Fotopoulou. They ask whether minimal selfhood is a built-in feature of our experiential life or a later, socioculturally determined acquisition, emerging in the process of social exchanges and mutual interactions. Building on empirical research on affective touch and interoception, Ciaunica and Fotopoulou argue in favor of reconceptualizing minimal selfhood so that it goes beyond such debates and their tacitly "detached," visuospatial models of selfhood and otherness. They trace the relational origins of the self back to fundamental principles and regularities of the human embodied condition, such as the amodal properties that govern the organization of sensorimotor signals into distinct perceptual experiences. Interactive experiences with effects "within" and "on" the physical boundaries of the body (e.g., skin-to-skin touch) are necessary for such organization in early infancy when the motor system is not yet developed. Therefore an experiencing subject is not primarily understood as facing another subject "there." The authors conclude that the minimal self is by necessity co-constituted by other bodies in physical contact and proximal interaction.

In the following chapter, Dan Zahavi responds to the critique of the concept of the minimal self by Ratcliffe and Ciaunica and Fotopoulou. Zahavi acknowledges that the discussion of the minimal self has entered a new phase with the foregoing chapters, not only because they engage with the recent arguments of *Self and Other* (Zahavi 2014) but also because their criticisms differ from the criticism offered in the past, for example, by advocates of a no-self view, narrativists, and phenomenal externalists. Rather than denying the existence of the minimal self, the critiques published here are concerned is with its proper characterization and interpersonal constitution. Zahavi maintains, however, that the minimal self is not interpersonally constituted. He argues that it can coherently be defined more thinly and independently of interpersonal aspects of the self, for which it is the condition of possibility.

Mark Bickhard presents a model of persons as emergent dynamic forms of sociocultural agency. Such a model requires a metaphysical framework that makes sense of the normative dynamic emergence of agents, which in turn requires a metaphysics of process. He also briefly addresses how this model of persons as interactive agents relates to persons as moral agents. Moreover, Bickhard turns against radical enactivism and holds that even organisms like frogs need normative truth-valued representational capacities, and that such capacities are incorrectly captured in the traditional encoding models of representation. He maintains that representational capacities are needed to understand the functional normative level of emergence of organisms in evolution and, a fortiori, persons in sociocultural settings. Decisive for the concept of persons, so Bickhard, is that they are individually constituted in interactive processes and potentialities as sociocultural agents.

Part 3, "Cultural Affordances and Social Understanding," explores the basis of social understanding, including the social understanding of significance in the context of its cultural conditions. The contributors here ask questions such as: What does it mean to feel ashamed of somebody else; how can we account for the social dimension of hetero-induced shame? What is the nature of affordance, how can we conceive of joint affordances, and how could they be produced by social interaction? What does it mean to pretend something? How can we understand cultural significance if most of the available evidence is limited to a few material items, such as scattered bones and traces of ornaments? How can the concept of radical enactive cognition (cf. Hutto and Satne, this vol.) be applied to evolutionary anthropology?

Shaun Gallagher argues in the first chapter that the distinction between *significance* and *meaning* made in debates about the nature of interpretation in hermeneutics is relevant to contemporary discussions of social cognition. He reviews the debates about interpretation, focusing on Gadamer's hermeneutics. Then, within the framework of a pluralist approach to social cognition, he discusses some problems with mind-reading approaches that attempt to get to the "inner" meaning of the other. He defends a Gadamerian view of social cognition that models our encounter with others on a dialogical interaction and the emergence of significance.

Within the literature, shame is generally described as a self-conscious emotion, meaning that shame is about the self that feels that emotion. But how can this account accommodate cases in which I feel ashamed of someone else? Alba Montes Sánchez and Alessandro Salice's chapter pursues two goals. The first is to vindicate the phenomenological credentials of what might be called "hetero-induced shame" and to resist possible attempts to reduce its specificity. The second goal is to show how the standard account of shame as self-directed can be made hospitable to cases of hetero-induced shame. They argue that a promising way to do this is by supplementing the standard account by a theory of group identification.

John Elias holds that in virtue of our sociability and plasticity we are especially open to altering and developing our capacities and abilities, thereby expanding the scope of available affordances. The distinctively dynamic and extensive nature of human abilities, however, raises questions concerning the ontology of affordances, given their relativity to abilities, their *being* relative to abilities. These questions are particularly pressing, since much of the power of the concept comes from the claim that affordances are *real*, that they *exist* in some sense. Resolving these issues, Elias suggests, involves taking the temporal dimension of abilities and affordances seriously, particularly in terms of interaction across multiple temporal scales. Such a temporal perspective encompasses the modulating role of motivation, as well as questions concerning the presence and salience of affordances. He ends by addressing abilities as they extend into, and are extended by, social interaction and coordination, and introduces the notion of *joint affordances* specifically, in contrast to the more general sociality of affordances.

Pretending is often conceptualized as an imaginative and symbolic capacity, explained in terms of mental representations. Zuzanna Rucińska proposes an alternative way to explain pretending by using affordances, instead of mental representations, as explanatory tools. Rucińska shows that a specific notion of affordance has to be appropriated for affordances to play the relevant explanatory roles in pretense. Her analysis opens up a discussion of the nature of affordances, clarifying how, in various conceptions, the environment and the animal play a role in shaping affordances. She then clarifies which notion best explains pretending, and suggests that a particular conception of affordances as dispositional properties of the environment (à la Turvey 1992) can make affordances explanatorily useful. Rucińska shows how environmental affordances together with animal effectivities, placed in the right context (formed by canonical affordances or other people), could form an explanation of basic kinds of pretend play. The idea is that some forms of cognitive activity, such as basic pretense, can be explained by embodied and enactive theorists without the need to posit mental representations. Social and cultural factors provide a crucial aspect of a coherent explanation of basic pretense.

Duilio Garofoli writes that evidence of feather extraction from scavenging birds by late Neanderthal populations, supposedly for ornamental reasons, has recently been used to bolster the case for Neanderthal symbolism and their cognitive equivalence with modern humans. This argument resonates with the idea that the production and long-term maintenance of body ornaments necessarily require a cluster of abilities defined here as the material symbolism package. This implies the construction of abstract meanings, which are then mentally imposed on artifacts and socially shared through full-blown mind reading, assisted by a metarepresentational language. However, a set of radical enactive abilities, above all direct social perception and situated concepts, suffices to explain the emergence of ornamental feathers without necessarily involving the material symbolism package. The embodied social structure created by body ornaments, augmented through behavioral-contextual narratives, suffices to explain even the long-term maintenance of this practice; no recourse to mentalism is needed. Costly neurocentric assumptions that conceive the material symbolism package as a homuncular adaptation are eschewed by applying a nonsymbolic interpretation of feathers as cognitive scaffolds. Garofoli concludes that the presence of body adornment traditions in the Neanderthal archaeological record does not warrant Neanderthals' cognitive equivalence with modern humans, for it does not constrain a metarepresentational level of meaning.

Part 4, "Embodiment and Its Cultural Significance," addresses the concept of the body in relation to culture. Here the contributors reevaluate the body in the light of studies in neurophysiology, cultural neurophenomenology, biopsychology, phenomenology, anthropology, and psychopathology. They argue that the human body is shaped by cultural forms of experiencing and conceptualizing. It is a carrier of collective memories, and its experience is an integral part of the world shared by the members of a culture. Furthermore, the body is malleable by cultural products such as movies and is dependent on culture.

Vittorio Gallese addresses the notion of embodiment from a neuroscientific perspective, by emphasizing the crucial role played by bodily relations and sociality in the evolution and development of distinctive features of human cognition. Gallese accounts for the neurophysiological level of description in terms of bodily formatted representations, and he replies to criticisms recently raised against this notion. The neuroscientific approach that Gallese proposes is critically framed and discussed against the background of the evo-devo focus on a rarely explored feature of human beings in relation to social cognition: their neotenic character. Neoteny refers to the slowed or delayed physiological and somatic development of an individual. Such development depends largely on the quantity and quality of interpersonal relationships the individual is able to establish with her or his adult peers. It is proposed that human neoteny further supports the crucial role played by embodiment, here spelled out by adopting the explanatory framework of embodied simulation, in allowing humans to engage in social relations and make sense of others' behaviors. This approach can fruitfully be used to shed new light on nonpropositional forms of communication and social understanding and on distinctive human forms of meaning making, such as the experience of artificial fictional worlds.

Thomas Fuchs holds that the concept of *body memory* comprises all forms of implicit memory that are mediated by the body and actualized without explicit intention in our everyday conduct—for example, habitual patterns of movement and perception, instrumental skills, or behavioral and cultural habits. The lifelong plasticity of body memory enables us to adapt to the natural and social environment, in particular to become entrenched and feel at home in the social and cultural environment. Fuchs introduces the concept of *collective body memories* that develop in social groups through recurrent shared experiences and lead to spatial and temporal patterns of joint group behavior. Examples of such memories are the formation of a well-attuned football team and its fluent interplay, the habitual ways of interacting that characterize a family, or the enactment of social ceremonies and rituals. In such situations, the intercorporeal memories of the individuals unite to form overarching procedural fields. Moreover, the interactive processes develop an autonomous or emergent dynamic involving the individuals in behavior they would not exhibit outside the formation. Once the group joins again in a similar configuration and situation, the resulting collective body memory is reactualized. Fuchs analyzes these phenomena mainly from a phenomenological, but also from a dynamical systems, perspective.

Joerg Fingerhut and Katrin Heimann explain that, over the last decade, the role of the spectator's body has become considerably more important in theoretical as well as experimental approaches to film perception. However, most positions focus on how cinema has adapted to the spectator's body over time, that is, on the basic principles of human perception and cognition, in developing its immersive power. This chapter presents the latest contributions to this topic while also providing a new stance regarding the relationship between the mind and movies. Drawing on selected research from embodied approaches to cognition and picture perception, the authors suggest that humans learn to see film by integrating filmic means into their body schemata and, through this process, develop a "filmic body," available to them during film watching and, possibly, also offscreen. Film language and film cognition are plastic products of mutual influence between films and embodied agents and thereby move the medium toward novel filmic means and us toward novel experiences. The authors propose a number of research designs for further exploring these claims.

Peter Henningsen and Heribert Sattel present and interpret data on significant cultural influences on pain-related psychosocial workplace conditions, one of the core issues of psychosomatic medicine, and discuss consequences for a cultural neuroscience of pain. Chronic pain encompasses the experience of the pain sensation itself and a whole universe of related emotions, thoughts, behaviors, and suffering; tissue damage is no necessary precondition for it. A biopsychosocial view of risk factors typically concentrates on the intra-individual level and includes genetic dispositions or injuries, for example, whereas an embodied approach emphasizing the "body being in the world" and integrating cultural perspectives seems more appropriate. However, recent epidemiological work has demonstrated the relevance of group-level psychosocial risk factors for chronic pain. Lack of social support at work, injustice, high levels of job stress, and effort-reward imbalance are important factors. Nevertheless, even these perspectives do not capture all relevant differences: studies in different societies reveal significant cultural influences, both in an "etic" and in an "emic" perspective. The link between culture and pain involves various factors. Culturally shaped ways of world making influence the interpretation, labeling, and treatment of distress. New knowledge on the relational biology of pain shows how culture determines differences in the neural processes underlying emotion and pain.

In the book's final chapter, Laurence Kirmayer and Maxwell Ramstead investigate the field of cultural psychiatry, which is concerned with understanding the implications of human cultural diversity for psychopathology, illness experience, and intervention. The emerging paradigms of embodiment and enactment in cognitive science provide ways to approach this diversity in terms of variations in bodily and intersubjective experience, narrative practices, and discursive formations. This chapter outlines an approach to cultural neurophenomenology and psychopathology through metaphor theory, which examines the interplay of culturally shaped developmental processes of embodied experience and narrative practices structured by ideologies of personhood and social positioning. The new paradigm has broad implications for psychiatric theory, research, and practice, and these are illustrated with examples from the cross-cultural study of delusions.

Acknowledgments

Christoph Durt's work on this book was supported by the DFG excellence initiative Cultural Dynamics in Globalised Worlds and by the European Union's Horizon 2020 research and innovation program under the Marie Skłodowska-Curie Individual Fellowship no. 701584.

References

Bourdieu, P. 1977. *Outline of a Theory of Practice*. Cambridge University Press.

Bourdieu, P. 1990. *The Logic of Practice*. Stanford University Press.

Campbell, R. 2009. A process-based model for an interactive ontology. *Synthese* 166:453–477.

Carroll, N. 1996. *Theorizing the Moving Image*. Cambridge University Press.

Casey, E. 1984. Habitual body and memory in Merleau-Ponty. *Man and World* 17:279–297.

Clark, A. [2008] 2011. *Supersizing the Mind: Embodiment, Action, and Cognitive Extension*. Oxford University Press.

Clark, A., and D. Chalmers. 1998. The extended mind. *Analysis* 65:1–11.

Colombetti, G., and E. Thompson. 2008. The feeling body: Toward an enactive approach to emotion. In *Developmental Perspectives on Embodiment and Consciousness*, ed. W. F. Overton, U. Müller, and J. L. Newman, 45–68. Erlbaum.

Csordas, T. J. 1990. Embodiment as a paradigm for anthropology. *Ethos* 18 (1):5–47.

Dawkins, R. [1976] 2006. *The Selfish Gene*. Oxford University Press.

De Jaegher, H., and E. A. Di Paolo. 2007. Participatory sense-making: An enactive approach to social cognition. *Phenomenology and the Cognitive Sciences* 6:485–507.

De Sousa, R. 1987. *The Rationality of the Emotions*. Cambridge University Press.

Di Paolo, E. A. 2009. Extended life. *Topoi* 28:9–21.

Di Paolo, E. A., M. Rohde, and H. De Jaegher. 2010. Horizons for the enactive mind: Values, social interaction, and play. In *Enaction: Toward a New Paradigm for Cognitive Science*, ed. J. Stewart, O. Gapenne, and E. A. Di Paolo, 33–87. MIT Press.

Donald, M. 1991. *Origins of the Modern Mind: Three Stages in the Evolution of Culture and Cognition*. Harvard University Press.

Dumas, G., J. A. S. Kelso, and J. Nadel. 2014. Tackling the social cognition paradox through multi-scale approaches. *Frontiers in Psychology* 5:882. doi:10.3389/fpsyg.2014.00882.

Froese, T., and T. Fuchs. 2012. The extended body: A case study in the neurophenomenology of social interaction. *Phenomenology and the Cognitive Sciences* 11:205–236.

Fuchs, T. 2013. Depression, intercorporality, and interaffectivity. *Journal of Consciousness Studies* 20:219–238.

Gallagher, S. 2008. Direct perception in the intersubjective context. *Consciousness and Cognition* 17 (2): 535–543.

Gibson, J. J. 1979. *The Ecological Approach to Visual Perception*. Houghton Mifflin.

Hmelo-Silver, C. E., R. G. Duncan, and C. A. Chinn. 2007. Scaffolding and achievement in problem-based and inquiry learning: A response to Kirschner, Sweller, and Clark (2006). *Educational Psychologist* 42 (2): 99–107.

Hornik, R., N. Risenhoover, and M. Gunnar. 1987. The effects of maternal positive, neutral, and negative affective communications on infant responses to new toys. *Child Development* 58 (4): 937–944.

Husserl, E. [1913] 1976. *Ideen zu einer reinen Phänomenologie und phänomenologischen Psychologie: Allgemeine Einführung in die reine Phänomenologie* (Text der 1–3 Auflage). Martinus Nijhoff.

Husserl, E. 1966. *Analysen zur passiven Synthesis*. Martinus Nijhoff.

Jonas, H. 1966. *The Phenomenon of Life: Toward a Philosophical Biology*. Harper & Row.

Kelso, J. A. 1995. *Dynamic Patterns: The Self-Organization of Brain and Behavior*. MIT Press.

Laland, K. N., J. Odling-Smee, and M. W. Feldman. 2000. Niche construction, biological evolution, and cultural change. *Behavioral and Brain Sciences* 23:131–175.

LeDoux, J. [1998] 1999. *The Emotional Brain: The Mysterious Underpinnings of Emotional Life*. Simon & Schuster.

Lewis, M. D. 2000. Emotional self-organization at three time scales. In *Emotion, Development, and Self-Organization: Dynamic Systems Approaches to Emotional Development*, ed. M. D. Lewis and I. Granic, 7–69. Cambridge University Press.

Lewis, M. D. 2005. Bridging emotion theory and neurobiology through dynamic systems modeling (target article). *Behavioral and Brain Sciences* 28:169–194.

Malafouris, L. 2013. *How Things Shape the Mind: A Theory of Material Engagement*. MIT Press.

Malafouris, L., and C. Renfrew. 2010. Introduction: The cognitive life of things; Archaeology, material engagement and the extended mind. In *The Cognitive Life of Things: Recasting the Boundaries of the Mind*, ed. L. Malafouris and C. Renfrew, 1–12. McDonald Institute Monographs.

Menary, R. 2007. *Cognitive Integration: Mind and Cognition Unbound*. Palgrave Macmillan.

Menary, R. 2010a. Introduction: The extended mind in focus. In *The Extended Mind*, ed. R. Menary, 1–25. MIT Press.

Menary, R. 2010b. Cognitive integration and the extended mind. In *The Extended Mind*, ed. R. Menary, 227–243. MIT Press.

Merleau-Ponty, M. [1945] 2005. *The Phenomenology of Perception*. Trans. C. Smith. Taylor and Francis e-Library.

O'Regan, J. K., and A. Noë. 2001. A sensorimotor account of vision and visual consciousness. *Behavioral and Brain Sciences* 24:939–1011.

Ratcliffe, M. 2008. *Feelings of Being: Phenomenology, Psychiatry, and the Sense of Reality*. Oxford University Press.

Robinson, J. [2005] 2007. *Deeper Than Reason: Emotion and Its Role in Literature, Music, and Art*. Clarendon Press.

Stephan, A. 2012. Emotions, existential feelings, and their regulation. *Emotion Review* 4 (2): 157–162. doi:10.1177/1754073911430138.

Stern, D. N. 1985. *The Interpersonal World of the Infant: A View from Psychoanalysis and Developmental Psychology*. Basic Books.

Sterelny, K. 2006. The evolution and evolvability of culture. *Mind and Language* 21 (2): 137–165.

Sterelny, K. 2010. Minds: Extended or scaffolded. *Phenomenology and the Cognitive Sciences* 9:465–481.

Thompson, E. 2005. Sensorimotor subjectivity and the enactive approach to experience. *Phenomenology and the Cognitive Sciences* 4 (4): 407–427.

Thompson, E. 2007. *Mind in Life: Biology, Phenomenology, and the Sciences of Mind*. Belknap Press of Harvard University Press.

Thompson, E., and M. Stapleton. 2009. Making sense of sense-making: Reflections on enactive and extended mind theories. *Topoi* 28 (1): 23–30.

Thompson, E., and F. J. Varela. 2001. Radical embodiment: Neural dynamics and consciousness. *Trends in Cognitive Sciences* 5 (10): 418–425.

Tomasello, M. 1999. *The Cultural Origins of Human Cognition*. Harvard University Press.

Tomasello, M., M. Carpenter, J. Call, T. Behne, and H. Moll. 2005. Understanding and sharing intentions: The origins of cultural cognition. *Behavioral and Brain Sciences* 28:675–691.

Trevarthen, C. 1989. Development of early social interactions and the affective regulation of brain growth. In *Neurobiology of Early Infant Behaviour*, ed. C. von Euler, H. Forssberg, H. Lagercrantz, and V. Landin, 191–216. Macmillan Education UK.

Trevarthen, C., and K. J. Aitken. 2001. Infant intersubjectivity: Research, theory, and clinical applications. *Journal of Child Psychology and Psychiatry, and Allied Disciplines* 42 (1): 3–48.

Turvey, M. T. 1992. Affordances and prospective control: An outline of the ontology. *Ecological Psychology* 4 (3): 173–187.

van Gelder, T. 1998. The dynamical hypothesis in cognitive science. *Behavioral and Brain Sciences* 21:615–665.

Varela, F. J., E. Thompson, and E. Rosch. 1991. *The Embodied Mind: Cognitive Science and Human Experience*. MIT Press.

Williams, L. E., J. Y. Huang, and J. A. Bargh. 2009. The scaffolded mind: Higher mental processes are grounded in early experience of the physical world. *European Journal of Social Psychology* 39:1257–1267.

Zahavi, D. 2014. *Self and Other: Exploring Subjectivity, Empathy, and Shame*. Oxford University Press.

Zajonc, R. B. 1984. On the primacy of affect. *Psychologist* 39 (2): 117–123.

I Phenomenological and Enactive Accounts of the Constitution of Culture

1 Intercorporeality and Intersubjectivity: A Phenomenological Exploration of Embodiment

Dermot Moran

1 Phenomenology as an Eidetic Description of *Ineinandersein* and the Life of Spirit

The phenomenological movement—especially as originally developed by Husserl and elaborated by Scheler, Heidegger, Sartre, and Merleau-Ponty—has been responsible for the radical reconception of human existence that revolutionized philosophy in the twentieth century and is still being assimilated more generally in philosophy of mind and action, as well as in the cognitive sciences (see, e.g., Varela, Thompson, and Rosch 1999; Thompson 2010; Shapiro 2014). Edmund Husserl, in particular, focused intensively on a number of central themes, such as the intentionality of consciousness and the constitution of sense, the essential description of the a priori structures of consciousness ("the ABC of consciousness"), the essential structures of lived embodiment (*Leiblichkeit*), the nature of empathy (*Einfühlung*) and the experience of the foreign (*Fremderfahrung*), and finally the wider a priori structures of intersubjectivity (*Intersubjektivität*) and "sociality" (*Sozialität*).

The mature Husserl aimed at nothing less than a holistic phenomenological description of the entire "life of consciousness" (*Bewusstseinsleben*) or "life of spirit" (*Geistesleben*), including human sociality, communalization, historicality, generativity, and life in culture and tradition. He wanted to describe human life in its rich intersubjective concreteness. Regrettably, phenomenologists who concentrate narrowly on the early Husserl of the *Logical Investigations* (1900–1901; Husserl 2001a) and *Ideas I* (Husserl 1977a)[1] often overemphasize his focus on the individual life of intentional consciousness as reconstructed from within (and even on the structure of individual, atomistic lived experiences [*Erlebnisse*]) and tend to overlook Husserl's original, radical, and fundamentally groundbreaking explorations of intersubjectivity, sociality, and the constitution of historical cultural life (which would later influence Heidegger and Schütz, among others).

---

1. Hereafter the work will be cited as *Ideas I*, followed by the paragraph number (§), page number of the translation (2014b), and then the *Husserliana* volume number and page.

In respect of this individualist misinterpretation, Husserl is often his own worst enemy, since he repeatedly and very publicly, for example, in his *Cartesian Meditations* (Husserl 1950, 1967; hereafter *CM*), compared his phenomenological breakthrough to subjectivity with Descartes's discovery of the *ego cogito* and modeled his phenomenological *epoché*, albeit with important changes of emphasis, on Descartes's radical doubt. As a result, Husserl's phenomenology has too often been designated a methodological solipsism that proceeds through individualistic introspection of conscious experiences, and Husserl's wider explorations of social and cultural life have been passed over (and many of his original discoveries have been attributed to others, e.g., Heidegger and Gadamer). It is worth reminding ourselves, therefore, of the originality of Husserl's meditations on the nature of the self, its embodiment, and its intercorporeal, intersubjective communal relations with others.

In this chapter,[2] then, I want to focus on Husserl's mature reflections (i.e., as specifically found in his writings of the 1920s and 1930s) on the intentional constitution of culture, particularly as he understood it to relate to lived embodiment and, especially, the specific relations that hold between lived bodies, their *Ineinandersein*, *Füreinandersein*, or what Husserl calls in *Cartesian Meditations* "a mutual being-for-one-another" (*ein Wechselseitig-für-einander-sein*; *CM*, 129; *Hua I*, 157). As he puts it elsewhere, in the *Crisis of European Sciences* (Husserl 1970, 1962), Husserl approaches human subjects not only as having "subject being for the world" (*Subjektsein für die Welt*) but also as possessing "object being in the world" (*Objektsein in der Welt*; *Crisis*, 178; *Hua VI*, 182). How humans can be both in the world and for the world is, for him, the riddle of transcendental subjectivity.

Human beings not only have a sense of the individual identity and continuity of the flow of consciousness but also have a sense of being involved with one another. Human beings unite in many forms of social and collective intentionality. But Husserl also writes of humans as possessing a "world-consciousness" (*Weltbewusstsein*). This world-consciousness is a very real and complex phenomenon. It allows human subjects not only to experience things from their individual points of view but also to participate in evolving historical cultural life. Yet few phenomenologists have explored Husserl's conception of *Weltbewusstsein*, although it is, for him, a central component to human "being-in-the-world."

Human beings are, in the parlance of contemporary cognitive science, embodied, enactive, embedded, and "enworlded."[3] Human subjectivity, moreover, is always a cosubjectivity (*Mitsubjektivität*) with others in the shared context of an evolving, historical world. For Husserl, that consciousness must be embodied is an eidetic law. He speaks broadly

2. An earlier version of this paper was presented at the conference *Enacting Culture: Embodiment, Interaction and the Development of Culture*, October 15–17, 2014, TESIS, University of Heidelberg, Germany, on Wednesday, October 15, 2014. I am grateful to Thomas Fuchs, Christoph Durt, Stefano Micali, Sarah Heinämaa, and Dan Hutto for their comments.

3. Gerda Walther, one of Husserl's students, explicitly speaks of "embedding" (*Einbettung*) in her *Zur Phänomenologie der Mystik* (1923).

of embodiment as *Leiblichkeit* and *Verleiblichung* (1973e; *Hua XV*, 289),[4] but also as "incorporation" (*Verkörperung*; 1973c; *Hua XIII*, 139), and even of the "enworlding" (*Verweltlichung*) or the "humanization" (*Vermenschlichung*; 1973e; *Hua XV*, 705) of the transcendental subject.

While contemporary cognitive science has come to see the human being as essentially intersubjective and social, it has not yet fully taken on board the dynamics of human temporal life—a life lived in the context of a historical community. Phenomenology still has much to contribute in this regard.

It is this world-consciousness that allows each conscious subject to feel a belonging with other subjects in a shared world. Humans live necessarily in various forms of community (from family to state), as well as participating in a generation and having a sense of other generations (father, mother, forefathers, etc.; see Husserl 1973e; *Hua XV*, 178)—part of what Husserl speaks of as the life of "generativity" (*Generativität*, a concept he seems to have developed entirely separately from Heidegger; see *Crisis*, 188; *Hua VI*, 191).[5] Generativity is, for Husserl, just one of the many a priori structures (Heidegger will call them "existentialia") that go to constitute communal, social life (for a discussion of the world as it appears to acting, practical subjects, for example, the way the death of a child can break with the normal course of life, something Husserl and his wife Malvine experienced when their son was killed in action in the Great War, see Husserl 1973e; *Hua XV*, 212).

Furthermore, Husserl very early established the interrelation between objectivity and intersubjectivity, between the intentional experience of something as having an identity and a sense that transcends our distinctive points of view and the sense that we have that our points of view are partial and that others bring different perspectives to bear on situations. Husserl saw that the undeniable primary experience of the objective world is of the *one shared world* that all of us experience ("for all," *für jedermann*, Husserl 1974, 1969; *Hua XVII*, 244). The sense of objectivity is co-constituted by us, and we are constituted as living beings in relation to this backdrop of world. The objective world means simply the world that is the object of our intersubjective intentions and attitudes. The constitution of objectivity is

4. Husserl's terminology is flexible; he also speaks of *organische Leiblichkeit* (1968; *Hua IX*, 109), *physische Leiblichkeit* (1968; *Hua IX*, 129), and even of the body as *Leibding*.

5. Husserl develops his own terminology for talking about history and tradition. His concept of "generativity" is one such example. David Carr translates the term—somewhat misleadingly—as "genesis." The term usually has a medical meaning of "concern with the next generation," e.g., in rearing children, but is broadened by Husserl to mean the overall process by which cultural meaning is creatively filtered and transmitted from one generation to another. The concept was later developed by the psychoanalyst Erik Erikson (1902–1994) to cover all kinds of ways in which traditions may be passed on or inhibited, e.g., a child deciding to stand up to familial abuse. Husserl discusses "generativity" in greater detail in texts associated with the *Crisis*; see especially the 1934 supplement, *Verschiedene Formen der Historizität* (*Different Forms of Historicity*, in Husserl 1992).

closely related to the experience of embodiment. One's own lived body is, in an important sense, the first object that one experiences.

2   The Centrality of Embodiment for Husserl: *Leib* and *Körper*

In line with Husserl's meditations on cultural life, the centrality of embodiment in his phenomenology is also not yet fully appreciated, primarily because these reflections were carried out in *Ideas II* (Husserl 1952),[6] the draft manuscript of which Husserl felt unable to complete and publish in his lifetime. As a result, Maurice Merleau-Ponty, who had access to the *Ideas II* manuscript from as early as his visit to Leuven in 1939, is usually credited with inaugurating the phenomenology of embodiment in his *Phenomenology of Perception* (Merleau-Ponty 1945) and in subsequent publications.[7] Despite Merleau-Ponty's popularization of embodiment, especially perceptual and motor embodiment, Husserl's own phenomenology is, in many ways, almost totally and even exclusively a "phenomenology of embodiment" (*Phänomenologie der Leiblichkeit*), a phrase he uses in his *Phenomenological Psychology* lectures (1977b, 153; 1968; *Hua IX*, 199) of 1925 (which are themselves an extended reflection on embodiment). Embodiment forms the basis of perception and agency but also lies at the root of the human "existential" experience of being-in-the-world most generally, of "acting and suffering" (*Tun und Leiden*), as Husserl says.

The phenomenological tradition generally—including not just Husserl but also Max Scheler, Edith Stein, and later Helmuth Plessner—begins its analysis of embodiment from the key distinction, found originally in Fichte and the earlier German idealists, between the animate, "lived body" (*Leib*), that is experienced first-personally and the physical, material, objective "body" (*Körper*) that is subject to the laws of physics, causation, gravity, and so on (see Waldenfels 2000; Grätzel 1989). It must be emphasized that *Leib* and *Körper* are on a continuum; in fact, Husserl speaks of a *Verflechtung* between *Leib* and *Körper*. The body can be experienced as a living, responsive organism (when I run and jump) or as a physical thing (when I fall and bang my knee). Indeed, Sartre points out in *Being and Nothingness* that an anesthetic can render the organic body as mere physical thing by numbing the limb (1995, 304; 1943, 366), thereby rendering the touching-touched circle inoperative.[8] Normally,

---

6. Hereafter cited as *Ideas II*, followed by paragraph number (§), page number of the translation (1989), and then the *Husserliana* volume number and page.

7. There are exceptions; see, in particular, Behnke 2010; Bernet 2013; Heinämaa 2011; and Taipale 2014.

8. In contrast to Merleau-Ponty, however, Sartre claims that the phenomenon of double sensation does not reveal something essential about embodiment. For Sartre, the double sensation is simply a contingent feature of our embodied existence and is not a significant or exemplary phenomenon. Sartre claims that the double sensation can easily be removed by morphine, which makes my leg numb and insensitive to being touched (1995, 304; 1943, 366). The intertwining of touching and touched is

however, the body-subject or organically experienced body is prior in that it is the primary way we experience our own bodies.

Normally and primarily, for Husserl, we experience our bodies as living organisms, and the body as purely physical entity is grasped only by abstraction from everything psychic. As Husserl writes:

Thus the body [*der Leib*] is seen at this level ... and seen in its two-sidedness, at the same time in its physical externality [*Äusserlichkeit*] and its animating internality [*Innerlichkeit*]. Only if we abstract from this internality co-existing in every normal experience of the body [*Leibeserfahrung*] do we have the purely physical body-thing [*Leibding*]. Thus, e.g., the foot as material spatial body [*Körper*]. But as my foot in the experience of my body, it is more; the field of touch-sensations and contact sensations governs it, is localized in it. (1977b, 100; *Hua IX*, 131)

Husserl writes about our own peculiar relation to our bodies in the following research note from 1921:

Thus the body has in itself the most original character of being mine [*den ursprünglichsten Charakter des Meinen*]; belonging to me, it contrasts with the foreign, in which I am not involved, i.e., not practically [*nicht praktisch*]. ... My body is among all things the closest [*Mein Leib ist mir unter allen Dingen das Nächste*], the closest in perception, the closest in feeling and will. And so I am, the functioning I, before all other worldly objects united with it [the body] in a special way. (Husserl 1973; *Hua XIV*, 58, my translation)

The lived body, in Husserl's account, is experienced from within as a center of orientation, sensations, movement, action and affection, as a series of "I can's." But the body is not just a set of abilities and habitualities. It also engages in its own self-constitution and is expressive of the ego-subject that inhabits it. The bodily subject's self-presence, moreover, is permeated by gaps and absences. As Husserl says, the body is "a remarkably imperfectly constituted thing" (*Ideas II*, §41(b), 167; *Hua IV*, 159). This allows the body to become continually expressive and to develop new significations that eventually become sedimented into its very embodied condition.

3   The Embedded and Enworlded Body-Subject

In his later works, Husserl tries to explore this embeddedness and enworldedness of the human subject in terms of his original concept of "horizons," zones of genuine significance

---

not revelatory of our being-in-the-world. Rather, for Sartre, to touch and be touched reflect different "orders" or "levels" of being. When one hand touches the other hand, I directly experience the hand that is being touched first. In other words, I am intentionally directed at the *object*. It is only because of the possibility of *a certain reflection* that I can turn back and focus on the sensation in the touching hand. This reflection is not inbuilt in the primary act of intending. Sartre maintains that this constitutes ontological proof that the body-for-me and the body-for-the-other are entirely separate intentional objectivities.

that cannot be objectified in the manner of the objects of perceptual experience but serve to contextualize in fundamentally different ways the objects in experience.[9] Horizons have various levels of indeterminacy. In §27 of *Ideas I*, for instance, he speaks of the horizons of inactualities that surround the perceived actualities:

What is currently perceived, what is determinate (or at least somewhat determinate) and co-present in a more or less clear way is in part pervaded, in part surrounded by a horizon of indeterminate actuality, a horizon of which I am dimly conscious. I am able to direct the illuminating focus of my attention on it with varying success. (*Ideas I*, §27, 49; *Hua III/1*, 49)

In later years, Husserl speaks of the need to examine in particular the nature of what he calls "horizon intentionality" (*Horizontintentionalität*) that is quite distinct from the more usual object intentionality that has been the focus of most philosophical discussion. Horizon in everyday parlance is primarily a spatial notion. It marks the limits of the visible. But in Husserl, horizon is primarily a temporal notion, marking the way in which the present emerges from the past and projects toward the future.

Every lived experience (*Erlebnis*) is not just built on the past, layered with sedimented meaning and surrounded by environmental affordances, but also projects a "horizon" of emptiness, of possibility, anticipation, and futurity. Furthermore, conscious experiences are not just first-personal decisions and position taking (*Stellungnehmen*); the conscious subject emerges from and builds on unconscious living experiences that are passively experienced. The subject can simply be drawn toward certain sounds as attractive or can like certain colors or tastes. It is conditioned and motivated by these passive allures: "I choose the fabric for the sake of its beautiful color or its smoothness" (*Ideas II*, §34, 148; *Hua IV*, 140). The self "sinks its tap roots into nature," as Edith Stein says (2000, 115). It is affected by what Husserl speaks of as "psychophysical conditionalities," whatever has been given to us in terms of height, strength, basic health, and all those other factical features of our embodied experience that might be considered as "the given."

The self, moreover, is never experienced in a complete self-disclosure (in the immediacy of the *cogito*) but experiences itself as mediated through interaction with other selves (e.g., parents, siblings, teachers, friends, colleagues, and what Alfred Schütz calls, more generally, "consociates" [1967, 109]). Human subjectivity in its temporal flow of experiential consciousness can only be understood if its horizons (temporal, motivational, senseful), especially its overlapping horizons with other subjects, can definitively be charted. For Husserl, the intertwining and overlapping of sensory modalities in the embodied subject give us the place to start reflecting on the experience of otherness and especially the other's lived body.

9. In the *Crisis* (264; *Hua VI*, 267) Husserl praises William James for his notion of "fringes," which Husserl takes to be a version of his own concept of "horizons." Husserl's notion of horizon includes not just spatial and temporal horizons but all kinds of experiences of boundaries, zones, and shades of indeterminacy that surround what is determinate in our experience.

4   Reversibility, the "Double Sensation" (*Doppelempfindung*), and the Flesh

What makes Husserl's phenomenology radically original lies not only in its extensive elabo-ration of the concept of "lived embodiment" (*Leiblichkeit*) and its perceptual, willing, and active life, but also in its exploration of the "intertwining" (*Verflechtung*; see Moran 2014) that already occurs *within* the body (e.g., hand-touching-hand, which produces "double sen-sations") and, crucially, prefigures embodied relations *between* subjects in the constitution of the "we-world" (*Wir-Welt*) of intersubjectively shared culture. It was indeed Husserl who first took over a traditional trope from nineteenth-century psychology, namely, the exploration of the phenomenon of self-touching, our ability to touch ourselves, and made it indicative of the very essence of embodiment.[10]

Merleau-Ponty, inspired by Husserl's discussion of the "double sensation," seized and expanded on this insight to make the touching-touched relation central to the experience of embodiment, or what he termed "the flesh" (*la chair*). Deeply influenced by Husserl's manu-script of *Ideas II* (which Merleau-Ponty continued to read creatively in later years after his first encounter with the manuscript in 1939, stimulated especially by Marly Biemel's *Husserliana* edition that appeared in 1952), Merleau-Ponty (1968, 1964a) in *The Visible and the Invisible* (hereafter *VI*) speaks primarily of "incarnation" (*incarnation*) or simply just "flesh" (*la chair*) (*VI*, 31; 51), a term he could also have encountered in Sartre's *Being and Nothingness*, which has an important chapter on embodiment (see Moran 2011). "Flesh," as Merleau-Ponty puts it in *The Visible and the Invisible*, is essentially characterized by "reversibility" (*réversibilité*), "the finger of the glove that is turned inside out" (*VI*, 263; 311) and "the doubling up of my body into inside and outside" (*VI*, 264; 311). Flesh is one entity that is sensitive on both sides: inward and out.

It was Merleau-Ponty, too, who recognized that the phenomenon of touching-touched cannot be separated from the wider issue of the relation between subjects in empathy and, indeed, the overarching issue of the intersubjective constitution of the objective world as such. In his late essay commemorating Husserl, *The Philosopher and His Shadow* (*Le philosophe et son ombre*), written in 1959 and published in *Signs* (Merleau-Ponty 1960, 1964b), Merleau-Ponty writes that "the problem of *Einfühlung*, like that of my incarnation, opens on the meditation of sensible being, or, if you prefer, it betakes itself there" (*Le problème de l'Einfühlung comme celui de mon incarnation débouche donc sur la méditation du sensible, ou, si l'on préfère, il s'y transporte*; *Signs*, 171; 215). In other words, the apprehension of the other in its various forms (from the other of my body to other bodies) is itself based on the under-standing of incarnation, whose very essence is revealed through the touching-touched relation. The touch opens up one's own body, and that of the other, as a living presence: "Es wird Leib, es empfindet," as Merleau-Ponty often repeats, echoing a passage in Husserl's *Ideas II*, §36:

10. For a discussion of the phenomenology of touch and vision, see Moran 2015.

If I speak of the *physical* thing, "left hand," then I am abstracting from these sensations (a ball of lead has nothing like them and likewise for every "merely" physical thing, every thing that is not my Body [*Leib*]). If I do include them, then it is not that the physical thing is now richer, instead *it becomes Body, it senses* [*es wird Leib, es empfindet*]. ... The sensation is doubled. Hence the Body is originally constituted in a double way. (*Ideas II*, 152–153; *Hua IV*, 145)[11]

As Merleau-Ponty insists, the touching-touched relation totally transforms our ontology and moves us beyond the dualism of subject and object. He writes in his famous essay *The Philosopher and His Shadow*:

It is imperative that we recognize that this description also overturns our idea of the thing and the world, and that it results in an ontological rehabilitation of the sensible. For from now on we may literally say that space itself is known through my body. ... When we say that the perceived thing is grasped "in person" or "in the flesh" (*leibhaft*), this is to be taken literally: the flesh of what is perceived, this compact particle which stops exploration, and this optimum which terminates it all reflect my own incarnation and are its counterpart. (*Signs*, 166–167; 210–211)

Perceptual presence of the intentional object, the experience of subjective incarnation, and the distinctive experience of empathy with the other (in all its gradations) are all interrelated in both Husserl's and Merleau-Ponty's phenomenology.

5   The Openness to the Other in Empathy

Empathy, as Husserl conceives it (and as Merleau-Ponty quite brilliantly recognizes), always includes a recognition of the other's lived body as expressive of his or her subjectivity that is perceived and lived through, but cannot be inhabited fully from within, the first-person perspective (see Moran 2004; Zahavi 2014). I apprehend you in your unique subjectivity, and I live through your emotional and cognitive states as perceived directly by me, but I cannot inhabit them in the precise manner you inhabit them. Empathy, according to the Husserlian and Merleau-Pontian approaches, involves a direct perceptual relation with the other. When the phone rings, I *hear* John's voice; I do not hear a set of electronic sounds and *infer* that John is the cause of these noises and conclude, therefore, that John is speaking to me. Empathy is not "inference" (*Schluss*), as Husserl often proclaims. It does not come about, under normal conditions, as the conclusion of a chain of reasoning (although clearly such chains of inference can occur, as when Sherlock Holmes infers the suspect's motive for the crime). In everyday life, empathy is direct and is experienced as immediate (even if it is in fact mediated through the cues or expressivity of the lived body).

Empathy has to incorporate or (to use Husserl's term *Fundierung*) *be founded on* the perception of the living body of the other person, but it never stops there (unless one is explicitly

11. Merleau-Ponty (*Signs*, 166; 210) invokes Husserl's phrase "es wird Leib, es empfindet" in his essay *The Philosopher and His Shadow* to express the peculiar manner in which touching brings one's own flesh to animate life.

intentionally directed to the body's movements as a director might direct an actor). Empathic perception *sees through* the living body, as it were, to the intentional situation, to the matter that is in play. I enter a room and intuitively grasp that an argument is going on or that the relation between the people in the room radiates awkwardness. Similarly, I am attuned to others (even to consociates I do not personally know) when I negotiate public spaces, step around people in the street, stand behind people in a queue (although the acceptable distances or "body space" involved is highly culturally specific). This *attunement* (Husserl speaks of *Paarung*, "pairing") is taking place primarily in the intentional space of sense or meaning, but not necessarily explicitly in the sphere of linguistic meaning; it is incorporated in bodily expression. The smile *reveals* happiness, the grimace pain. As we move to the higher levels of communal and social organization, empathy requires us to deploy our resources as fully engaged, acting, embodied subjects who are responsive to motivations and reason. From this point of view, the communication between subjects takes place at the level of *persons* (as Husserl will emphasize) who relate to one another in an interpersonal way that prioritizes people as free, reason-responsive agents.

6   Intercorporeity and Intersubjectivity

In his many private meditations on intersubjectivity and, of course, in his published *Fourth* and *Fifth Cartesian Meditations*, Husserl deftly tried to articulate the full complexity of the manner in which human subjects negotiate this tricky intertwining of first- and second-person perspectives in the "personalistic attitude." As is well known, Husserl never felt he had resolved precisely the manner that the other persons are given to the subject precisely as other, as another subject with his or her distinctive point of view. Husserl was not able to fully characterize how it is that one lived-body subject experiences another. Merleau-Ponty, despite his many advances over Husserl, also could not deliver a sophisticated account of empathy. Even in his mature reflections in the posthumously published *The Visible and the Invisible*, Merleau-Ponty never develops an account of empathy that could rival Husserl's in its detail and complexity, but he does offer a new name for the embodied subject's intercourse with other embodied subjects when he proposes the name "intercorporeity" or "intercorporeality" (*intercorporéité*). In fact, Husserl himself never uses the equivalent German word for "intercorporeality," *Zwischenleiblichkeit*, but it is undoubtedly the case that he explores the phenomenon without explicitly naming it. By explicitly naming it, Merleau-Ponty foregrounds it as a key element of intersubjectivity. Moreover, for Merleau-Ponty, intercorporeality has many different forms, and, indeed, he even speaks of "interanimality" (*interanimalité*), widening the net of intercorporeal relations beyond the interhuman and including various forms of interspecies encounters (an area of research that is only now being recognized in the philosophy of embodiment and the cognitive sciences). Interanimality is to be found in petting an animal, milking a cow, riding a horse, doing "doga" (yoga with a pet dog as partner), or even in more extreme cases, such as the practice of zoophilia, having sex with animals. For thousands of years, the lifeworld of farmers has involved living in shared relationships with animals.

In relation to his understanding of intercorporeal relations, Husserl is initially, as Merleau-Ponty also acutely recognizes, concerned with the interplay between, and indeed fusion of, the different corporeal senses (primarily sight and touch) in constituting the one, public, spatiotemporal, material world. It is especially the running together of sight and touch that preoccupies Husserl, for instance, in his *Ding und Raum* (*Thing and Space*) lectures of 1907 (1973a, 1997) and thereafter in *Ideas II*. In *Thing and Space*, Husserl asserts that the lived body's modality of occupying space is constituted by sensory experiences being objectivated and combined with sensations of bodily self-movement: "A body is constituted as a sensuous schema by the sense of touch and the sense of sight, and every sense is a sense through an apperceptive conjunction of the corresponding sense-data with kinaesthetic data" (1997, 257; 1973a; *Hua XVI*, 298).

In these lectures, Husserl spends a great deal of time examining how vision itself is constituted from the intertwining of the sensory experiences of each eye, combined with eye and head movements, but then he combines this with the analysis of the role played by touch and self-awareness of the body's location through this sense. Touching, having proprioceptive experiences, and self-movement are all essential to the body's self-constitution. There can be no sensory experience without bodily movement. As Merleau-Ponty will later proclaim: "*Wahrnehmen* and *sich bewegen* are synonymous" (*VI*, 255; 303), and "I am that animal of perceptions and movements called a body" (*Signs*, 167; 211). Husserl emphasizes that the eye cannot see itself, so it is excluded from the circularity involved in the touching relation. Although the eye in one sense "touches" the object it sees (alights on it), the eye itself does not appear as a component in its own vision. Thus in *Ideas II*, §37, Husserl attests that this "double sensation" (*Doppelempfindung*) or "double apprehension" (*Doppelauffassung*) belongs exclusively to touch and is *not* found in vision. He declares: "In the case of an *object constituted purely visually* we have *nothing* comparable" (*Ideas II*, §37, 155; *Hua IV*, 147). On this basis, he concludes that touch is more basic than sight in the body's sensory constitution of spatiality. Merleau-Ponty will take over and expand on Husserl's analysis of the double sensation to claim that all five senses have a distinctive degree of reciprocity, reflectivity, and doubling over. In what seems to be more than pure metaphor for him, he speaks of our vision "palpating"—touching—the visible (*VI*, 131; 171) and of the seer as "incorporated" into the visible in a genuine incarnation. Quoting the experiences of artists in particular (painters and poets), Merleau-Ponty talks about vision as taking place in a world that is not just visible but in its own way looks back at the seer. There is doubling over or recoil of the seen on the seer, just as in the case of touch.

Merleau-Ponty in particular emphasizes not just the body's self-constitution, motility, reflexivity, and incarnation but also the experience of the fleshly subject encountering a world that responds to it also as flesh: "It is already the flesh of things that speaks to us of our own flesh" (*VI*, 193; 243). But Husserl, equally, does not neglect the experience of one person perceiving the other through the mediation of the other's animate body and the combined intersubjective experiences contributing to the experience of being in the one, shared world. These reflections are found primarily in Husserl's *Nachlass*, especially in the three volumes of *Zur Phänomenologie der Intersubjektivität: Texte aus dem Nachlass* (*On the Phenomenology of Intersubjectivity*) (1973). Husserl emphasizes that intercorporeal cooperation is itself the foundation for objectivity: "The body, the living body of the other, is the first intersubjective thing" (1973d; *Hua XIV*, 110, my translation). Similarly, in his published work *Formal and Transcendental Logic*, Husserl talks about the need to explore phenomenologically the extraordinary phenomenon that I experience not only the other person as currently present to me (in empathy) but also have a sense that we share one and the same world:

Moreover it must be made understandable that I necessarily ascribe to someone else (in his lived experiences, his experiences and the rest, which I attribute to him as processes other than mine), not a merely analogous experienced world [*Erfahrungswelt*], but the *same* world [*dieselbe*] that I experience; likewise, that I mean him as experiencing me in the world and, moreover, experiencing me as related to the same experienced world to which he is related; and so forth. (1974, §96, 239, translation modified; *Hua XVII*, 246)

For Husserl, it is an absolute and undeniable fact that each of us experiences ourselves and others as participating in and belonging to the one, *same* world, the world that is "there for everyone" (*für Jedermann daseiend*; 1974, §96, 240; *Hua XVII*, 247), an absolute fact of the "intersubjective knowledge community" (*intersubjektive Erkenntnisgemeinschaft*; 1974; *Hua XVII*, 247). For Husserl, as we have seen, a key to the constitution of objectivity is the relation of one body-subject with another in empathy. It is through empathy and the communion with another that the genuine sense of the objective world in itself is born. As Husserl writes:

Communication creates unity [*Kommunikation schafft Einheit*]. Separated things remain external [to each other], they can be put beside one another and touch, they can never have a common *identity*. Consciousness, however, really coincides with consciousness; consciousness, when it understands another consciousness, is constituted as the same, what is constituted in one; both are one in the same way. (1973d; *Hua XIV*, 198, my translation)

This is a really important insight. The fusion of consciousness with consciousness forms an intentional unity, and the intentional object of these fused consciousnesses equally has a high degree of unity. Thus the "world" (the horizon of significations) as the common object of our conscious experiences is constituted by us as *one and the same*. For Husserl, it simply makes no sense to posit a plurality of worlds; all regions belong to the one overall world, as the "horizon of horizons."

7 Intersubjectivity, Intercorporeality, and the Constitution of World

For Husserl, one of the most important tasks for phenomenology is to chart how the *sense* of an objective world intersubjectively available to all comes to be constituted. This is doubly important, not just to understand our dwelling in the lifeworld but also because it is precisely this shared objective world that becomes purified and formalized in the scientific concept of the "true world" (*die wahre Welt*) or "world in itself" (*Welt an sich*; *Hua VI*, 119), as Husserl elaborates it in his *Crisis of European Sciences*. Everyone must be able to experience *the* object (from his or her perspective) within an already given common world. For each one of us, to be a subject is to occupy a perspective, to have a slant on things, as it were, but it also means that each of us recognizes that our particular perspective or slant is just one of many and the objective world is precisely the object of all our perspectives.[12] Husserl makes both embodiment and experience of the other in empathy necessary conditions for the possibility of intersubjective agreement and validity. Part of this intersubjective communication involves, as we have seen, what Merleau-Ponty calls "intercorporeality." There are many puzzling forms of intercorporeality, but it includes, at the higher level, two people communicating by language with each other or, to use Max Scheler's example, being oriented emotionally (emotions are surely the most embodied of feelings) toward the same shared values. Intercorporeal communication does not have to be merely the intimate, for example, whispering in the ear of a loved one. The martial arts and dance are exemplary forms of intercorporeality.

Medicine has many forms of intercorporeal practice, from various forms of massage to the manipulation of joints. Indeed, intercorporeality takes many forms, from the double body of pregnancy, through the caress, the kiss, the embrace with the loved one, sexual intercourse, and the handshake, to corporal punishment, wrestling, martial arts, and team sports, even singing together, joint chanting, and other forms of intercorporeal blending. Each offers its own unique mode of attunement to the other through the lived body and develops its own universe of signification.

Husserl's discussion of the phenomenon of intercorporeality is limited, but it focuses primarily on the mother-child relation. His analysis of the connections between subjects leads him also to discuss, albeit briefly, the "most primordial genetic continuity" between mother and child, which he refers to, in a fragmentary text from 1927, as being a part of *Ich-Du-Leben* (1973d; *Hua XIV*, 504) and a special kind of "pairing." The mother is oriented to the "child-within," and the child-within (even before birth) is attuned to the voice, mood, and movements of the mother. Husserl's remarks here are regrettably brief. He is interested in exploring different kinds of intentional "fulfillment" (*Erfüllung*) that occur when one person seeks to interpret the other. Husserl continues this reflection in another text from June 22, 1933

12. For a discussion of how modern science elaborated on the idea of both perspective and pure perspectiveless objectivity (the "God's eye"), see Harries 2001.

(1973e; *Hua XV*, 582), where he observes that a mother has already within her the experience of being a child, herself a "child-of-a-mother." In other texts (collected in Husserl 2014a; *Hua XLII*), he talks of the "motherly instinct," which he sees as a primitive but highly complex interrelation. He asks, for instance, how a mother would feel if her child were not well (2014a; *Hua XLII*, 27; see also p. 357).

Husserl's reflections on the mother-child relation are sparse but suggestive of how he might develop the notion of the particular attunement of subjectivities that comes about with the intertwining of bodies between mother and the baby she is carrying. In various texts, Merleau-Ponty discusses the child's relationship with others, but he too has been criticized (see Olkowski 2006) for downplaying and even ignoring the prebirth relations between mother and the child in the womb. In fact, Merleau-Ponty is interested, as in his essay *The Adult's View of the Child*, in Jacques Lacan's notion of the mirror stage, which Merleau-Ponty invokes under the figure of Narcissus. The child recognizing itself in the mirror is performing a form of integration, an overcoming of the splitting of the self as earlier experienced. "For the infant," Merleau-Ponty writes, "the event of the mirror image signifies a certain recuperation of his own body" (2010, 86). Merleau-Ponty likes to invoke the image of Narcissus fascinated by his own image reflected in the pool. As Merleau-Ponty, in his *Course Notes*, puts it: "The flesh is a mirror phenomenon and the mirror is an extension of my relation with my body" (*VI*, 255; 303).

Merleau-Ponty, following Lacan—and Sartre—emphasizes that human relationships combine both love and hatred. He discusses phenomena such as jealousy where there is both identification with the other and also the desire to kill the other. Merleau-Ponty drew heavily on the psychoanalytic and anthropological literature of his day (including Lacan, Malinowski, Lévi-Strauss, and others).

Recently, the British psychologist Colwyn Trevarthen has introduced the concept of "primary intersubjectivity," which he regards as taking place already in the womb when the mother and baby are in symbiotic communication. Trevarthen's work certainly offers a useful supplement to the earlier discussions of Husserl and Merleau-Ponty. According to Trevarthen's observations, intimate reciprocal communications occur between mother and baby in the womb. The mother hums to the child, and at the same time, the child in the womb has been observed to move or wriggle in time to the music (see, e.g., Trevarthen 2011; Nagy 2011; Trevarthen and Aitken 2001). The child in the womb responds already to the mother's voice, to external sounds, to music, and so on. There is the mother's intercorporeal experience of the child kicking in the womb (i.e., there is one subject experiencing another's kinesthetic movements), or there is just the sense of another subject being present, who is listening, who is aware, another person sharing one's body. In early pregnancy, the child is first aware through touch and can be observed (in ultrasound) reaching and touching itself. By twenty-five or twenty-six weeks, the child is moving in the womb and responding to sounds. The baby will gradually show a particular adaption to the rhythm of the mother's language.[13] This whole area of intercorporeal communication needs a great deal of further exploration, but it certainly supports the view that human embodiment is intercorporeal from the outset.

8 Merleau-Ponty on the Phenomenon of Intercorporeality

Let us spend a little longer examining Merleau-Ponty's concept of *intercorporéité*, a term that occurs only a few times and in his late work. In his exploration of intercorporeality, Merleau-Ponty, following Husserl, gives primacy to the experience of one hand touching another—so there is already within the individual an intercorporeal experience—with a self–other dimension and an experienceable "gap" (*écart*) between them. But there are other such experiences—the baby suckling a breast, or the baby sucking its thumb, or even the act of masturbation—that are forms of the *touchant/touché*, each of which has its own essential structure and sense.

In *The Visible and the Invisible*, Merleau-Ponty initially introduces his notion of intercorporeity as part of his larger meditation on the reflexivity of the flesh:

If we can show that the flesh is an ultimate notion, that it is not the union or compound of two substances, but thinkable by itself, if there is a relation of the visible with itself that traverses me and constitutes me as a seer, this circle which I do not form, which forms me, this coiling over of the visible upon the visible, can traverse, animate other bodies as well as my own. And if I was able to understand how this wave arises within me, how the visible which is yonder is simultaneously my landscape, I can understand a fortiori that elsewhere it also closes over upon itself and that there are other landscapes besides my own. If it lets itself be captivated by one of its fragments, the principle of captation is established, the field open for other Narcissus, for an "intercorporeity." (*VI*, 140–141; 183)

In *The Philosopher and His Shadow*, he speaks of "carnal intersubjectivity" as producing the sense of objectivity, but as then being forgotten by this objectivity (*Signs*, 173; 218). Furthermore, this intercorporeal objectivity is obliterated once we move to the notion of what he calls "logical objectivity":

Logical objectivity derives from carnal intersubjectivity on the condition that it has been forgotten as carnal intersubjectivity, and it is carnal intersubjectivity itself which produces this forgetfulness by wending its way toward logical objectivity. Thus the forces of the constitutive field do not move in one direction only; they turn back upon themselves. Intercorporeality goes beyond itself and ends up unconscious of itself as intercorporeality; it displaces and changes the situation it set out from, and the spring of constitution can no more be found in its beginning than in its terminus. (*Signs*, 173; 218)

For Merleau-Ponty, intercorporeity or intercorporeality (*l'intercorporéité*)—the dynamic encounter of one lived body with another—is part of what he calls the "pre-objective world"

---

13. See also the work of Heidelise Als on early experiences of children in the womb, e.g., Als and Butler 2008.

(*l'être préobjectif*), that is, the layer of the world that is first encountered in sensory experience and is the ground and condition for the "objective" world (which is always the outcome of intersubjective constitution) but paradoxically is also reconstituted by the objective world. Merleau-Ponty is inspired by Husserl to think through this notion of the reversibility of the flesh and its relationship to the constitution of a sense of a unique, common, shared world accessible to all. He writes in *The Philosopher and His Shadow*:

The relationship between logical objectivity and carnal intersubjectivity is one of those double-edged relationships of *Fundierung* Husserl spoke about in another connection. Intercorporality culminates in (and is changed into) the advent of *blosse Sachen* without our being able to say that one of the two orders is primary in relation to the other. The preobjective order is not primary, since it is established (and to tell the truth fully begins to exist) only by being fulfilled in the founding of logical objectivity. Yet logical objectivity is not self-sufficient; it is limited to consecrating the labors of the pre-objective layer, existing only as the outcome of the "Logos of the esthetic world" and having value only under its supervision. (*Signs*, 172–173; 218)

What Merleau-Ponty is implying when he says that "intercorporality culminates in (and is changed into) the advent of *blosse Sachen*" is that the experience of things as "mere things" or as "sheer objects" of our natural attitude is a product of intercorporeal constitution. The phrase *blosse Sachen* (mere things) comes from Husserl's *Ideas II*, §51 (*Hua IV*, 190). Husserl is contrasting the scientific manner of treating animals and humans as "mere things" in contrast to the personalistic (moral, legal) perspective that sees animals and human beings as participating in our common community, as "subjects in our common surrounding world" (*Subjekte einer gemeinsamen Umwelt*; *Ideas II*, §51; *Hua IV*, 190). Merleau-Ponty, following Husserl, uses the phrase to mean things understood as entities in physical nature and insists that "we do not live naturally in the world of *blosse Sachen*" (*Signs*, 163; 206). Rather, we live communally in a world of other subjects, a world of "spirit" (*Geist*).

Merleau-Ponty speaks of circles of reversibility and reflexivity, both within the body and between bodies, that widen out into the experience of organic nature. He writes about "an intercorporeal being, a presumptive domain of the visible and the tangible [*un être intercorpo-rel, un domaine présomptif du visible et du tangible*], which extends further than the things I touch and see at present" (*VI*, 143; 186).

Merleau-Ponty is speaking about the communication of the body with itself across the different media of touch and sight, as well as the communication of one body with another. For him, these relations cannot be construed as occurring on the level of consciousness and thought but somehow are corporeal and fleshly: "There is a body of the mind, and a mind of the body and a chiasm between them" (*VI*, 259; 307).

Husserl maintains something very similar. For Husserl, it is important that we begin with the "natural attitude" (*die natürliche Einstellung*), which he always insists is, for us as acting, constituting subject, the world of persons as experienced in the "personalistic attitude" (*die personalistische Einstellung*; see esp. *Ideas II*, §49). It is the specific interaction of persons (embodied subjects freely acting, willing, and deciding, who are also, at the highest level, motivated by reasons) that constitutes the fully concrete lived and shared cultural lifeworld that we inhabit. Husserl speaks about the cultural world as a world interwoven by the intentional activities of persons, and in *The Vienna Lecture* in particular he speaks of the specific durability and timelessness of scientific acquisitions.

Persons bound together [*Miteinander ... verbundende Personen*] in direct mutual understanding [*in aktueller Wechselverständigung*] cannot help experiencing what has been produced by their fellows in similar acts of production as being identically the same [*als identisch das Selbe*] as what they themselves produce. (*Crisis*, 278; *Hua VI*, 323)

For Husserl, moreover, the interpersonal dimension is always set up against the backdrop (horizon) of a shared world: "We could not be persons for others if there were not over against us a common surrounding world [*eine gemeinsame Umwelt*]. The one is constituted together with the other" (*Ideas II*, 387; *Hua IV*, 377). Furthermore, participating in the cultural world means being a subject who can be motivated and have goal-oriented behavior. As Husserl writes in his 1920 research notes *Problem der Apperzeption*: "If I had never even brought to realization object-goals, nor ever even tried something and accomplished it, then I could never understand a work or an implement, etc., not any cultural object" (1973c; *Hua XIII*, 358–359, my translation).[14] Identifying something as a cultural object (a tool, a road sign, a work of art, a piece of speech or writing) requires entering into a world of intentional meaning apprehension and meaning constitution, and this in turn requires the presumption that subjects are persons with teleological, goal-oriented comportment (*Verhalten*) or behavior; Husserl sometimes employs the English word (see *Hua VI*, 251).

Husserl clearly states that cultural objects can be recognized as cultural (as opposed to natural objects) only by taking intentional constituting factors (motivations) into account:

Descriptively, I can describe from the purely psychological point of view even cultural products, such as science, art, etc., namely, analyze them with regard to their conscious motivations, in which they are instantiated as the result of actions. The things that have the form of culture, that previously could be grasped as nature, as natural objects, as objects of physics and psychophysics, are not recognized as the things, they come into play only as intentional objects of consciousness. (1973c; *Hua XIII*, 89–90, my translation)

Given that much of Husserl's project is to investigate the kind of objectivity possessed by science, his analysis of the human or cultural sciences (*Geisteswissenschaften*) is often passed over, even though the role and status of the human sciences occasioned a particularly lively debate in Germany in the work of Dilthey, Windelband, Rickert, and Weber, with whose work Husserl was generally familiar.[15] Cultural objects are intentional objects par excellence

14. "Hätte ich nicht selbst Zweckobjekte verwirklicht, nicht selbst schon etwas versucht und zustande gebracht, so könnte ich kein Werk und kein Werkzeug etc., kein Kulturobjekt verstehen."

15. Indeed, Husserl regularly lectured on the relation between the sciences of nature and those of spirit; see Husserl 2001b.

(although Husserl later argued that constituting things as objects of nature is itself a very special form of cultural constitution).

Husserl does not think all cultural formations are of the same kind. He makes a special exemption of political institutions and the law, which is not merely an artifact in the manner in which other cultural objects are.

Law [*Das Recht*] is not a cultural formation [*Kulturgebilde*], like language, literature, art, that emerges as a mere result of the working together of communicating humans [*des Zusammenwirkens miteinander verkehrender Menschen*], as a "communal accomplishment" [*Gemeinschaftsleistung*], but rather it is a solid communal bond [*ein festes Gemeinschaftsband*], creating unity, in which a unity of willing consciousness [*Einheit des Willensbewusstseins*] is installed, unity of obligations and rights, etc. The state, Plato said, is the man writ large. (1973c; *Hua XIII*, 106, my translation)

Cultural objects are, strictly speaking, "idealities" in Husserl's terminology. They have a unique form of transtemporal existence; that is, they exist as identically the same across time. There is, for Husserl, only one *Mona Lisa*, only one *Hamlet*, and so on. Later Husserl will speak of them as "bound idealities" or "irrealities." They are bound to certain material conditions; for example, the *Mona Lisa* has just that amount of oil paint, canvas, brush strokes, and other material components. He writes in *Experience and Judgment* (published posthumously from Husserl's research manuscripts in 1939):

To be sure, an ideal object like Raphael's Madonna can *in fact* have only one mundane state and in fact is not repeatable in an adequate identity (of the complete ideal content). But *in principle* this ideal is indeed repeatable, as is Goethe's *Faust*. Another example of an irreal objectivity which will lead us to an important distinction in the domain of irrealities is a civil constitution. A state (a nation) is a mundane reality, at once unitary and plural. (1999, §65, 266; 1973b, 319–320)

I do not have the time here to discuss all the complexities that emerge in trying to understand the ideality of a work of art (Husserl's student Roman Ingarden made a significant contribution in this regard). It is obvious that it is not always possible to identify a work of art uniquely (there may have always been several master copies of a film); it is not always the case that there is a unique original or that that unique individual is the sole bearer of artistic value. But the point here is to recognize that intentional constitution is responsible for generating a whole new domain of objectivities on various levels that come to be the objects of culture. Human beings live primarily in this cultural world.

9 Life in Tradition

For Husserl and others of the phenomenological tradition (including Max Scheler and Edith Stein), the empathic encounter is primarily a lived encounter with a copresent living subject. Husserl speaks of "the interconnected caring between I and you" (*die Ineinandergeborgenheit von Ich und Du*; 1973d; *Hua XIV*, 172). In some important but often neglected remarks (some taken up by Gadamer), Husserl elaborates on the nature of communication. It can be reciprocal, as in the case of living subjects communicating with each other, or it can be one-sided, as when subjects in the present read or respond to subjects in the past. This is the nature of historical communication. As Husserl writes:

We have one-way and two-way relations of communication with constitutive function: (1) All unity of historical spirit as historical is a one-sided relation. My life and that of Plato are one. I continue his life-work, the unity of his accomplishments is a partner in the unity of my accomplishments, his striving, his willing, his formations are continued in mine. Scientific knowledge as historical unity is a correlate of a unity of accomplishment. (1973d; *Hua XIV*, 198, my translation)

This is the kind of historical appropriation of other subjects that Husserl develops in his notion of the "poeticizing of history" (*die Dichtung der Geschichte*), discussed in fragmentary supplements to the *Crisis* (394; *Hua VI*, 512). History is experienced (not necessarily consciously) by subjects as a continuity to which they themselves belong. Humans edit out the bits with which they do not identify and highlight the bits they do. Notoriously, for instance, Heidegger emphasized the Greek origins of European thought and neglected the Jewish contribution. Poets, famously, feel themselves influenced only by some that they admire from the past. Taking as his example the kind of dynamic interaction between subjects that is found in the history of philosophy, Husserl explains how dead philosophers of the past are joined with living thinkers of the present as a single "community of philosophers" (*Philosophengemeinschaft*; *Hua VI*, 444), or "community of thinkers" (*Denkergemeinschaft*; *Hua VI*, 444). Philosophical problems and their discussion are kept alive from generation to generation by the activity of thinkers reflecting on earlier thinkers, in a highly particular one-sided "generativity" that is typical of the manner cultural tradition evolves, is sedimented, forgotten, revived, and transformed (*Crisis*; *Hua VI*, 444). The dynamics of constituting what has to be experienced as "belongingness to a tradition" is something Husserl only briefly interrogates, but this adds greatly to his mature understanding of human beings as being-in-the-world.

The experience of the social world, then, inevitably has a transgenerational and historical component. The world has to be experienced as having come to be as it is from the labors of anonymous others: the city was built by others, others travel the roads, and so on. Tradition, moreover, requires the sense of a *shared* present and a shared past. For Husserl, this cultural interaction with others requires a material or corporeal basis. Human beings are being-with-others-in-the-common-world through being embodied.

One interesting point is that Husserl often places great emphasis on the body's activity and motility (self-movement) and its sense of governing in its own sphere; freely engaging in movement seems to be a major characteristic of how the body inhabits the world. But Husserl also emphasizes the *passivity* of experience, a point that Merleau-Ponty picks up and develops in his *Passivity Lectures* (*L'institution et la passivité*, 2003). I am located firmly in my body and thereby in the room. It is this passivity of the body that in some respect guarantees the stability and hence objectivity of the surrounding world. As Husserl writes in *Leib-Ding-Einfühlung*: "In any case, embodiment is the condition of possibility of a passivity in the subject, through which an intersubjective world is passively constituted and can be actively controlled" (1973d; *Hua XIV*, 73, my translation).[16] It is precisely because the body is in a sense passively inserted in the world that the body has the capacity for activity. One can only step if one is pushing off from the stable situation of standing firmly on the ground. Husserl's conception of the complex interweaving of passivity and activity is often seen as related to the synthesis of perception and judgment, but it applies much more deeply in the experience of embodiment itself. The world has a stability, materiality, and "thereness" precisely because of my embodied experience. Husserl speaks somewhat confusingly of "passive synthesis," implying that an activity of constitution is going on even to produce the experience of passivity. It is better perhaps to think of passive synthesis not as implying "synthesis" but as giving the sense of stability and permanence. In this sense, although our bodies age and change, we have the strong sense that they anchor us in the world of stable things.

As I have been emphasizing, for Husserl the constitution of the social world is deeply connected with the constitution of the body and the constitution of others in empathy. How precisely the sense of the common world (both the social world of persons and the objective world of scientific "mere things") is constituted is an enormous and extremely complex challenge for Husserl. Husserl's conception of the constitution of sociality through intersubjective intercorporeality surely deserves further exploration.

References

Als, H., and S. Butler. 2008. Die Pflege des Neugeborenen: Die frühe Gehirnentwicklung und die Bedeutung von frühen Erfahrungen. In *Der Säugling: Bindung, Neurobiologie und Gene; Grundlagen für Prävention, Beratung und Therapie*, ed. K. H. Brisch and T. Hellbrügge, 44–87. Klett-Cotta.

Banerjee, R., and J. Herbert, eds. 2011. The intersubjective newborn. Special issue, *Infant and Child Development* 20 (1): 1–2.

Behnke, E. 2010. Edmund Husserl's contribution to phenomenology of the body in *Ideas II*. In *Issues in Husserl's Ideas II: Contributions to Phenomenology*, ed. T. Nenon and L. Embree, 135–160. Springer.

Bernet, R. 2013. The body as a "legitimate naturalization of consciousness." *Royal Institute of Philosophy* 72 (supplement): 43–65.

Grätzel, S. 1989. *Die philosophische Entdeckung des Leibes*. F. Steiner.

Harries, K. 2001. *Infinity and Perspective*. MIT Press.

Heinämaa, S. 2011. Embodiment and expressivity in Husserl's phenomenology: From *Logical Investigations* to *Cartesian Meditations*. *SATS: Northern European Journal of Philosophy* 11 (1): 1–15.

16. "Jedenfalls, Leiblichkeit ist Bedingung der Möglichkeit einer Passivität im Subjekte, durch die sich eine intersubjektive Welt passiv konstituieren und sich aktiv beherrschen lassen kann."

Husserl, E. 1900–1901. *Logische Untersuchungen*. 2 vols. Verlag von Viet.

Husserl, E. 1950. *Cartesianische Meditationen und Pariser Vorträge. Husserliana I*. Ed. S. Strasser. Nijhoff. Abbreviated in the text as *Hua I*.

Husserl, E. 1952. *Ideen zu einer reinen Phänomenologie und phänomenologischen Philosophie*, vol. 2: *Phänomenologische Untersuchungen zur Konstitution. Husserliana IV*. Ed. M. Biemel. Nijhoff. Abbreviated in the text as *Hua IV*.

Husserl, E. 1962. *Die Krisis der europäischen Wissenschaften und die transzendentale Phänomenologie: Eine Einleitung in die phänomenologische Philosophie. Husserliana VI*. Ed. W. Biemel. Nijhoff. Abbreviated in the text as *Hua VI*.

Husserl, E. 1967. *Cartesian Meditations*. Trans. D. Cairns. Nijhoff. Abbreviated in the text as *CM*.

Husserl, E. 1968. *Phänomenologische Psychologie: Vorlesungen Sommersemester 1925. Husserliana IX*. Ed. W. Biemel. Nijhoff. Abbreviated in the text as *Hua IX*.

Husserl, E. 1969. *Formale und transzendentale Logik: Versuch einer Kritik der logischen Vernunft; Mit ergänzenden Texten. Husserliana XVII*. Ed. P. Janssen. Nijhoff. Abbreviated in the text as *Hua XVII*.

Husserl, E. 1970. *The Crisis of European Sciences and Transcendental Phenomenology: An Introduction to Phenomenological Philosophy*. Trans. D. Carr. Northwestern University Press. Abbreviated in the text as *Crisis*.

Husserl, E. 1973a. *Ding und Raum: Vorlesungen 1907. Husserliana XVI*. Ed. U. Claesges. Nijhoff. Abbreviated in the text as *Hua XVI*.

Husserl, E. 1973b. *Experience and Judgement: Investigations in a Genealogy of Logic*. Trans. J. S. Churchill and K. Ameriks. Revised and edited by L. Landgrebe. Routledge & Kegan Paul.

Husserl, E. 1973c. *Zur Phänomenologie der Intersubjektivität: Texte aus dem Nachlass; Erster Teil, 1905–1920. Husserliana XIII*. Ed. I. Kern. Nijhoff. Abbreviated in the text as *Hua XIII*.

Husserl, E. 1973d. *Zur Phänomenologie der Intersubjektivität: Texte aus dem Nachlass; Zweiter Teil, 1921–1928. Husserliana XIV*. Ed. I. Kern. Nijhoff. Abbreviated in the text as *Hua XIV*.

Husserl, E. 1973e. *Zur Phänomenologie der Intersubjektivität: Texte aus dem Nachlass; Dritter Teil, 1929–1935. Husserliana XV*. Ed. I. Kern. Nijhoff. Abbreviated in the text as *Hua XV*.

Husserl, E. 1974. *Formal and Transcendental Logic*. Trans. D. Cairns. Nijhoff.

Husserl, E. 1977a. *Ideen zu einer reinen Phänomenologie und phänomenologischen Philosophie*, vol. 1: *Allgemeine Einführung in die reine Phänomenologie. Husserliana III/1*. Ed. K. Schuhmann. Nijhoff. Abbreviated in the text as *Hua III/1*.

Husserl, E. 1977b. *Phenomenological Psychology: Lectures, Summer Semester 1925*. Trans. J. Scanlon. Nijhoff.

Husserl, E. 1989. *Ideas pertaining to a Pure Phenomenology and to a Phenomenological Philosophy, Second Book*. Husserl Collected Works III. Trans. R. Rojcewicz and A. Schuwe. Kluwer. Abbreviated in the text as *Ideas II*.

Husserl, E. 1992. Verschiedene Formen der Historizität. In *Die Krisis der europäischen Wissenschaften und die transzendentale Phänomenologie: Ergänzungsband; Texte aus dem Nachlaß, 1934–1937, Husserliana XXIX*, ed. R. N. Smid, 37–46. Springer. Abbreviated in the text as *Hua XXIX*.

Husserl, E. 1997. *Thing and Space: Lectures of 1907*. Husserl Collected Works VII. Trans. R. Rojcewicz. Kluwer.

Husserl, E. 1999. *Erfahrung und Urteil: Untersuchungen zur Genealogie der Logik*. Ed. Ludwig Landgrebe. 7th ed. Felix Meiner.

Husserl, E. 2001a. *Logical Investigations*. 2 vols. Trans. J. N. Findlay. Ed. D. Moran. Routledge.

Husserl, E. 2001b. *Natur und Geist: Vorlesungen Sommersemester 1927. Husserliana XXXII*. Ed. M. Weiler. Kluwer. Abbreviated in the text as *Hua XXXII*.

Husserl, E. 2014a. *Grenzprobleme der Phänomenologie: Analysen des Unbewusstseins und der Instinkte; Metaphysik; Späte Ethik. Husserliana XLII*. Ed. R. Sowa and T. Vongehr. Springer. Abbreviated in the text as *Hua XLII*.

Husserl, E. 2014b. *Ideas for a Pure Phenomenology and Phenomenological Philosophy. First Book: General Introduction to Pure Phenomenology*. D. O. Hackett Publishing. Abbreviated in the text as *Ideas I*.

Merleau-Ponty, M. 1945. *Phénoménologie de la perception*. Gallimard.

Merleau-Ponty, M. 1960. *Signes*. Gallimard.

Merleau-Ponty, M. 1964a. *Le visible et l'invisible*. Ed. C. Lefort. Gallimard.

Merleau-Ponty, M. 1964b. *Signs*. Trans. R. C. McCleary. Northwestern University Press. Abbreviated in the text as *Signs*.

Merleau-Ponty, M. 1968. *The Visible and the Invisible*. Trans. A. Lingis. Northwestern University Press. Abbreviated in the text as *VI*.

Merleau-Ponty, M. 2003. *L'institution et la passivité: Notes de cours au Collège de France (1954–1955)*. Ed. D. Darmaillacq, C. Lefort, and S. Ménasé. Preface by C. Lefort. Éditions Belin.

Merleau-Ponty, M. 2010. *Child Psychology and Pedagogy: The Sorbonne Lectures, 1949–1952*. Trans. T. Welsh. Northwestern University Press.

Moran, D. 2004. The problem of empathy: Lipps, Scheler, Husserl and Stein. In *Amor Amicitiae: On the Love That Is Friendship: Essays in Medieval Thought and Beyond in Honor of the Rev. Professor James McEvoy*, ed. T. A. Kelly and P. W. Rosemann, 269–312. Peeters.

Moran, D. 2011. Sartre's treatment of the body in *Being and Nothingness*: The "double sensation." In *Jean-Paul Sartre: Mind and Body, Word and Deed; A Collection of Essays*, ed. J.-P. Boulé and B. O'Donohue, 9–26. Cambridge Scholars Press.

Moran, D. 2014. The phenomenology of embodiment: Intertwining (*Verflechtung*) and reflexivity. In *The Phenomenology of Embodied Subjectivity*, ed. D. Moran and R. T. Jensen, 285–303. Contributions to Phenomenology 71. Springer.

Moran, D. 2015. Phenomenologies of vision and touch: Between Husserl and Merleau-Ponty. In *Carnal Hermeneutics*, ed. R. Kearney and B. Treanor, 214–234. Fordham University Press.

Nagy, E. 2011. The newborn infant: A missing stage in developmental psychology. *Infant and Child Development* 20 (1): 3–19.

Olkowski, D. 2006. Only nature is mother to the child. In *Feminist Interpretations of Merleau-Ponty*, ed. D. Olkowski and G. Weiss, 49–70. Penn State University Press.

Sartre, J.-P. 1943. *L'être et le néant: Essai d'ontologie phénoménologique*. Gallimard.

Sartre, J.-P. 1995. *Being and Nothingness: An Essay on Phenomenological Ontology*. Trans. H. Barnes. Routledge.

Schütz, Alfred. 1967. *The Phenomenology of the Social World*. Trans. G. Walsh and F. Lehnert. Northwestern University Press.

Shapiro, L., ed. 2014. *The Routledge Handbook of Embodied Cognition*. Routledge.

Stein, E. 2000. *Philosophy of Psychology and the Humanities*. ICS Publications.

Taipale, J. 2014. *Phenomenology and Embodiment: Husserl and the Constitution of Subjectivity*. Northwestern University Press.

Thompson, E. 2010. *Mind in Life: Biology, Phenomenology, and the Sciences of Mind*. Harvard University Press.

Trevarthen, C. 2011. What is it like to be a person who knows nothing? Defining the active intersubjective mind of a newborn human being. *Infant and Child Development* 20 (1): 119–135.

Trevarthen, C., and K. J. Aitken. 2001. Infant intersubjectivity: Research, theory, and clinical applications. *Journal of Child Psychology and Psychiatry, and Allied Disciplines* 42 (1): 3–48.

Varela, F. J., E. Thompson, and E. Rosch, eds. 1999. *The Embodied Mind: Cognitive Science and Human Experience*. MIT Press.

Waldenfels, B. 2000. *Das leibliche Selbst: Vorlesungen zur Phänomenologie des Leibes*. 3rd ed. Suhrkamp.

Walther, G. 1923. *Zur Phänomenologie der Mystik*. Niemeyer.

Zahavi, D. 2014. *Self and Other: Exploring Subjectivity, Empathy, and Shame*. Oxford University Press.

2   We Are, Therefore I Am—I Am, Therefore We Are: The Third in Sartre's Social Ontology

Nicolas de Warren

The aim of this chapter is to incite you to read an unreadable book. Jean-Paul Sartre's *Critique of Dialectical Reason* ([1960] 2004) unquestionably ranks as one of the most unreadable books of twentieth-century philosophy, matched only, if not exceeded by its parallel project, the equally unfinished Flaubert study *The Family Idiot*. The *Critique*, in spite of being the crowning masterpiece of Sartre's political thinking, fell stillborn from the press, largely due to its unforgiving attitude toward its readers. It is an imposing work that from its opening pages does not seem to want to be read. Such indifference toward any readership is commensurate with the peculiar character of Sartre's thinking, which arguably attains in the *Critique* its most consummate form in fusing together unbridled self-absorption (a thinking entirely given to itself) with unsurpassed generosity (a thinking that gives itself entirely for thought). This generosity reflects that indefinable quality of Sartre's thinking as a pedagogy of philosophical creativity. In Gilles Deleuze's eloquent homage, Sartre was perhaps the last great "teacher" (*maître*: untranslatable into any single English equivalent) of philosophy in the twentieth century, who not so much instructs *what* to think, but more fruitfully displays *how* to think with an audacity and verve rarely equaled since (Deleuze 2002). The apparent contradiction in Sartre's philosophical writing of how an unreadable book could at all teach us anything is only resolved as a productive *tension* within a thinking that tirelessly remained uncompromising toward the gifts of its own genius.

As I propose here, the originality of the *Critique* consists in a sophisticated treatment of central problems in sociology and social ontology.[1] The *Critique* offers a social ontology that challenges entrenched views regarding how to understand the constitutive interaction between the individual (the "I") and the collective (the "we"). How are social groups and collective actions constituted? What is the relationship between individual and collective

---

1. Although Sartre's *Critique of Dialectical Reason* has received attention as a work of political philosophy, it has not—to my knowledge—yet been appreciated as a sophisticated social ontology. For an exegetical presentation of Sartre's work, see Catalano 1986. For an analysis of Sartre's unfinished vol. 2, see Cambria 2009. For an interpretation of Sartre's *Critique* that places its Marxism against the background of his earlier existentialist ethics, see Flynn 1984.

agency? Sartre's innovative approach to this cluster of problems is based (in part) on the introduction of a novel concept: the third (*le tiers*).[2] As with other Sartrean creations, the concept of the third operates in both a critical and a productive register. Sartre conceived of the social ontology in the *Critique* as advancing a critique of sociology and, more generally, targets an assumption that arguably continues to shape numerous philosophical approaches to social ontology today. As he observes: "It is a common error of many sociologists to treat the group as a binary relation (individual–community), whereas, in reality, it is a ternary relation" (Sartre [1960] 2004, 374). If the question of how to understand the constitution of collective agency and action turns on understanding the relationship between an individual subject (a "first-person" structure of intentionality) and a collective subject (a "we-intentionality"), Sartre proposes to question any strict opposition between first-person singular and first-person plural. In thinking of intersubjectivity beyond the stricture of binary terms, the concept of the third underpins a compelling account of how social relations are forged in collective action, or *praxis*, through which social agency is constituted on an individual *and* collective plane. With this critique, Sartre rejects a variety of sociological approaches to the constitution of groups: the theory of social contract popular within the French Republican tradition; theories of moral sentiment (natural sympathy for social existence); and Comte's positivistic notion of society. Sartre is, however, especially critical of Durkheim's conception of collective consciousness and its lasting imprint on sociological thought. As Sartre remarks in his 1961 Rome Conference address:

Ce qui nous offre la possibilité de comprendre en quoi la subjectivité est indispenable pour la connaissance dialectique du social. C'est parce qu'il n'ya que des *hommes*, qu'il n'y a pas de grandes formes collectives, comme Durkheim et d'autres l'ont imaginé, et que ces hommes sont obligés d'être la médiation entre eux de ces formes d'extériorité qu'est, par exemple, l'être de classe. (Sartre 1993, 35)

Whether one begins from the "I" to arrive at the third person or begins from the "we" to arrive at the first person, both approaches—in Sartre's account—fail to realize what is essential about human *freedom*, namely, that the freedom of an individual can only be genuinely realized in praxis (i.e., action) in concert with other self-realizing individuals. Sartre's guiding intuition here is that individual human freedom *requires* a form of collective existence, forged in reciprocal praxis, for its genuine realization; but, likewise, the realization of genuine collective agency *must* in turn be founded on the praxis of individual agents.[3] In this manner, Sartre argues that there are two fundamental ways of speaking about agency: in the third-person singular ("the group acts") and in the first-person singular ("I am acting in concert with a group"). As he declares: "We must know how to say 'we' in order to say 'I'—that is

2. For an analysis of the centrality of the third within the project of the *Critique of Dialectical Reason*, see Flynn 1997.

3. For the important relation between Sartre and Durkheim, see Cormann 2000, 77–110.

beyond question. But the opposite is also true" (Sartre 1989, 33). My aim in this paper is to probe the meaning and significance of this fundamental insight.

1   Enter the Third

My argument centers on the claim that the philosophical originality of Sartre's concept of the third consists in a twofold thesis. Sartre's rejection of treating social groups in binary terms critically hinges on his philosophical insight that individual human freedom can only become genuinely realized in collective praxis. Only under specific (and historical) forms of social existence can human freedom be *fully* realized (a claim already proposed in Hegel's *Philosophy of Right*). The realization of individual freedom is inseparable from *reciprocity* with other individuals in freedom. As Sartre insists, however, the realization of genuine collective agency (i.e., genuine reciprocity) is founded on the praxis of individuals. As I shall suggest, this complex relation between collective praxis and individual praxis—where the former conditions the realization of the latter while the latter founds the realization of the former— can best be understood in terms of the dual function of participation and solidarity for genuine collective agency in which an individual's freedom *to act* is realized.

Sartre's principal insight into the social realization of *individual* freedom in collective praxis has as its converse the claim that social forms of coexistence that inhibit or prevent the realization of individual action in concert with others represent the reification, or alienation, of human freedom. The central problem of social ontology is thus twofold: to understand how collective agency is constituted on the basis of the reciprocal praxis of individuals, *as well as* how individual agency can become reified or alienated in collective forms of existence; or, in other words, how social agents come to forfeit their individual freedom in collective forms of existence. In such cases, as Sartre brilliantly demonstrates, material objects play a critical role of mediation. The paradox of the social world is that it is by the very means of its human fabrication that human agency becomes itself displaced. The world of our making is equally the world of our unmaking. In this account, the locus of social agency becomes centered within material objects that organize and shape fields of possible interaction. The materiality of social relations and, more specifically, the materiality of tools and technology effectively function as social agents, since such material objects mediate and determine social interactions among individuals.

As I shall explore, the concept of the third allows for the perspicuous description of how material objects attain the status of what we might call "social quasi agency." It is not just that such objects *extend* and *enhance* human agency (an established view among Russian sociologists such as Vygotsky and French anthropologists such as Leroi-Gourhan) but more emphatically (and more problematically) that such material objects—necessary for any social world—can also usurp and displace genuine human agency. Sartre's social ontology is in this manner committed to the view that human praxis is always mediated by the praxis of other social agents, as well as always situated within a material field of action (i.e., a field of tools, natural resources, everyday objects, technology, etc.). Over and above this stress on the material conditions of social existence, this twofold thesis articulates more broadly Sartre's essential problem in the *Critique*: human beings make history, yet history makes human beings.[4]

Sartre's concept of the third not only allows for a consideration of the constitutive function of materiality (tools, technologies, institutions) for collective existence. It also allows for an understanding of how *ideas* come to function as organizing structures of social agency. The central insight here is that ideas can mediate social agency in the specific sense that, as Dostoyevsky made into the theme of *Demons*, ideas shape an individual's orientation toward the world and field of social action. As vividly portrayed in Dostoyevsky's novel, "the demons" represent various ideas and ideologies (in the context of *Demons*: atheism, liberalism, Westernization, etc.) from which individuals are *born* as social agents. In the apt words of one of the novel's characters, Verkhovensky (addressed to Kirillov): "It was not you who ate the idea, but the idea that ate you." This emphasis on the social function of ideas marks the way in which Sartre incorporates a critique of ideology into the social ontology of the *Critique*. There is indeed a sense for Sartre's thinking in which the most urgent problem for social ontology is not so much how to understand collective agency and action as how to understand the forms of social existence in which we neither truly act nor think. Although I cannot pursue this angle further, Sartre's social ontology must ultimately be grasped in its *political* significance. The point of philosophy is not just to understand social reality but to envision how to change social reality, given that most of our social forms of existence are predicated on inaction and unthinking.

2 Scenes from Everyday Life

For evident reasons, I cannot hazard a summary of Sartre's massive *Critique of Dialectical Reason*. Nor would it be feasible to present an overview of the elaborate methodology (the regressive-progressive method of dialectic reason) that Sartre employs in the *Critique*—a discussion further complicated by the relationship between *Search for Method* (ostensibly, the methodological prolegomena to the *Critique*) and the *Critique*. Equally impractical would be to begin with an inventory of Sartre's distinctive vocabulary (*practico-inert*, *totalization*, etc.), without which one cannot understand the *Critique*, but an understanding of which is nearly impossible without having first worked through the *Critique* as a whole.

Since I understand my present purpose as principally didactic, that is, as motivating the reading of an unreadable book, my aim cannot be to discuss the conceptual universe of the

4. The material conditions of human praxis include nature, but contrary to the dialectical materialism of Engels, Stalin, and Tran Duc Thao, Sartre considers any dialectical structuring of material nature *independent* of human mediation, or significance, to be philosophically untenable. Nature is always mediated through human significance, much as human praxis is always mediated by materiality.

Critique or even the concept of the third with any measure of adequacy. To understand any given concept concretely (and not merely abstractly or as mere definition) would demand that one follow how a given concept *behaves* over the course of its employment in the *Critique*. Concepts are in this sense *dialectical*: the work of a concept (i.e., its production of intelligibility) is inseparable from how a concept operates within a universe of other concepts in motion.

This dialectical significance of concepts relates directly to another important feature of Sartre's thinking. Sartre's genius was very much the genius of examples. Examples, however, are not images that arrest the movement of a concept but anchoring points for a concept's dialectical unfolding. By the same token, examples in Sartre's thinking are not "instances" or "tokens" for general or abstract notions but rather *exemplifications* of complexity. A Sartrean example renders perspicuous a degree of conceptual complexity through an intuitive portrayal that is neither flat nor lifeless. All of Sartre's philosophical examples, whether in the *Critique* or in other writings, are more akin to manners of staging a philosophical concept, and it is by virtue of such a staging of concepts that he perfectly enacts a thinking that is at once theoretical and theatrical.

This distinctive employment of philosophical examples lends itself to a heuristic solution to the apparently insoluble challenge of how to discuss an unreadable book. Given the purpose of this paper, I organize my discussion around three examples taken from the *Critique*: "waiting for the bus," "the pledge," and "goalkeeper on a soccer team." Each example allows me to enter in medias res into the thick of Sartre's concept of the third without forfeiting its complexity while at same time economizing on its deployment. In this manner, by tracking different variations of the concept of the third across these three examples, I would consider myself successful in motivating the reading of an unreadable book if the curiosity piqued in the reader at the paper's beginning becomes by the end a full-fledged interest in Sartre's social ontology.

3   Waiting for the Bus

Let me begin with a truism, the significance of which remains surprisingly obscure in much of the contemporary debate surrounding questions of social ontology. If one takes into account the rich empirical material from anthropology and sociology, and especially if one encompasses non-Western cultures, it is clear that human social groups exist—and have existed historically—in immensely different forms. The diversity of human social life would seem to preclude the feasibility of any robust social ontology or else reveal most social ontologies as severely limited by unstated assumptions regarding any claims to have illuminated the presumptive elementary forms of social existence. The question of how theoretically to countenance the diversity of social life while striving for conceptual intelligibility is not easily resolved. Sartre's own effort in the *Critique* is explicitly conditioned by the historical development of modern Western society, yet methodologically mindful that "only experience can indicate the internal relation of a definite group and as a definite moment of its interior dialectic" (Sartre [1960] 2004, 254). For my immediate concern, what this acknowledgment signals is that Sartre's own elementary concepts must be understood *dialectically*, that is, as structures of intelligibility that must be grasped across different configurations of social forms of existence. Ideally, the immense diversity of social life would have to be described *concretely* to understand how in each instance the "internal relations of the group" are constituted. And yet, since it would defeat the purpose of developing a theory of social ontology if one *had* to undertake an exhaustive classification of social forms of existence, elementary concepts within a social ontology must be both flexible and rigorous. They must allow for varieties of descriptions *within* the intelligibility of a form or structure. They must, in other words, incorporate dialectical intelligibility with a phenomenological return to the things themselves.

Such is the case for one of Sartre's more elementary concepts: *serial collectives* or *seriality*. As with other elementary forms of social existence, the concept of seriality is anchored in the *Critique* with a set of examples, the simplicity of which belies the complexity and variability of the form under description. According to Sartre, this form of social existence pervades modern societies and, indeed, represents the kind of social existence that populates and determines our most banal, everyday interaction with others. In the simplest form, serial collectives are exemplified in the common experience of waiting in lines, whether when caught in traffic, moving on an escalator, or boarding a plane. Sartre's own example in the *Critique* is a series of individuals waiting for a bus. This apparently inconspicuous example encapsulates the principal characteristics of a serial collective: a collection of individuals who act together without acting in concert with and concern for each other, and whose collective agency is thus "atomized" without any genuine reciprocity and investment in a common end. Serial collectives range from insubstantial, fleeting forms of social interaction circumscribed to the particular time and place of encounter (i.e., waiting for a bus) to politically more charged forms of social organization in which each individual is permanently incorporated into a seamless totality.

In Sartre's stylized description, each person waiting for the bus does not speak with others in the line; each remains isolated within his or her own singular purpose (to go home, to go to the store, etc.) in ignorance of the aims of the others. Yet insofar as each person is waiting for the *same* bus, each acts in relation to the others, and this reciprocity of action is structured according to a set of social norms and conventions: giving right of way to the elderly, not cutting in line, having a valid ticket of transportation, and so on. In Sartre's vocabulary, serial collectives are structured forms of social isolation: each individual is isolated from the others, and yet each isolated individual nonetheless acts in reciprocity with others, albeit "negatively"—without the internalization of a common objective to which individual actions strive. The stress here is on the meaning of "internalization." Collective agency and common objectives (e.g., not allowing a certain person to cut in line) structure seriality; however, the coordination of such actions is mediated through a reciprocity that is not yet genuinely internalized. When the bus arrives, each person shuffles forward in step with the others; each person animates his or her own body in relation to the moving bodies of others with the consequence that bodies are set in motion in a serialized manner. The mundane action of shuffling one's feet forward to board the bus highlights the *embodied* manifestation of action in relation to other agents. As the line advances, the bodily movements of each individual are determined by an individual intention (to get home, etc.), but the bodies move collectively as determined by an interest of getting on the bus. This reciprocal action represents the integration of social norms into the body, as well as the mediation of recipro-cal action by *material objects* (in this example: the bus). The insight here is double: the human body is constituted through its incorporation of social norms, and the actions of the human body are mediated by material objects. Social relations are thus materialized in the body as the intersection of norms and technology.

In serial collectives, individuals relate to each other "in exteriority": individuals act in reciprocity with each other without acting in concert with and concern for the other. This does not mean, however, that individuals do not freely choose to act according to the social conventions regulating proper bus-waiting and bus-taking behavior. It does mean, however, that the exercise of voluntary decisions is not *mediated* through collective realization of everyone's freedom (in the group). Even though—as with the example of the bus—individu-als are united by a shared interest, and hence determined by the third, namely, what "we" want to do, individuals have not "interiorized" reciprocity: each still acts for the sake of a private interest under which the shared aim is subsumed. Moreover, as Sartre stresses with evident political implication, each member in a serial collective is in principal *interchange-able* and hence, in this sense, only *quantitatively* distinct. Expressed in Sartre's abstruse ter-minology: "In the series, everyone becomes himself (as Other than self) in so far as he is other than the Others, and so, in so far as the Others are other than him" (Sartre [1960] 2004, 262). When standing in a line, each individual is other than his or her neighbor, yet to the extent that each individual shares the goal of getting on the bus, each individual is the *same* as the other: acting collectively in unison. Indeed, if an individual decides to "cut in line" or prevent the passage of a pregnant women, other individuals within the queue speak up in unison and hence speak *for the third*. The third is here spoken for through an individual: "One doesn't act that way," "It is not proper to do so," "The rules state that pregnant women have priority."

In Sartre's handling of this example, it is the bus that organizes the "practico-inert ensem-ble" (the field of possible actions). Individuals move in reciprocity with regard to each other *in view of the bus*, or, in other words, the actions of individuals are collective to the extent that such actions are mediated by the bus (i.e., you get on, then I get on; I give my seat to the old lady; I move my hands so as not to brush up against your shoulder, etc.). It is the material object (the bus) that *effectively* provides the *locus* of social agency. The bus is here a quasi-social agent to the extent that, as Sartre writes, it is a "material object that realizes the unity of interpenetration of individuals as beings-in-the-world-outside-themselves to the extent that it structures their relations as practical organisms" (Sartre [1960] 2004, 256). The *absence* of a genuine common end among individuals allows by default the unity of the serial collective (as *one* action, i.e., the line advancing) to be placed, as it were, in the hands of the material object, the bus, that structures and organizes the discrete actions of the collective. The bus allows for ("mediates") the constitution of the third, that is, the *we* who get on the bus.

It is in virtue of waiting for the bus that the actions of individuals are related to each other in a reciprocal manner. The bus is a social object: it is a material object produced by human intention and for human purpose. Moreover, the bus materializes a set of social relations among human beings: how the seats are arranged, how the bus operates within a system of public transportation, and so on. The world of human artifice is a world made for human doings, yet the paradox is that it is precisely such a ready-made world that unhinges human agency from its own locus. In the example of the bus, human agency is mediated through material objects: we act with regard to each other in patterns determined by the material objects that facilitate and structure our social interaction. In turn, the bus and its function within a system of public transportation are mediated (i.e., embedded) by a larger (and more complex) network of social relations and their material realizations. The significance of the modern city for social forms of life in modern mass societies and their overwhelming serial forms of existence operates tacitly in the background of Sartre's thinking. As he writes:

And, through the medium of the city, there are given the millions of people who are the city, and whose completely invisible presence makes of everyone *both* a polyvalent isolation (with millions of facets) *and* an *integrated* member of the city. (Sartre [1960] 2004, 257)

Sartre's emphasis on the modern city further exemplifies how the material organization of social space *determines* social (collective) agency. In the nineteenth century, Baron Haussmann's reconstruction of Paris with its grand boulevards had the clear political purpose of establishing channels for the rapid deployment of troops while also ensuring a channelization of potential unrest and revolt into isolated, easily surrounded centers.

Two further examples—not from Sartre's *Critique*—may serve to further illustrate the diverse phenomena of serial collectives. In Jacques Tati's *Playtime*, the iconic Monsieur Hulot epitomizes Sartre's dual insight into embodiment and technology in modern social life. Collective seriality is on full display throughout the film: cars inching forward in traffic jams, passengers sitting on a bus, the bustle of crowded streets, and so on. Although Tati's vision of modern city life and its alienating architecture contrasts with a nostalgic vision of French village life (colorful, spontaneous, and joyous) that is absent from Sartre's own thinking, both Tati and Sartre give expression to the ways in which the material organization of modern cities and our dependency on modern technology *organize* forms of collective agency, such that the veritable "subject" of social agency is the anonymous functioning of a collective rather than individual human agents or the genuine social agency of a group.

In both Tati and Sartre, the robust connection between the embodiment of social norms and the mediation of social reciprocity through material objects is understood through the basic metaphor of the "mechanization" of social life and—with brilliant comic effect in Monsieur Hulot—the body.[5] With evident Bergsonian accents, the *physicality* of Tati's burlesque comedy turns on the multiplicity of ways in which the *lived* body becomes imbued within collectivized movements not of its own (individual) initiation; it is only through such a mechanization (serialization) of the body that serial collective agency (waiting in line for the bus, going up an escalator) is constituted.[6] Monsieur Hulot's sympathetic, carefree, and light-footed gait is his recognizable signature; his fumbling and general physical awkwardness with modern gadgets and bureaucratic procedures represent in turn the effacement of his singularity (in fact: his identity is his idiosyncrasy). The raucous explosion of dance in the brilliant concluding chaos of the restaurant in Tati's movie marks the final counterpoint of spontaneous collective *freedom* against the drabness and inertness of serial collective existence.

Beyond Sartre's mundane example of waiting for a bus resides a more profound critique of modern capitalist society, as well as modern totalitarianism. Serial collective existence corresponds to what the German cultural writer Siegfried Kracauer understood as "the mass ornament," as exemplified in the dance spectacle of the Tiller Girls.[7] This spectacle of dancers moving (kicking legs, etc.) in strict synchronization constructs a collective agency of formal rationalization. It is the *form* of such a collective display of technique and precision that appeals to the eye of the beholder. The audience is enchanted by the human fabrication of an ornament. In a Kantian vein, the autonomous form of aesthetic beauty has been displaced by an ornamental advertisement of charm. The latter gives the *appearance* of living movement while in fact reducing lived bodies to inert objects arrayed in a serialized motion indistinguishable from the motion of material objects. In Sartre's terminology, the dance line of the Tiller Girls is an instance of the practico-inert. As Kracauer argues:

5. "Jacques Tati: *Playtime* Clip," YouTube, https://www.youtube.com/watch?v=Wj6l5kEzPhw.

6. It is important not to misunderstand the point made here. This description of "mechanized" serial agency does *not* imply an absolute separation between collective movements and an individual's voluntary control over her own body, decision to move in such a manner, and so on. The way, however, in which collective schemas of movement (dance steps, etc.) are inscribed within, and hence dictate *how*, an individual moves her body requires an analysis of habit and habitualization, or what Sartre calls *hexis*. For lack of space, however, I cannot develop Sartre's account of hexis here. Much as with Bergson, an individual's agency can never be fully "mechanized," if only because an individual's freedom can never be fully extinguished; it can only become "petrified" and "habitualized" (i.e., constrained and shaped) in greater and greater degrees. Hence the importance of Sartre's general term *practico-inert*.

7. "Tiller Girls of London 1958," YouTube, https://www.youtube.com/watch?v=0XQ17OZ4mwU.

The ornament, detached from its bearers, must be understood rationally. It consists of lines and circles like those found in textbooks on Euclidean geometry, and also incorporates the elementary components of physics, such as waves and spirals. Both the proliferations of organic forms and the emanations of spiritual life remain excluded. The Tiller Girls can no longer be reassembled into human beings after the fact. Their mass gymnastics are never performed by the fully preserved bodies, whose contortions defy rational understanding. Arms, thighs, and other segments are the smallest component parts of the composition. (Kracauer [1927] 1995, 77–78)

The modern social organization of labor and production is manifested in the line of dancers acting together in machinelike precision. In a more aggressive form, as a form of social agency that epitomizes totalitarian societies, the mass synchronization of dancers in the Mass Games in North Korea exhibits the complete suppression and incorporation of individual freedom into a collective agency that achieves the perfection of a machine.[8] In both the Tiller Girls and the Mass Games, a serial collective *becomes* a machine, or material object, such that individuals are physically inscribed into the third ("the collective") while at the same time establishing the relation of each individual to others as individuals of a collective. Freedom is here short-circuited on both the individual *and* the collective plane of existence.

In Sartre's analysis of seriality, it is not only material objects that operate as quasi-social agents (the practico-inert) by virtue of their mediating function for the constitution of serial collectives. As a social-ontological category, seriality allows Sartre to understand the constitutive function of *ideas* within the formation of social existence. Social prejudices— for example, racist ideas—as well as ideological forms of collective consciousness, are social objects that mediate and constitute serial collectives. Whereas in the case of waiting for a bus, a serial collective is a social interaction among individuals in which individuals do not genuinely act in concert with each other, in cases of serial collectives formed through an idea, individuals are brought together in a shared form of *unthinking*. In the *Critique*, Sartre points to two contrasting forms of serial unthinking: public opinion and racism. In both cases, a collective consciousness is forged in which the possibility of genuine thought for an individual is short-circuited under the *appearance* of a shared form of thinking. Ideas are social realities to the extent that ideas mediate social relations and reciprocity; an idea is materialized as a structure of individual and collective habit, or what Sartre calls *idea-hexis*. Ideas are thus embedded in concrete forms of social life, and indeed, ideas can only change along with corresponding changes in social conditions and forms of collective existence. Sartre's trenchant insight here is to have identified how the bond of identification and adhesion within a social collective is mediated through a common "idea" or set of beliefs. Against such an entrenched idea, there is no possible argument, since the social function of the idea is not to convince or, indeed, *to be true* but to be socially

---

8. AsiaObscura, "Insanely Synchronized Children Dancing at the Mass Games in North Korea," YouTube, https://www.youtube.com/watch?v=WrQ0bJpx7D4.

effective as a mediator of social interaction and group membership. We believe so as to be together so as to not expose our common beliefs to the possibility of falsehood or failure. In Sartre's inimitable prose, the idea approaches the impenetrability of stone: "The Idea is a process; it derives its invincible strength from the fact that nobody thinks it" (Sartre [1960] 2004, 300). The idea is here characterized as a "process" insofar as Sartre's point is to stress the idea as a social process: the idea gives form and meaning to a collective consciousness. Within such a serialized consciousness, each individual is bound to the other through a shared set of beliefs and opinions; each individual does not think for himself or herself but finds apparent verification in recognizing the same idea with the others in the group. The words of others in the collective are a mirror in which each member sees the presupposed depth, or presumptive "truth," of his or her own words without recognizing the profound superficiality and, indeed, stupidity of the words we share in common that bind us to each other.

Racism is another, more pointed example of collective unthinking. In Sartre's account, racism cannot be understood as a set of beliefs or ideas that represent legitimate claims to truth. On the contrary, as he argues in his essay *Anti-Semite and Jew* (*Réflexions sur la question juive*, 1946), racism undermines the meaningfulness of language insofar as racism immunizes itself from any accountability for what one says and responsibility for the truth of what one professes. Racism gains its power less from its putative semantic content than from its social function of mediation, that is, as constituting social collectives. The social collective of those bound together in the common project of racism (protesting against immigrants, expressing racist views, engaging in violence, etc.) is mediated through a set of ideas and beliefs that are mutually reinforced and, in this sense, "validated" through social interaction. Within a group of racists, the ideas that bind the members together are both validated and immunized. Each member of the group verifies the purported "truth" of his or her racist ideas through other members in the collective. Social collectives bound by radical views, such as extreme political groups or religious cults, maintain a high degree of disciplined social cohesion as a function of the immunization of their founding ideas from any accountability or verifiability to reality. Sartre insists that what characterizes racism is that racism is in fact "not a thought at all," by which he means that it is a thought that *no one thinks* in thinking that everyone in the collective thinks it. This claim does not ignore that ideology and racism are loquaciously expressed, articulated, and—in some sense—"argued" by various ideologues and spokespersons (Goebbels, for instance, in the case of Nazism). The point is rather that the social function of such "arguments" and "reasonings" is to act essentially as a form of collective unthinking; what matters is not the content of the view but its social function of bonding individuals to each other. In a manner that anticipates the recent work of the French anthropologist Scott Atran (2010) and his work on religious fundamentalism and extremism, the "moral logic" of fundamentalism is at odds with rational sequences of thinking and choosing, yet entirely "logical" from the perspective of social integration and allegiance.[9] In Sartre's expression, the "I think" becomes subsumed and displaced by the "one thinks," which in turn masks that "no one thinks." The logic of social cohesion trumps adherence to the standards and ethics of rationality while masking the nonrationality of the sequence of beliefs that compose the structure of ideology.[10] If, as Kant famously argued, a genuine thought for a rational subject must be a thought that is accompanied by the possibility of saying "I think," then racism represents a form of thought that forecloses the possibility of saying "I think." There is literally *no subject of thinking*, but only a social collective bound in unthinking. As Sartre insightfully writes:

In fact, the affirmative force of this opinion [i.e., racism] derives from the fact that, in and through everyone, it is the inevitable stubbornness of others; and the certainty of the person who affirms it rests on his (cheerfully accepted) inability to occasion any doubt about this subject in any other member in the series. The Idea as a product of the common object has the materiality of a fact because no one thinks it. (Sartre [1960] 2004, 301)

4   The Pledge and Rituals of Initiation

In a variety of forms, from the innocuous waiting for a bus to the Tiller Girls to the politically shrill North Korean Mass Games, serial collectives are a form of social agency in which individual agency—freedom—becomes entirely absorbed and reified into a collective action that in turn lacks the essential character of genuine group agency, namely, the realization of individual freedom in concert with others. Serial collectives lack a reciprocity that is both internalized within the action of each member and, through each member, externalized into a robust sense of group agency. What serial collectives lack, in other words, is a robust sense of individual participation in a group action and the robust cohesion of the group for an individual's sense of solidarity with the group and its members. The passage to such a form of social existence in which individuals act in concert with each other, so as both to realize their own individual freedom (praxis) and to realize a robust sense of group agency, critically depends—in Sartre's analysis in the *Critique*—on the constitutive moment of pledging (taking an oath of allegiance, swearing to uphold norms of a group, signing a contract, etc.) and the formation of what he calls a "statutory group." Sartre understands the category of

9. In Atran's pithy formulation of this very Sartrean thesis: "It [Atran's research] tries to answer the question, 'why do people believe in a cause, and why do some die and kill for it?' The answer in a nutshell is that people don't simply kill and die for a cause. They kill and die for each other." For an analysis of Sartre's understanding of this phenomenon, see Nicolas de Warren 2015. My thanks to Christian Tewes for helping me formulate this point more clearly.

10. The situation is, in fact, more complex, since Sartre argues in *Anti-Semite and Jew* that anti-Semitism, for example, is based on a bad faith with regard to language itself. It is an attempt to speak without holding oneself to the responsibility of accountability for what one says, or, in other words, without accepting such a kind of speech as a claim to truth.

"pledging" broadly and diversely. In its essential function, the act of pledging binds individuals to each other in such a manner that each individual becomes bound to the group *through* other individuals and bound to other individuals *through* the group. In pledging, each individual constitutes herself as the "third" for each other, much as the "third itself"—the group as such—becomes constituted through each pledged individual who thus becomes incorporated into the group. It is in the act of pledging that (within the group) I am, because we are; and we are, because I am.

In Sartre's account, the pledge has the fundamental form of a performative speech act that constitutes mediated reciprocity among individuals in the group. The act of pledging institutes (in the phenomenological meaning of *Stiftung*) a normative social order that becomes, through the investment of individuals in the subsistence of a group identity, a self-regulating form of social existence. When an individual pledges to adhere to the norms and values that constitute a group, he or she makes this pledge before others in the group, who function as witnesses and recipients of the pledge. Other members of the group to whom I pledge my fidelity stand before me as *both* an individual and as the third. As Sartre writes: "I give my pledge to all the third parties, as forming the group of which I am a member, and it is the group which enables everyone to guarantee the statute of the permanence to everyone" (Sartre [1960] 2004, 421). Each member becomes the objective guarantee for the pledge of an individual, who thus becomes bound to certain obligations and norms in a dual sense: each individual is bound to the group as such (and hence to other members of the group); but also, each individual is *bound to herself* in terms of her group identity and membership. It is in this latter sense that the individual's relation to herself becomes mediated by the third (the group), while it is in the former sense that the group becomes constituted as group through the relation of the individual to other members. In Sartre's terminology, the pledge "interiorizes" a permanent community within each individual so as to give each individual social substance (as a constituent of the group), as well as "exteriorizes," or objectifies, the praxis of each individual into the existence, or social fact, of the group. Significantly, the *ontological* achievement of the pledge consists in establishing an objective guarantee from others in the group that protects and inhibits each individual from becoming *Other*, that is, from exiting or betraying the group. A robust social group is thus delimited by points of entry and exit, and hence by mechanisms of inclusion and exclusion. To know and be bound to others in the group, I must know what it would mean to become other than the group and, in this knowing, know what sanction and violence might befall me should I become other than the group through betrayal or disloyalty.

In many social groups, the intake of new members through the act of pledging is connected to rituals of initiation for the uptake of the pledge by the group. The pervasive phenomenon of initiation rites commonly involves violence against the bodies of individuals and an incorporation of bodies through dance or other forms of synchronized collective actions. As in the examples of premodern warrior dances or the synchronized drills and marches of the modern military, the incorporation of the body into a social group requires the ritualized submitting of the body to a norm of movement and bodily affectivity that binds individuals to the group as such, "the third" (McNeill 1995). The voluntary submission of an inductee's body to pain and shame (for example, a hazing ritual in which new recruits must crawl in mud, etc.) marks the uptake of a pledge by the group. Humiliation is pivotal: shame before others in the group represents a mechanism for the inclusion of the individual into a group through the internalization of the gaze of the Other, the third. In submitting oneself to the humiliation of such rituals of initiation, an individual *freely* allows for her complete incorporation into the body of the group. The submission to an identification with the group is thus both total and voluntary.

The social incorporation of the body, in the twofold sense of the physical and normative inscription of an individual's body into the group and the social incorporation of the group as such, is not the only function of the pledge and rituals of initiation. The pledge is not only a performative act (*praxis*, in Sartre's terminology) that institutes the group (intake of individuals into the group and uptake by the group of individuals); it is also an act that gives voice: it is the first act of speech that gives voice to the individual as member of the group. To have a voice in a group is to be recognized by others in the group as an individual who both can address other members *in the name of the group* and can be addressed as an individual by the group, namely, by others who speak for the group. In this manner, each individual is both "the first" and "the third." It is by virtue of this function of speaking for the third—for the group—that an individual exercises the normative function of speaking from the position of the "we": an individual may tell another member to abide by the rules, exhibit behavior befitting the group, and so on. The group regulates itself through a dialectical interplay between its constituent members as shifting between "the first" and "the third," between the position of an individual in the group and the position of the group itself. In its most pristine form, namely, as an ideal of *democratic* communities, this toggling between speaking from the position of the individual and from the position of the group would allow for a group dynamic in which confrontations and conflicts are resolved through a community of speakers (members) without any *hierarchy* and hence establishment of segmented forms of authority within the group.

5   The Goalkeeper

Especially in Western societies, team sports enjoy a superlative degree of cultural prestige, sociological value, and psychological significance. The virtues of team sports are widely celebrated as exhibiting the perfect balance of individual freedom and initiative with group agency and devotion. It is therefore unsurprising that team sports hold such an exalted status in democratic societies; team sports exemplify in perfect pitch the ideal of the individual as plural and the group as singular. In the words of Bill Belichick, lionized coach of the New England Patriots, "There's no mystery here, it's about trusting each other and everybody doing their job."[11] The mantra "Do Your Job" expresses the premium placed on individual praxis (i.e., each individual performing perfectly his or her function within the team) in conjunction with a premium of solidarity and trust that are essential for genuine (and successful) team agency, that the team plays as one.

In the *Critique*, Sartre turns to the example of a soccer team to illustrate the constitution of an organization in which individual freedom is genuinely realized in acting in concert with others while in turn the genuine social agency of the group is constituted through individual actions. A soccer team is an example of what Sartre calls "an organization" (as distinct from the two other forms of social groups I have discussed, serial collectives and statutory groups). As in the example of a soccer team (and Sartre has in view the team of eleven players acting on the field, not the wider organization as such), an organization is composed of individuals with differentiated functions (midfielder, goalie, defender, etc.), who act together in view of a common objective that determines the actions of each player on the field. Realizing the team's objective (winning the match) is achieved through the discrete actions of individuals in such a manner that, as Sartre writes, "the individual *praxis* is a *self-suppressing mediation*, or a mediation which negates itself for the sake of being transcended by a third party" (Sartre [1960] 2004, 459). This "self-suppression" of individual praxis is not, however, its oppression or coercion, nor either its negation in the sense of "elimination" (as perhaps Sartre's own use of the term "self-suppression" might inadvertently suggest). It is through individual praxis that the common objective is achieved, and this individual praxis becomes "negated" in the dialectical sense of "interiorized mediation"—"mediation" in the sense that individual praxis enables group agency, and in this manner, individual praxis is transcended in a twofold sense: the individual praxis is irreducible to the team agency, and yet individual praxis becomes, or constitutes, the team agency.

Each player's scope of actions is defined by a determined function and hence prescribed by the team (the third). The role of goalkeeper, for example, is defined by her duty to defend the goal, use her hands within the goal box, and so on. To the extent that she decides what to do in light of her prescribed function and in view of the common objective of winning, her actions are mediated by the third. In fact, the goalkeeper must situate her own actions within an understanding of the functions of other players, and this entails not only the mediation of her actions through the functions of other players but equally the mediation of her actions through the game plan or strategy that the coach has prescribed before the match. Each individual player must operate according to the demands and expectations of the third. In the words of the legendary European football coach José Mourinho: "I would rather play with ten men than wait for a player who is late for the bus." Yet each player must execute his or her prescribed function in view of a constantly changing landscape of possible actions

11. NFL, "'Inside the NFL': Super Bowl XLIX Second-Half Highlights," YouTube, https://www.youtube .com/watch?v=urYnkhU5t3w.

(with regard to other team members and opposing players) where individual initiative and intelligence are paramount. As Sartre observes:

Even before passing the ball to another member of his team and seeing the outcome of his decision gradually emerging, the individual must have triumphed, *through his personal qualities*, over another individual, in the other team, who has the same function. (Sartre [1960] 2004, 459)

The exercise of a player's functions thus demands "personal qualities" or, in other words, the realization of individual praxis that is irreducible and, indeed, in a sense, unprescribed by the team. What makes a good player is precisely the intelligence and creativity to mediate the team's agency through her own individual praxis. What makes a good team is the dual mediation of individual agency and the common agency of the team, or what Sartre simply calls *team spirit*: "the interdependence of [individual] powers in connection with the common objective" (Sartre [1960] 2004, 460). Genuine social agency is here constituted through individual participation and group solidarity. It is through the experience of participating in a common activity that individual freedom is genuinely realized, hence the pleasure each individual experiences in playing on the team; it is in terms of group solidarity that each individual action constitutes a common agency, namely, a unified team effort. In this manner, each player is a "common individual" or, in other words, an "I" who is the group in an organization of individuals who form the group. Each player is both "the first" and "the third." In Sartre's definition:

The common individual is defined as a historical and concrete individual in so far as his action was an unforeseen moment of the common undertaking—or, to put it differently, of the reshaping of the group by the group. What is revealed as *common* by the group is the *individual* particularity of his action. (Sartre [1960] 2004, 461)

Sartre's insight into the dual constitution of group agency in terms of the third, where individual freedom is genuinely realized in the same motion as genuine group agency, becomes consolidated, as it were, in his reflections on what makes a "good goalkeeper." By contrast, in the serial collective of the Tiller Girls, a *good* dancer is an individual whose freedom is entirely absorbed into the anonymity of a collective action that resembles the motion of a machine. The good dancer is the dancer who becomes indistinguishable from her peers, thus allowing for the visibility of the spectacle—the mass ornament—to appear in seamless unity and perfection. On the other hand, a *good* goalkeeper is an individual who must make existential choices for "the himself" *of* the team and the team *in* himself (one thinks here of Wim Wenders's film *Die Angst des Tormanns beim Elfmeter*). In making such choices, the goalkeeper "transcends himself as common individual in order to lose himself in the common objectification"; or, in other words, the goalkeeper must exercise his individual talent (he must "do his job") to constitute himself as the means for realizing the common objective of the team (Sartre [1960] 2004, 459). The beautiful game is here the beauty of the realization of human freedom in concert with others, where each can say to the other: I am, therefore we are; we are, therefore I am.

References

Atran, S. 2010. *Talking to the Enemy*. Penguin Books.

Cambria, F. 2009. *La material della storia: Prassi e conoscenza in Jean-Paul Sartre*. ETS.

Catalano, J. 1986. *A Commentary on Jean-Paul Sartre's "Critique of Dialectical Reason,"* vol. 1: *"Theory of Practical Ensembles."* University of Chicago Press.

Cormann, G. 2000. Le problème de la solidarité: De Durkheim à Sartre. In *Etudes sartriennes X: Dialectique, littérature, avec des esquisses inédites de la Critique de la Raison dialectique*, ed. F. Caeymaex, G. Cormann, and B. T. Denis, 77–110. Éditions Ousia.

Deleuze, G. 2002. *L'Ile déserte et autres textes (1953–1974)*. Les Éditions de Minuit.

de Warren, N. 2015. Brothers in arms: Fraternity-terror in Sartre's social ontology. In *The Phenomenology of Sociality: Discovering the 'We'*, ed. T. Szanto and D. Moran, 313–326. Routledge.

Flynn, T. 1984. *Sartre and Marxist Existentialism*. University of Chicago Press.

Flynn, T. 1997. *Sartre, Foucault, and Historical Reason*, vol. 1. University of Chicago Press.

Kracauer, S. [1927] 1995. *The Mass Ornament: Weimar Essays*. Harvard University Press.

McNeill, W. H. 1995. *Keeping Together in Time: Dance and Drill in Human History*. Harvard University Press.

Sartre, J.-P. 1946. *Réflexions sur la question juive*. Editions Morihien.

Sartre, J.-P. 1989. Preface to *The Traitor*, ed. A. Gorz and trans. R. Howard, 1–34. Verso.

Sartre, J.-P. 1993. La Conference de Rome, 1961: Marxisme et subjectivité. Ed. M. Kail. *Les Temps Modernes* 560 (March): 11–39.

Sartre, J.-P. [1960] 2004. *Critique of Dialectical Reason*. Trans. A. Sheridan-Smith. Verso.

3 Consciousness, Culture, and Significance

Christoph Durt

This chapter offers a new view on the relation between consciousness and culture by investigating their intertwinement with significance. I argue that consciousness discloses aspects of significance, while culture encompasses shared significance, as well as the forms of behavior that enact significance. Significance is linguistic or nonlinguistic meaning that is (partly) understood in intersubjective engagement and constantly reinstantiated in new contexts of relevance rather than belonging to single individuals (cf. Gallagher, this vol.). Significance is embedded in the shared world to which we relate through cultural forms of thinking and sense-making.

As in other chapters, several of which will be taken up here, the discussion that follows is inspired by thoughts from Edmund Husserl, without relying on his terminology. Furthermore, it integrates studies from two authors who would otherwise be underrepresented in this book—Gilbert Ryle and Ludwig Wittgenstein—and considers the implications of their thoughts for contemporary discussions of consciousness and culture. In spite of the fundamental differences between these three authors, they provide complementary insights into how consciousness and cultural forms of behavior accomplish significance.

Consciousness and culture are often defined in ways that obscure their relation to significance. Consciousness tends to be reduced to a limited concept of experience, such as in the debates around "phenomenal consciousness." Reflective aspects of consciousness are construed by what Ryle calls "thin description" (Ryle [1968] 2009, 501). Culture, by contrast, tends to be defined in very thick terms at the expense of thinner forms of cultural behavior. I will explain this by reference to Clifford Geertz, who made thick description the defining characteristic of culture, and argue that this definition neglects the foundational role of "thinner" cultural behaviors.

Significance is accomplished in embodied processes. At a basic level, these processes include the forms of behavior shared by most or all humans. Behavior is to be understood not thinly as lacking significance but thickly as itself resting on different levels of significance. At higher levels, forms of behavior become more complex, as well as more specific to groups of people, and allow for more complex manifestations of significance. The forms of behavior common to some or all people are cultural; culture accomplishes significance.

The claim that consciousness encompasses different levels of significance gives rise to the question of their connection. "Hybrid" concepts of cognition propose that there are two distinct parts of the mind, such as a motor-perceptual and a reflective part. Di Paolo and De Jaegher (this vol.) worry that hybrid concepts of cognition perpetuate dualism by seeing only "direct action-based mechanisms" as embodied and those involved in "more reflective tasks" as disembodied. Here I argue, with regard to Merlin Donald's concept of the "hybrid mind" (Donald 2001, 164), that the underlying issue should not be framed in terms of embodied versus disembodied systems of production; even computers are in some sense "embodied." Rather, there is a categorical difference between different levels of significance. Embodied consciousness not only accomplishes but also integrates different levels of significance.

1   Phenomenal and Reflective Consciousness and Significance

This section takes its departure from standard definitions of consciousness as phenomenal experience and argues that they do not suffice to explain reflective forms of consciousness such as understanding and thinking. These acts accomplish forms of significance that need to be accounted for by "thick description" (Ryle [1967] 2009, 489). While definitions of consciousness in experiential terms attempt to describe consciousness rather thinly, much of consciousness involves thicker levels that cannot be reduced to thin description.

1.1   Phenomenal and Reflective Consciousness

Since the early modern formulations of the distinction between primary and secondary qualities, philosophers have tried to come to terms with the paradox caused by (1) the assumption that ideas of secondary qualities must be produced by primary qualities, and (2) the claim that we cannot even conceive how this is possible. In Locke's words, we "can by no means conceive how any size, figure, or motion of any particles, can possibly produce in us the idea of any colour, taste, or sound whatsoever; there is no conceivable connexion between the one and the other" (Locke [1689] 1836, 419; cf. Durt 2012, 2–3). Today the distinction between ideas of primary qualities and ideas of secondary qualities tends to be glossed over,[1] and all are subsumed under the terms "phenomenal character," "phenomenal consciousness," "qualia," "phenomenal experience," or simply "experience." But the paradox lives on, now framed by the question of how experience can be accounted for by naturalism, a

---

1. The distinction is still implicitly at work in the tendency to define phenomenal character with reference to secondary rather than primary qualities, and it sometimes resurfaces explicitly, for example, in the claim that "we need qualia to make sense of secondary qualities" (Shoemaker 1990, 114). Understanding why ideas of primary qualities have traditionally not been seen as a challenge to a mechanistic account of the universe may be key to a more differentiated understanding of the challenge of "experience" to naturalism.

question that is supposed to constitute "the hard problem of consciousness" (Chalmers 1995, 202). Here again, the study of consciousness continues to be driven not by the richness of conscious phenomena themselves but by the troubles some aspects of consciousness seem to provide for naturalism's attempt to explain everything there is.

That the use of "consciousness" is driven by theoretical purposes explains why its discussion in these contexts is limited to *phenomenal* consciousness, but it doesn't justify the limitation. Framing problems of consciousness in terms of "qualia" and "intentionality" has even been called "extremely ethnocentric," because "very few non-Western cultures would view the matter in the way that Western consciousness researchers might conceive of it" (Throop and Laughlin 2007, 633). The discrepancy with ordinary views of consciousness, however, also exists also in relation to "Western culture," for "consciousness" in ordinary English has a much wider sense than its technical uses in contemporary philosophy. The discrepancy by itself is, of course, not an argument against framing consciousness in technical terms, but it indicates that a better understanding of consciousness needs to go beyond the very narrow use of "consciousness" within some currents of Western philosophy. This chapter does so, not by denying phenomenal and intentional aspects of consciousness but by reinterpreting them from a wider point of view that shows that both go together with reflective consciousness.

The ordinary concept of consciousness includes the intransitive aspect of "being awake rather than asleep or otherwise unconscious" and the transitive aspect of "being conscious of something or other" (Hacker 2002, 157). Defining consciousness in terms of the phenomenal character of experience considerably extends the concept to include "the whole domain of 'experience'—of 'Life' *subjectively understood*" (157). On the other hand, such definitions restrict the concept of consciousness to a limited notion of experience and tend to leave out reflective understanding and highly reflective activities such as thinking. This omission is ironic, since reflective understanding and thinking are prime examples of activities of consciousness in the ordinary sense of the word. Of course, not all consciousness is highly reflective, and there may be unconscious activities that can be called "understanding" and "thinking," but usually these are performed in consciousness.

The relatively recent discussion of "cognitive phenomenology" brings to the fore the idea that phenomenal consciousness and cognitive accomplishments are inherently connected.[2] Matthew Ratcliffe neatly summarizes the two basic questions of cognitive phenomenology as (1) whether "the phenomenal content of perception in one or more modalities incorporate[s] conceptual, propositional or other ingredients that are properly regarded as 'cognitive,'" and (2) whether "non-sensory cognitive states and processes have phenomenal content too, the focus being on 'thinking'" (Ratcliffe 2014, 355). Answering yes to either question would cast

---

2. Today the more common term is "achievement," which, however, carries the connotation of the completion of a difficult task or goal. Accomplishments are not necessarily difficult and do not need to involve goals.

doubt on the strict distinction between phenomenal consciousness and cognitive states. To make real progress at this point, moreover, we need to go beyond the dichotomy between phenomenal content and intentionality and consider how experience and understanding come together in consciousness.

The classical definitions of phenomenal experience try to refer to it deictically in terms of "what it is like." Thomas Nagel (1974), for instance, takes the "what it is like" to be a member of a species as the necessary and sufficient condition of consciousness. Other instances are the "what it is like" of seeing a certain shade of red or understanding the sentence "two plus two is four" (Searle 2000, 561). Higher-order theories of consciousness make a similar move by supposing that there are higher-order "states" that are supposed to make lower-order mental "states" conscious. The higher-order states are again usually defined by a "what it is like" to be in the lower-order state (Rosenthal 2004, 29). In the same vein, Strawson's concept of "understanding experience" (Strawson [1994] 2010, 5) or "meaning-experience" (7) concerns the thin experiential difference between, for example, understanding a French sentence and merely hearing the words.

The concept of "understanding experience" is clearly not enough to explain the difference between understanding and not understanding, since to understand something, it does not suffice to have an experience that goes along with understanding; Strawson himself admits that one can have such an experience without understanding ([1994] 2010, 7). But is the experience of understanding enough to account for the difference between what goes on *in consciousness* when one understands a French sentence and when one doesn't understand the sentence? An affirmative answer presupposes that consciousness is merely experiential consciousness in a thin sense of experience.

Against such a definition, I hold that consciousness can also comprise understanding in a fuller sense. "Understanding" is not meant to be independent of experience or to be another thing in addition to experience. Rather, it is a "thicker" level of experience. In consciousness, understanding always goes along with experience; we do not understand with a pure intellect but do so with the help of perceptions and imaginations. Claiming that understanding is exhausted by associations between perceptions and imaginations is psychologism, a position that, in contrast to other forms of reductionism, is not particularly common today. Instead, understanding is typically conceived as inessential to consciousness itself, and consciousness is restricted to experiential contents in a thin sense. I think that this conception disregards something extremely important: by becoming conscious of a thing or relation, we understand something about it.

1.2 Consciousness and Significance

When we consciously experience or understand something, we become aware of part of its significance. With the concept of significance, I take up the distinction between meaning and significance, which has been used to interpret Hans-Georg Gadamer as holding that even "if there actually is an unchanging meaning that belongs to a text, there is no access to it that doesn't go by way of significance" (Gallagher, this vol.). I here do not, however, limit the concept to texts or even language. The crucial point is that significance is not bound to individual consciousness but shared between conscious beings and always newly reinstantiated. This is so because each individual act of consciousness reveals only a certain significance, which can be developed further in additional acts. In some cases, this may lead to modifications to which the author or speaker can object "that's not what I meant," but that too is an interpretation of what she meant; it doesn't imply that meaning is fixed in one act of rigid designation.

I contend that disclosing significance is an essential accomplishment of consciousness. Consciousness is usually "about" something or "intentionally directed" toward something, which may be a perception of an object or the understanding of a state of affairs. What consciousness becomes aware of in both cases is an aspect of the significance of the object or relation it is about. In consciousness we disclose aspects of the significance of the things and the relations of which we are conscious. Significance is not just content in the sense of "the way [an intentional state] represents what it is about or directed on" (Crane 2013, 5), but it discloses aspects of the objects and relations intended. The disclosing of significance in consciousness is an accomplishment of consciousness in the verbal sense, and the disclosed significance is an accomplishment in the nominal sense.

The term "significance" is here understood as comprising not only the conventional meaning of signs and symbols or the importance of something for human life but also sense and meaning more generally. The concept avoids the—in my view—misguided dichotomy between intension and reference that goes back to a particular interpretation of Gottlob Frege's distinction between sense (*Sinn*) and meaning (*Bedeutung*). This interpretation conceives of meaning (reference) as being in the world and sense (intension) as somewhere else—either in Frege's "third world" (*drittes Reich*; Frege [1918] 2003, 50) or, as probably most adherents of the dichotomy today think, in the mind. In contrast to this distinction, I conceive of significance as, on the one hand, accomplished by culturally shared ways of interacting with one's environment; significance is not detached from the world. On the other hand, significance is accomplished by consciousness; it is not detached from the mind. Nor is it a third thing in between mind and world. Rather, it is the way the world is given to us due to our ways of making sense of it. While significance is embedded in activity, it is—unlike sense-making—not the activity itself; in the view presented here, sense-making is directed toward significance. Significance is accomplished by individuals in individual acts, but the accomplished significance is more than the accomplishing acts. Conceiving consciousness as disclosing significance thus stands radically opposed to purely intrinsic and solipsistic concepts of consciousness.

To say that consciousness discloses significance is not to say that consciousness discloses all at once the whole significance of what it is about. Rather, consciousness usually discloses only extremely limited parts of significance. In Husserl's expression, we only experience "adumbrations" (*Abschattungen*) of things. Analogously, in reflection we only understand certain aspects of the significance of things or relations. Consciousness is about things that are not all disclosed in a current state of consciousness or in any individual consciousness alone; it is open to features of the environment (McDowell [1994] 2002, 450). Significance concerns things and relations in the intersubjectively shared world, which Husserl also calls the "lifeworld" (1962, 48). Significance can reach into the sense-making of past and future generations, for example, through writing and inherited forms of interaction.

Those who know Husserl will readily recognize further connections to other key Husserlian concepts such as intentionality, *noesis* and *noema*, and accomplishment or achievement (*Leistung*). Husserl calls the accomplishing activity of "transcendental consciousness" in disclosing significance "constitution," and he uses this term to refer to, among other things, the constitution of objects by transcendental consciousness (cf. Husserl [1913] 1976, 344–359). A detailed study of the Husserlian terminology can contribute to sharpening the issues involved, but I will concentrate instead on bringing the previously outlined ideas together with insights from Ryle and Wittgenstein.[3] While important differences exist between phenomenology and ordinary language philosophy, including the emphasis each puts on egological consciousness and intersubjective behavior, Dermot Moran rightly states that those who "overemphasize [Husserl's] focus on the individual life of intentional consciousness as reconstructed from within … tend to overlook [his] original, radical, and fundamentally groundbreaking explorations of intersubjectivity, sociality, and the constitution of historical cultural life" (this vol., sec. 1). On the other hand, Ryle's and Wittgenstein's insights into significance and culture can complement the phenomenological study of consciousness, as I will show.

2   Thinner and Thicker Descriptions of Consciousness

This section reinterprets Ryle's distinction between thin and thick description of behavior as a distinction between levels of significance and considers how it can be applied to the concept of consciousness. Defining consciousness in terms of the phenomenal character of experience involves a relatively thin description of consciousness, and other activities of consciousness such as thinking require thicker descriptions.

2.1   Thinner Description of Consciousness
Ryle introduces the concept of "thick description" in his essay *Thinking and Reflection* (Ryle [1967] 2009, 489). He explains how it differs from "thin description" in *The Thinking of Thoughts: What Is "Le Penseur" Doing?* by way of example: "Two boys fairly swiftly contract the eyelids of their right eyes. In the first boy this is only an involuntary twitch; but the other is winking conspiratorially to an accomplice. At the lowest or the thinnest level of description the two contractions of the eyelids may be exactly alike" (Ryle [1968] 2009, 494). Thin

---

3. Another closely related author is Martin Heidegger, whose terminology I will not consider here. For a study on Heidegger in relation to language and significance, see Inkpin 2016.

description alone does not allow us to distinguish between a twitch and a wink; it consists only of observations of bodily movements, and in Ryle's example, the twitch and the wink look exactly alike. To distinguish them, we need thick description.

The lowest level of description of the two boys' behavior is not the twitch. A twitch is not just a contraction of an eyelid but an *involuntary* contraction of an eyelid, which also means it is not a wink. Because involuntariness has to be included in the description of the twitch, the description of a contraction of an eyelid as a twitch is a thicker description than just the description of the contraction of an eyelid, which Ryle calls "thin description." The description of the wink is even more obviously thick because it includes the fact that the winker was deliberately signaling to "someone in particular, without the cognizance of others, a definite message according to an already understood code" ([1968] 2009, 494). Recognizing a twitch or a wink entails thin description, at least implicitly; it depends on recognizing the movement of the eyelid. But it adds another layer that cannot be accounted for by thin description. Yet further layers can be added: Ryle's examples are the parodist mocking a clumsy wink, and the parodist practicing his parody. The series of layers of meaning could be extended indefinitely.

Ryle's metaphor for thick description, with its hierarchical structure and extendibility, "is a many-layered sandwich, of which only the bottom slice is catered for by that thinnest description" ([1968] 2009, 497). Ryle's term "bottom slice" sounds as if he thought of thin description as a foundation on which higher levels of description would have to be added. But the foundation of a sandwich is not like the foundation of indubitable knowledge that Descartes expects from his "je pense, donc je suis" (Descartes [1637] 1902 [*Discours*], 32). A bottom slice is neither an unanalyzable given nor a core; it may be necessary for the whole, but it is not its essence. Furthermore, the different slices or layers constitute a whole; there is only one action, which can be described in both thin and thick terms. The thin description of the wink may be perfectly true and relevant; the problem is that it doesn't suffice to distinguish a wink from a twitch. "Thin description" is to be understood not as an origin, essence, or unquestionable foundation but as something that is implicit and part of any thicker description, regardless of how many layers it has.

The distinction between thin and thick description is very different from Ryle's famous distinction between *knowing how* and *knowing that* because it does not categorically distinguish between two mutually exclusive types. The border between thick and thin is imprecise: just as there are thinner and thicker people, and people who are somewhere in between, so there are descriptions that are thicker than the thinnest possible description, yet not thick. It is not always possible to draw a sharp boundary between both, but usually we can distinguish pretty well between thin and thick. In other words, the difference between thin and thick description is not absolute but relative; thick description is relative to the thinner levels it builds on.

Ryle's distinction between thin and thick description is not a distinction between something that is observable and something that cannot be observed; both the twitch and the wink are observable behavior. Rather, thin description is an incomplete description in that it includes neither the cause nor the significance of the behavior. The observed contraction of an eyelid is not just a contraction of an eyelid; it is either a twitch or a wink, or it may indicate something else entirely. We may never know the cause of a particular contraction of an eyelid, but it certainly has some cause. Ryle doesn't say whether a description that includes the physiological causes of the behavior is still a thin description, and I will consider this question in section 4.1. Yet it is clear that when we can determine that the contraction of an eyelid is due to the attempt to signal something, we are giving a thick description.

I interpret Ryle's distinction between thin and thick description as a distinction between levels of significance. Thin description of an action has significance in its own right and can be understood if the respective observational terms are understood. But it abstracts from the full significance of the respective action and is in this regard underdetermined. While Ryle applies his concept of thin description only to behavior, we can apply it also to definitions of consciousness in terms of the phenomenal character of experience. Such definitions are thin because they make no use of significance beyond the level of observation.

In the case of the phenomenal character of experience, the observation is alleged to be possible from a first- rather than a third-person perspective. Both kinds of observation abstract from the full significance of the objects or relations of which we are conscious. For instance, a color experience is always part of the visual field. It usually discloses one aspect of an object, which exists in relations to other objects. Usually we become immediately aware of higher levels of significance, such as the significance of a green traffic light in its context of an empty crossing. I here leave open the question of how far it is even possible to describe the phenomenal character of experience independently of further significance. Even if it is at all possible, the phenomenal character is not completely void of significance in at least a minimal sense, for otherwise it would have no determinate character that is directed toward objects or relations in the world. In either case, the description is at least as thin as the description of a contraction of an eyelid. Attempts to define consciousness in terms of the phenomenal character of experience are attempts to give a thin description of consciousness.

2.2 Thicker Descriptions of Consciousness

One may say that in simple observation there is already understanding, for example, of what a wink means. Yet consciousness also includes and produces much more reflective forms of understanding. A highly reflective form of consciousness is thinking, which is worth considering a little further. As an example of a thinker, Ryle asks us to consider Rodin's bronze sculpture *Le Penseur*, which I here assume to stand for an actual person who really is thinking. The sculpture depicts a nude male sitting on a rock, resting his chin on the back of his right hand and his right elbow on his left thigh. His posture and muscular tension in masterly fashion reflect the dialectical tensions of thinking, and *Le Penseur* is quickly recognized as a thinker. It is obvious that a thin description alone does not get far in describing his activity; the description would be largely exhausted by detailing his posture. Perhaps he is also engaged in further observable activities, like Euclid when thinking about geometry, whom Ryle imagines "muttering to himself a few geometrical words and phrases, or scrawling on paper or in the sand a few rough and fragmentary lines" (Ryle [1968] 2009, 510). Such activities may be part of thin description, but at this level, the description would still be largely incomplete. The reasons why, however, may not be immediately obvious. Could *Le Penseur* not be thinking solely propositionally and at the same time speaking aloud his every thought? Wouldn't his thinking then be capable of being captured by thin description?

Ryle would have two answers to this objection. The first is that a precise description of all pitches of the thinker's voice or a complete recording of them would no more get to the essence of his thinking than the videotape of the wink would get to the core of the winking. On its own, the recording would grasp the meaning of the recorded voice no more than a parrot speaking words it doesn't understand. To be sure, thin description can be much more than just a recording. It may use terms such as "holding his chin" or "sitting in a crouched position." These expressions go well beyond the phonetic level and make sense only to those who understand English. Yet by themselves they don't enable us to understand what he is doing apart from sitting in that position. If we were also to include the propositions he is speaking, however, we would need to be a little more specific. Thin description can include some understanding, for example, the observational meaning of the action of closing one's eyelid. If that's all the thinker is thinking about, and he says everything he is thinking about, then this content of his thinking could be accounted for by thin description.

Nevertheless, that wouldn't suffice for thinking, which brings us to Ryle's second answer. Thinking goes beyond its contents by relating them to each other and reflecting on them. Thinking aloud is different from verbally explaining something. The thinker does not yet know the outcome of his thinking; he is a "pioneer" (Ryle [1968] 2009, 509). He does not yet know the way, and if the way he takes leads to a quagmire, the least he can learn is which way not to take. Ryle contrasts this with an expert explaining an issue, knowing already what she is trying to explain; thinking about something is not like explaining something. While thinking is not the same as speaking, it can be enacted by speaking, such as when one freely develops a thought in language.

According to Ryle, the thinking of *Le Penseur* is taken to stand on a high reflective level. We may say that he is "trying, by success/failure tests, to find out whether or not the things that he is saying would or would not be utilisable as leads or pointers" ([1968] 2009, 508). This is a very thick description because it includes several layers of reflection. In other cases, thinking may be less reflective. We can say of a person playing tennis that she is thinking while playing (cf. Ryle [1967] 2009, 480). While concentrating on the game, her thinking guides and directly expresses itself in her actions; she may, for instance, anticipate the strategy of her opponent and counter it with what she thinks is the best move. Unlike the reflective thinking of *Le Penseur*, her thinking is immersed in her environment, at least with regard to the features relevant to her game. The point about thinking is not whether it is detached from its environment but how it deals with the things it is about. The things may be in the environment, such as Tetris figures on a screen, or merely imagined. Thinking manipulates them in intelligent and creative ways that combine, for example, the characteristics of the things, purposes, and possibilities of change.

On the one hand, in thinking we understand the significance of what we think about, which is sometimes rather thin and sometimes rather thick. On the other hand, thinking brings the objects of thought together in ways that accomplish new significance, which stands on yet thicker levels. If we understand thinking as something that can be done consciously, as I think we should, then consciousness accomplishes significance not only on relatively thin but also on very thick levels. The more or less thick levels of significance may furthermore influence each other. An example of thicker significance potentially influencing the thinner perception of pain is given by Henningsen and Sattel (this vol.). They point out that the existence in Japanese of the medical concept *katakori* (insufficiently translated as "neck and shoulder pain") actually makes Japanese people more prone to experience the condition than speakers of other languages. Likewise, thick conceptions such as "medicinal beliefs" and theories can become part of a culinary tradition that influences food preferences, which in turn alter that tradition (Jain, Rakhi, and Bagler 2015, 5).

3   Thicker and Thinner Descriptions of Culture

In contrast to the preference for thin definitions of consciousness in contemporary philosophy, the academic discourse on culture tends to define "culture" in very thick terms. Edward Tylor famously defined culture as "that complex whole which includes knowledge, belief, art, morals, law, custom and any other capabilities and habits acquired by man as a member of society" (1871, 1). Kroeber, Kluckhohn, and Untereiner (1952) compiled a list of more than 150 definitions, a meticulous work continued by Baldwin et al. ([2006] 2008), who analyzed about 300 definitions and categorized them under the seven headings of structure, function, process, product, refinement, group, and power or ideology.

I lack the space here to even list the titles of the plethora of definitions, but it is worth pointing out that while they show a tendency to define culture in terms of reflective capabilities and their results, everyday habits and ways of perception are often included as well. Section 3.2 gives a good reason why: more reflective forms of significance are embedded in cultural behavior. Culture thus concerns all forms of significance that are common to groups of people and inherited by social rather than genetic means. The description of culture does not always have to be very thick, and less-thick description is also important for investigating the relation between culture and significance.

3.1 Thicker Descriptions of Culture

Clifford Geertz attempts to overcome the "theoretical diffusion" (1973, 6) in definitions of culture such as those of Kroeber, Kluckhohn, and Untereiner by defining culture in terms of thick description. In his seminal essay *Thick Description: Toward an Interpretive Theory of Culture*, Geertz defines anthropology as "an elaborate venture in, to borrow a notion from Gilbert Ryle, 'thick description'" (6). Geertz points out that the thin description of a behavior as "rapid contraction of his right eyelid" would not yet amount to a description of its cultural significance; "*as a cultural category*, [twitches] are as much nonwinks as winks are nontwitches" (7; italics in original). A twitch is not a twitch if it could be a wink, and distinguishing twitches from winks is decisive for the anthropological description of winking behavior.

Assuming that Geertz is right, we may still ask whether all thick descriptions of the customs of a foreign culture are equally thick, or whether it makes sense to distinguish between levels of thickness. The latter approach is suggested by Wittgenstein. He was an avid reader of James Frazer's classic text *The Golden Bough: A Study of Magic and Religion* ([1890] 2003). What Wittgenstein gained from his reading was not so much knowledge about practices of magic and religion at various times and in various cultures as insights into what can go wrong in studying them. Frazer's explanations of numerous dark and mysterious practices are as questionable as they are fascinating. For instance, Frazer explains the custom of the rain dance as a consequence of the superstition that the dance itself causes rain (last paragraph of chap. 4). When by chance it rains, this merely confirms the dancers in their superstition that their dance was the cause. Wittgenstein objects that it seems odd that "people don't realize earlier that sooner or later it's going to rain anyhow" (Wittgenstein 1993, 121). He thinks that Frazer's explanations are primitive, crude, and misleading. They are prone to make the "explained" behavior look more strange than understandable.

Wittgenstein makes an important distinction between two descriptions of a culture we know very little about. He asks us to distinguish between the superstitiousness of the belief described and the superstitiousness of the description. If we ascribe to the people of a culture the belief that their head will fall off when they have killed an enemy, we would describe a belief that one can arguably call superstitious. The description itself, however, would "contain nothing superstitious or magical in itself" (Wittgenstein 1993, 133). That the people really have the belief that their head falls off when they have killed an enemy seems unlikely, as it could be contradicted by experience in a relatively straightforward manner. If they really held that superstitious belief, however, we could describe it without recurring to superstitious beliefs.

Describing the beliefs of the people being studied in terms of "ghost" or "soul," by contrast, is likely to bring superstitions into the description: "I should like to say: nothing shows our kinship to those savages better than the fact that Frazer has on hand a word as familiar to himself and to us as 'ghost' or 'shade' in order to describe the views of these people" (Wittgenstein 1993, 133). The kinship may actually help in understanding the other culture, but in Frazer's examples, the kinship seems to be limited to his imagination of the other culture. Frazer gathered all his examples from the stories and reports of others, and we just don't know enough about the customs themselves to make good inferences about the beliefs of their practitioners. It is likely that what seems to be a profound belief behind the customs is in fact nothing but a projection of Frazer's or our own fantasies.

Wittgenstein's examples of "head" and "soul" show that it makes sense to distinguish between levels on which concepts are embedded in a framework of cultural behavior, beliefs, and knowledge. Even the concept of "head" is embedded in such a framework. There may be cultures that have a different concept of body parts or don't even have a concept of the body part "head." If the anthropologist and the native who points toward a rabbit and says "gavagai" share the same pointing behavior, "gavagai" can mean "rabbits, stages of rabbits, integral parts of rabbits, the rabbit fusion, and rabbithood" (Quine [1960] 2013, 51). It could even mean something like "there it goes" (Quine 1973, 44).

Furthermore, the ordinary concept of the physical object head relates to cultivations of the concept in religion and science. These develop the ordinary concept and embed it in further meaningful contexts, some of which may in turn become part of the ordinary concept. Ultimately, even the ordinary concept of "head" as a physical object is embedded in a rich cultural context, but that does not mean that the whole context of meaning is evoked in every use of the concept. With concepts such as "ghost," "spirit," or "soul," by contrast, it is much harder to find a description that does not use an intricate framework consisting of theory, worldview, superstition, religion, or other knowledge and belief that has to be described in very thick terms. Taking all description of a culture to be thick in the same sense erases an important difference between forms of behavior that are expressions of metaphysical, theological, or ideological beliefs and forms of behavior that do not directly depend on such beliefs. Culture is expressed not only in the former but also in the latter.

Of course, there are radical differences between Frazer and Geertz. Frazer believes in a developmental theory of knowledge from superstition through religion to his own secular "science," and his interpretations of the customs he is trying to explain are informed by such theories rather than careful observation. Geertz's anthropological work, by contrast, involves direct contact with his "informants" in other cultures; he reflects on his own presuppositions and tests and constantly revises his theories. Many of Wittgenstein's criticisms of Frazer's methodology would not apply to Geertz. Geertz would surely agree with Wittgenstein that interpreting rain dances and other customs as superstitions or stupidities is bad anthropology, at least if based only on secondhand reports. But his strong emphasis on the thickness of anthropological description is prone to lump together decisively different forms of interpretation and to miss the use of thinner description of a culture.

3.2 Thinner Descriptions of Culture
Geertz's claim that all cultural description is thick shows the need to interpret and think about interpretation when understanding cultures and has been groundbreaking for

anthropology. From a philosophical perspective, however, it is embedded in a problematic constructivist framework. Since every description is thick, even the "data" of anthropology are supposed to be constructions: "What we call our data are really our own constructions of other people's constructions of what they and their compatriots are up to" (Geertz 1973, 9). Geertz's "constructions of constructions" definition of the data of anthropology is echoed in his concept of culture: "The culture of a people is an ensemble of texts, themselves ensembles, which the anthropologist strains to read over the shoulders of those to whom they properly belong" (452). If this were literally so, we would be in a similar situation to Frazer, who only has recourse to texts describing the customs. Since Geertz does engage in fieldwork and spends time conversing with the people of specific cultures, his expression "ensemble of texts" must actually be somewhat metaphorical, indicating that even the most basic forms of interaction are embedded in "webs of significance" (5).[4] Yet one may wonder if restricting the concept of anthropological data to linguistic expressions suffices to describe what is going on in a culture.

There are things that don't make it into the words of a people, perhaps because the authors don't find them important enough, or find them too embarrassing, or perhaps don't even realize they exist. For instance, gender differences are often either unspoken or denied. Of course, such things are often expressed in written or oral speech. But trying to gather everything about a culture solely from what is said or written would disregard a lot of useful evidence, such as interactions between genders. Texts are further removed from ordinary behavior and experience. Living cultures, in contrast, allow observation of behavior, immersion in the culture, interaction beyond talking with the people of the culture, and the opportunity to study them in experimental settings. Geertz did use some of these important means for understanding; his actual way of doing ethnology is not in question here. The problem is his conception of culture as a text, a metaphor that neglects all of these ways of understanding. It thus makes sense that, as Laurence Kirmayer points out, anthropology has subsequently moved on from that metaphor, and "current medical anthropology emphasizes that culture is embodied as well as discursive" (this vol., sec. 1).

Differentiating between levels of thickness allows us to see the abundance of forms of cultural behavior that can be described relatively thinly before considering their relations to thicker concepts. These encompass not only less-thick concepts such as those of body parts, which are likely to be common to most or all cultures. One can also think of basic forms of behavior that vary considerably from culture to culture, for example, the rhythm and speed of life or bodily forms of interaction such as the frequency of touch and the distance that feels comfortable in personal communication. Basic forms of behavior may indirectly and sometimes directly be due to thick beliefs, but this is evidently not always the case. On their own, they don't have any more significance than the eye contraction that could signify

4. For Geertz's own reconsideration of the notion of culture as a text, see Geertz 1988.

either a twitch or a wink. Nevertheless, they may turn out to be more important than Geertz suggests.

Geertz calls his definition of culture "semiotic" (1973, 5, 24, 29–30), but what he means by this term is relatively narrow. Semiotics comprise syntax and semantics as well as pragmatics. This is so because language needs to be understood in the context of its use. Seen this way, thin and thick description do not merely require an understanding of the syntax and semantics of the interpretations of behavior or events by informants but also imply an understanding of the ways behavior gives meaning to the language of the culture and the interpretations by the members of the culture.

In contrast to Geertz's definition, Wittgenstein's writings express an extreme awareness of the role of behavior and interaction for significance. One can always define words with the help of other words, but significance goes beyond the relations of words to each other. Phonemes and symbols have no significance by themselves, but only when used in regular ways; without at least some regularity, language games would lose their point (cf. Wittgenstein [1953] 1997, §142). The regular uses of language can be quite specific, but Wittgenstein thinks that one can dig deeper and find a common ground beneath specific uses. The ground is not unquestionable or given once and for all; it may be more like a riverbed that changes much more slowly than the water running over it (cf. Wittgenstein 1997, §§96–99). As such a ground, it is likely to be common to all or most cultures. Wittgenstein writes that "the common behavior of mankind is the system of reference, with which we interpret a foreign language" (§206). The latter may be universal to all humans, but Wittgenstein's expression "mankind" should not be overstated (cf. Schulte 1990, 157); forms of behavior may be common to only certain groups of people (Durt 2005, 62–64). Such groups of people are cultures or subcultures; cultural behavior accomplishes significance.

Much of the "common behavior of mankind" may be described in relatively thin terms. However, I interpret Wittgenstein to mean that this is not always the case and that at least some of this common behavior needs to be described thickly. The history of the use of the concept of behavior in theories such as behaviorism makes the concept prone to be misunderstood as something that can always be described thinly. Behaviorism assumed a highly reductive concept of behavior, which often meant little more than bodily motion. But, to the contrary, behavior is often extremely complex, such as in the case of reflective thinking, and has to be described thickly. The more the behavior is interwoven with particular beliefs and ways of thinking, the more likely it is to be specific to groups smaller than the whole of humankind.

Together with the beliefs and ways of thinking with which behavior is interwoven, the thinner behaviors common to a culture provide a system of reference. Culture accomplishes significance through both specific and more common forms of behavior. This is the fundamental reason why Geertz's metaphor of text is insufficient: texts cannot be understood without understanding the underlying forms of behavior.

4   Embodied Accomplishments and the Integrative Nature of Consciousness

Behavior is an important topic for the study of embodiment because all behavior is embodied. But talk of different levels of accomplishment of significance may raise the question of whether they involve different kinds of embodiment. I argue that the categorical difference exists not between different levels of embodied accomplishment but between different levels of significance. Even very thick accomplishments such as thinking build on preconscious accomplishments as well as lower levels of significance. Consciousness is embodied at all levels of accomplishment, integrating different levels of significance.

4.1   Embodied Accomplishments and Accomplishments of the Body

Like Wittgenstein, Ryle sometimes thinks of behavior as being thick, as is obvious from the fact that he views even thinking as a form of behavior (cf. Ryle [1940] 2009, 197); behavior comprises externally observable thin behavior as much as complex forms of behavior. As forms of behavior, all levels of accomplishing activity are embodied. The higher levels of accomplishing activity build in turn on lower levels, preserving their embodied accomplishments.

Conscious accomplishments, such as the conscious recognition of a wink, build on preconscious accomplishments. We usually become aware of the wink immediately, without having to go through a conscious process of interpreting the contraction of the eyelid. Ryle does not undertake an analysis of the different accomplishments entailed in perception, but other philosophers have done so. Edmund Husserl, for instance, distinguishes between "passive" and "active synthesis," both of which disclose the significance of the intentional objects of consciousness. Passive synthesis concerns the subjective and intersubjective processes that prereflectively accomplish the constitution of the intentional objects, while active synthesis concerns more reflective accomplishments such as judgments (cf. Husserl 1966). One may furthermore attempt to go beneath experience altogether and try to determine the roots of significance in the biological activity of organisms. In Varela's interpretation, the activity of bacteria in swimming toward parts of a solution with a higher sugar gradient can be described as "food significance" (Varela 1991, 87); the bacteria make sense of their environment by interacting with it in meaningful ways. One may question whether biological significance is still the same kind of significance, but even if not, this would be no reason to deny that there are bodily accomplishments that go beneath consciousness and on which conscious accomplishments can build.

But how much can we explain by describing the unconscious activities going on in the body? For instance, in the act of thinking, the externally visible movements of the thinker's body play only a minor role. But beneath these externally visible movements we find a profusion of bodily goings-on. Suppose that it was practical to look at all the processes within the thinker's body, in particular his brain. To describe the brain processes, we can imagine either a thin or a thick description.

It is often assumed that the thinker's brain processes can be described in rather thin terms. One may naively imagine small wires connecting nodes and think of electric charges moving like lights between them. Such a description would not be on the same level as the description of the contraction of the eyelid of the twitcher or winker, but it may be close enough to call it "thin description." If such a thin description were possible, however, the described processes would clearly not be on the level of accomplishment of thick description, and no number of such descriptions, taken by themselves, could remedy this. Thick description is about a level of significance not entailed in thin description and is thus not reducible to thin description.

In fact, however, brain science gives a much thicker description of the brain. Scientific description generally has to do with conceptualizing and interpreting data by means of hypothetical assumptions and theories, and is in this regard always thick. But even in this regard, a thick description of brain processes alone would not amount to a thick description of the thinking process because, by itself, it would not relate these brain processes to the levels of accomplishment pertinent to thinking. The description of thinking not only has to be thick but also has to be thick in the *right way*. Its study would have to include the levels of accomplishments of the particular act of thinking, but that would be a science very different from today's brain science and maybe closer to today's human sciences.

In other words, we can coherently say that the thinker is thinking by using his brain, or even that he is thinking with his brain. But that does not mean that his thinking is *in* his brain, because the description of the brain by itself would not include the relations between the levels of accomplishments involved in thinking. Rather than being a place of thinking, the human brain is a "relational organ" that is not only part of an organism but also embedded in wider contexts, including those of cultural significance (cf. Fuchs 2008, 217).

It would be a misunderstanding to conceive these relations independently of bodily *experience*. The embodied being is more than a physical body. It is a body that is physical and at the same time experienced as a sensible body. Furthermore, it is experienced as a body that can initiate or control a range of movements. We may say that the embodied being is a living body or, with Husserl, that it is a *Leib* ([1913] 1976, 116). I think that as soon as there is an experience of one's body as a *Leib*, we can speak of a conscious being. Reflective consciousness is not required, and contra Descartes, animals experience their body as a *Leib* and thus have consciousness. Nevertheless, as I argued earlier, human consciousness also encompasses reflective levels of significance. The next section delineates the idea that embodiment of consciousness contributes to the integration of different levels of significance.

4.2 The Integrative Nature of Embodied Consciousness

The distinction between different levels of significance may sound to some like a hybrid concept of the human mind. Merlin Donald argues for such a view and claims that the human mind is torn between two modes:

We have hybrid minds. Like the monsters of Greek mythology, we are two creatures struggling within a single body. We are capable of operating within that same fuzzy analogue mode that constituted the whole of the cognitive universe for our ancestors, while another part of us operates like the symbolic machines we have made. But mostly we muddle through with various patched-together hybridized modes of thinking and feeling. Our conscious experience reflects this. These two sides of our being are engaged in a constant struggle of ownership or awareness. (Donald 2001, 164)

In this citation, Donald draws not only on ancient mythology but also on ancient Greek conceptions of competing parts of the soul, such as Plato's tripartite concept of the soul in his chariot allegory (Plato 1903, 246a–254e). Donald simplifies and modernizes the ancient conceptions with his claim that there are two competing parts of the human mind: the "analogue" and "symbolic" mode. This does not really contradict the notion of embodiment, though, since symbolic machines are material and could become part of a body, for example, by implanting a computer in someone's brain. But we are not cyborgs yet, and we have no neuroscientific evidence for a symbol-processing system separate from the "analogue" neuronal processes in the human nervous system. As with Donald's interpretation, the claim criticized by Di Paolo and De Jaegher (this vol.) that embodiment can only explain "direct action-based mechanisms" while "more reflective tasks" are disembodied makes little sense.

Donald's metaphoric claim makes more sense if we understand it to mean a categorical distinction not between different parts of the body but between different kinds of significance. With regard to significance, one can indeed draw a categorical distinction between "analogue" and "symbolic." On the one hand, generalizations comprise essential features of concretely experienced objects and may be called "analogue" because they are always vague or "fuzzy." On the other hand, idealizations and formalizations are limit objects that can be manipulated by syntactic operations (cf. Durt 2012, 164). This is the working mode of our "symbolic machines." Generalizations can be experienced, while idealizations and formalizations cannot be experienced in themselves, although they are accomplished by operations on experienceable objects such as measurement techniques (145–166).

Ryle's distinction between levels of accomplishment offers one way of understanding how higher levels of significance build on lower levels, preserving the embodied accomplishments of the lower levels and giving rise to new embodied accomplishments. Often the implied accomplishments are difficult to bring to light because they have developed over generations in cultural processes. Besides the symbolization techniques at the core of symbolic language, humans have mastered a large number of techniques that accomplish significance far beyond concrete experience. An important example is the ability to deal with ideal objects. Numbers are abstracted from concrete experience via techniques of idealization, which are then further developed by formalization techniques, allowing for new kinds of significance. Such techniques are cultural not in that they could exist in only one culture but in that they derive from specific forms of behavior that are intertwined with cultural systems of knowledge and belief. As techniques, they are just as embodied as any cultural forms of behavior.

These considerations tell against Donald's claim that the two modes of the human mind are opposed and in struggle. Of course, everything that grasps our attention has to prevail against other things that simultaneously demand our attention. But when this is experienced as a constant struggle and insurmountable conflict between two monsters, creatures, or actors in one consciousness, we come close to a psychopathological case. Dissociative identity or multiple personality disorder may be an obvious example, but one may also think of forms of "schizophrenic alienation" (*schizophrene Alienation*; Blankenburg 1971, 9). Patients with this condition suffer from what Sass calls "hyperreflexivity" (2001, 259); they feel the need to constantly reflect on even the most mundane actions, which can prevent them from doing those very things.

In healthy persons, by contrast, more or less reflective forms are combined in productive ways that allow the person to act coherently. Of course, there are also conflicts in the consciousness of healthy beings between different levels of reflection. But in this chapter I have given reasons for that these levels are not intrinsically contradictory and rather build on each other. To conclude, let us consider an example of a conscious activity of a healthy person that suggests that consciousness unites perception and movement into a kinesthetic system.

Consider a person who is psychologically and physiologically in good shape learning to perform a handstand. While practicing, she will suppress thoughts about dinner in favor of concentrating on holding her feet in the air. Some of that suppression will already be done subconsciously, and later, when she is able to do the handstand "automatically," she will be able to think about her dinner at the same time. But while she is learning, she needs to focus on the right things on pain of falling. She may have a teacher giving her verbal instructions such as "activate your core," "clasp the floor with your fingers," "look in front of your hands," "move your hip up as high as possible," and "carefully raise your feet." But such propositions are only scaffolds she is learning to apply in the right way. Ultimately she will be able to do the handstand without them. For now, they (hopefully) help her guide her actions.

This is so because she knows or learns their significance for her action and knows or learns how to combine them into one action. Consciousness integrates thinner and thicker forms of significance in experience, reflection, understanding, desire, and action. This integrative nature is what Merleau-Ponty calls the "intentional arc" (cf. Tewes, Durt, and Fuchs, introduction to this vol.). That consciousness is embodied is not a hindrance to integration but rather enforces the integration of different streams that otherwise may veer off indefinitely. In the case of the learner of a difficult practice, there is simply no time for consciousness to wander off.

These considerations are more compatible with the ancient chariot allegory than with Donald's concept of the digital and analogue parts of the mind. In contrast to Donald's concept, the chariot allegory had a placeholder for the integrative nature of consciousness, namely, the charioteer. As said, part of the integration of levels of significance is done subconsciously or automatically, and consciousness only controls the performance. But, as in the case of the learner, other parts will have to be done by conscious effort. Rather than being a mere passive stream of subjective experience, consciousness actively integrates the various experiences, thoughts, desires, and actions. The "stream of consciousness" or "stream of thought" (James [1890] 1918, 224–290; Husserl 1973, 171–191) can itself be guided by consciousness. Consciousness always has a character of mineness (cf. Zahavi 2014, 19), and already here we may be able to speak of a "self." Even with respect to thin self-experience in this sense, one may distinguish between the sense of ownership and the sense of agency and ask whether the minimal self includes a sense of agency (cf. Gallagher 2000). Agency in connection with thicker levels of significance, such as in thinking, is another important topic that has to be left to future investigation. For now, I hope it has become plausible that culture, consciousness, and significance are intrinsically intertwined, and that their relation is important for the investigation of each of the three.

Conclusion

Going beyond the common restriction of consciousness to phenomenal experience, this chapter has contended that, on the one hand, consciousness discloses aspects of significance in experience and understanding. Culture, on the other hand, systematically comprises shared significance and accomplishes significance through shared forms of behavior. Consciousness discloses aspects of the significance that is established and expressed in cultural behavior. Cultural behavior, in turn, builds on conscious experience and understanding and is guided by consciousness. Consciousness and culture are closely interdependent through their accomplishment of significance.

Acknowledgments

Many thanks to Christian Tewes, Christian Spahn, Thomas Fuchs, Aloisia Moser, Anita Galuschek, Tilman Staemmler, and Ruiming Zhang for their insightful and inspiring comments on drafts of this chapter. Work on this chapter was supported by the DFG excellence initiative Cultural Dynamics in Globalised Worlds and by the European Union's Horizon 2020 research and innovation program under the Marie Skłodowska-Curie Individual Fellowship no. 701584.

References

Baldwin, J., S. L. Faulkner, M. L. Hecht, and S. L. Lindsley, eds. [2006] 2008. *Redefining Culture: Perspectives across the Disciplines*. Taylor & Francis e-Library.

Blankenburg, W. 1971. *Der Verlust der Natürlichen Selbstverständlichkeit: Ein Beitrag zur Psychopathologie Symptomarmer Schizophrenien*. Ferdinand Enke.

Chalmers, D. 1995. Facing up to the problem of consciousness. *Journal of Consciousness Studies* 2 (3): 200–219.

Crane, T. 2013. *The Objects of Thought.* Oxford University Press.

Descartes, R. [1637] 1902. Discours de la méthode pour bien conduire sa raison, et chercher la vérité dans les sciences. In *Œvres de Descartes: Discours de la méthode & essais VI,* ed. C. Adam and P. Tannery. Librairie Philosophique J. Vrin.

Donald, M. 2001. *A Mind So Rare: The Evolution of Human Consciousness.* Norton.

Durt, C. 2005. Wittgensteins interkulturelle Perspektive: Auf der Suche nach gemeinsamen Handlungs- weisen. In *Wahr oder Tolerant: Religiöse Sprachspiele und die Problematik ihrer globalen Koexistenz,* Wittgen- stein-Studien 11, ed. W. Lütterfelds, T. Mohrs, and D. Salehi, 57–76. Peter Lang.

Durt, C. 2012. The paradox of the primary-secondary quality distinction and Husserl's genealogy of the mathematization of nature. PhD diss., University of California at Santa Cruz. http://www.durt.de/diss/ Paradox.html.

Frazer, J. G. [1890] 2003. *The Golden Bough: A Study of Magic and Religion.* Project Gutenberg eBook. https://www.gutenberg.org/files/3623/3623-h/3623-h.htm (accessed August 30, 2004).

Frege, G. [1918] 2003. Der Gedanke: Eine logische Untersuchung. In *Logische Untersuchungen,* ed. G. Patzig, 35–62. Vandenhoek & Ruprecht.

Fuchs, T. 2008. *Das Gehirn—Ein Beziehungsorgan: Eine phänomenologisch-ökologische Konzeption.* Kohlhammer.

Gallagher, S. 2000. Philosophical conceptions of the self: Implications for cognitive science. *Trends in Cognitive Sciences* 4 (1): 14–21.

Geertz, C. 1973. *The Interpretation of Cultures.* Basic Books.

Geertz, C. 1988. *Works and Lives: The Anthropologist as Author.* Stanford University Press.

Hacker, P. 2002. Is there anything it is like to be a bat? *Philosophy* 2 (4): 157–174. doi:10.1017/ S0031819102000220.

Husserl, E. [1913] 1976. *Ideen zu einer reinen Phänomenologie und phänomenologischen Philosophie. Erstes Buch: Allgemeine Einführung in die reine Phänomenologie 1. Halbband: Text der 1–3 Auflage. Husserli- ana III/1.* Ed. K. Schuhmann. Martinus Nijhoff.

Husserl, E. 1962. *Die Krisis der europäischen Wissenschaften und die transzendentale Phänomenologie: Eine Einleitung in die phänomenologische Philosophie. Husserliana VI.* Ed. W. Biemel. Martinus Nijhoff.

Husserl, E. 1966. *Analysen zur passiven Synthesis: Aus Vorlesungs- und Forschungsmanuskripten, 1918–1926. Husserliana XI.* Ed. M. Fleischer. Martinus Nijhoff.

Husserl, E. 1973. *Zur Phänomenologie der Intersubjektivität: Texte aus dem Nachlass Erster Teil, 1905–1920. Husserliana XIII.* Ed. I. Kern. Martinus Nijhoff.

Inkpin, A. 2016. *Disclosing the World: On the Phenomenology of Language.* MIT Press.

Jain, A., N. Rakhi, and G. Bagler. 2015. Spices form the basics of food pairing in Indian cuisine. *Physics and Society*. arXiv:1502.03815v1 (accessed January 2, 2016), 1–30.

James, W. [1890] 1918. *The Principles of Psychology*. Dover.

Kroeber, A., C. Kluckhohn, and W. Untereiner. 1952. *Culture: A Critical Review of Concepts and Definitions*. Peabody Museum of American Archaeology and Ethnology.

Locke, J. [1689] 1836. *An Essay Concerning Human Understanding*. Tegg.

McDowell, J. [1994] 2002. The content of perceptual experience. In *Vision and Mind: Selected Readings in the Philosophy of Perception*, ed. A. Noë and E. Thompson, 443–458. MIT Press.

Nagel, T. 1974. What is it like to be a bat? *Philosophical Review* 83 (4): 435–450.

Plato. 1903. Phaedrus. In *Platonis Opera*, ed. J. Burnet. Oxford University Press. http://data.perseus.org/texts/urn:cts:greekLit:tlg0059.tlg012 (accessed March 12, 2015).

Quine, W. [1960] 2013. *Word and Object*. MIT Press.

Quine, W. 1973. *The Roots of Reference*. Open Court.

Ratcliffe, M. 2014. Some Husserlian reflections on the contents of experience. In *Philosophical Methodology: The Armchair or the Laboratory?* ed. M. Haug, 354–378. Routledge.

Rosenthal, D. 2004. Varieties of higher-order theory. In *Higher-Order Theories of Consciousness: An Anthology*, ed. R. Gennaro, 17–44. John Benjamins.

Ryle, G. [1940] 2009. Conscience and moral convictions. In *Collected Essays, 1929–1968: Collected Papers*, vol. 2, 194–202. Routledge.

Ryle, G. [1967] 2009. Thinking and reflecting. In *Collected Essays, 1929–1968: Collected Papers*, vol. 2, 479–493. Routledge.

Ryle, G. [1968] 2009. The thinking of thoughts: What is "Le Penseur" doing? In *Collected Essays, 1929–1968: Collected Papers*, vol. 2, 494–510. Routledge.

Sass, L. 2001. Self and world in schizophrenia: Three classic approaches. *Philosophy, Psychiatry, and Psychology* 8 (4): 251–270.

Searle, J. 2000. Consciousness. *Annual Review of Neuroscience* 23:557–578. doi:10.1146/annurev.neuro.23.1.557.

Schulte, J. 1990. *Chor und Gesetz: Wittgenstein im Kontext*. Suhrkamp.

Shoemaker, S. 1990. Qualities and qualia: What's in the mind? *Philosophy and Phenomenological Research* 50 (4): 109–131.

Strawson, G. [1994] 2010. *Mental Reality*. 2nd ed. MIT Press.

Throop, J., and C. Laughlin. 2007. Anthropology of consciousness. In *The Cambridge Handbook of Consciousness*, ed. E. Thompson, 631–669. Cambridge University Press.

Tylor, E. 1871. *Primitive Culture: Researches into the Development of Mythology, Philosophy, Religion, Art, and Custom*. Vol. 1. John Murray.

Varela, F. 1991. Organism: A meshwork of selfless selves. In *Organism and the Origin of Self*, ed. A. Tauber, 79–107. Kluwer Academic.

Wittgenstein, L. 1993. Bemerkungen über Frazer's *Golden Bough*: Remarks on Frazer's *Golden Bough*. In *Philosophical Occasions, 1912–1951*, ed. J. Klagge and A. Nordmann, 118–155. Hackett.

Wittgenstein, L. [1953] 1997. *Philosophical Investigations*. Blackwell.

Wittgenstein, L. 1997. *Über Gewissheit*. Suhrkamp.

Zahavi, D. 2014. *Self and Other: Exploring Subjectivity, Empathy, and Shame*. Oxford University Press.

4   Neither Individualistic nor Interactionist

Ezequiel Di Paolo and Hanne De Jaegher

Enactive approaches to social understanding have been the subject of much development and debate over the last few years. We think that these debates are fruitful, but sometimes it is useful to take stock and clarify what aspects of these discussions may point to gaps in the theory or needs for clarification, and what aspects may be rooted in misunderstandings and misinterpretations.

In this essay, we provide an overview of the various claims defended by the enactive approach to intersubjectivity, from how social interactions can be defined operationally to how social understanding is rooted in participatory sense-making, even when we are not interacting with others. In contrast to prevailing views, the enactive approach does not put all the emphasis on individual capabilities to explain forms of social understanding and social action. This has often been interpreted as adopting an interactionist stance on intersubjectivity. It is indeed the case that social interaction patterns have not played a prevalent role in cognitive science and social neuroscience until recently, and in the light of this, a participatory as opposed to spectatorial stance still needs to be presented and defended.

One common misreading of this interactive emphasis, however, is to consider the enactive perspective as downplaying the role of individual processes and subjectivity. But in truth, the claims made position social interaction and embodied agency as equiprimordial loci of scientific and philosophical inquiry. The realm of intersubjectivity is animated by a force that is *neither* what goes on in people's brains or in their self-affective bodies *nor* what occurs in social interaction processes—if we consider each alternative on its own. On the contrary, intersubjective phenomena emerge only as a *dynamic relation* between these two broad domains: the personal and the inter-personal. Any emphasis on either side of this relation at the expense of the other fails to capture the complete picture.

To see the enactive approach as defending an interactionist position is to repeat the mistakes of methodological individualism we criticize in others. This interactionist interpretation is the cousin of another common misunderstanding of enactive ideas, namely, the idea that meaning is somehow "generated" in the agent-world coupling, and hence that enactive explanations are limited to concrete cognitive targets present in the here and now of interaction, for example, coordinating movements in a joint action task. Enactivism, so the criticism goes, cannot deal with anything beyond what is concurrently present to the cognizer. Thinking, planning, and imagining are supposedly out of reach for enactive explanations of cognition. This criticism is often invoked to defend some updated version of representationalism, since how else could we think, plan, or imagine, if not by using representations to supplement what is seemingly absent in our immediate couplings? We discuss the roots of this misunderstanding and explain why we can indeed propose that agent-world (and agent-agent) coupling is partly constitutive of sense-making—partly, because on its own it is not sufficient. Constitutive as well are the agent's multiple forms of autonomy, her history, and the broader values and norms that matter to her.

Similarly, the claim that the relation between persons and social patterns—embodied intersubjectivity—is primordial makes the enactive approach suspicious of hybrid solutions. These hybrids emerge as manufactured middle-ground positions between agonistic conceptual frameworks and hardly ever as the overcoming of their tensions. The theory of participatory sense-making instead offers dialectical tools for the self-deployment of the tensions that give rise to partial frameworks (individualist and interactionist). In the process, it offers accounts of increasingly complex forms of social agency, from bodily coordination to languaging.

1   Participatory Sense-Making

Let us review some of the enactive claims made regarding the study of social cognition and intersubjectivity and highlight how they always involve both interactive and individual elements.

1.1   Social Interaction

The enactive approach has seen an important amount of development over recent years. It offers a nonreductionist, naturalistic perspective on cognition and aims to answer questions that have largely been ignored by traditional functionalist approaches (Varela, Thompson, and Rosch 1991; Di Paolo, Rohde, and De Jaegher 2010; Di Paolo 2005, 2009; Froese and Di Paolo 2011; Thompson 2007). It offers a conception of autonomy and individuality of the embodied cognizer based on the precarious organization of its material, ongoing self-constitution (Di Paolo and Thompson 2014). This conception of the embodied cognizer, or sense-maker, serves to ground notions of interiority, agency, and normativity (Barandiaran, Di Paolo, and Rohde 2009). Unlike functionalism, the enactive perspective is concerned with what makes cognitive systems individual subjects with their own experience and perspective. It offers the conceptual categories for approaching this question in a naturalistic manner. Few other frameworks put so much emphasis on the individual cognizer; functionalist (computationalist and representationalist) approaches dodge these questions.

An issue that is raised when conceiving of cognitive agents in this way is whether their engagements with the world show any qualitative difference when the world is in part constituted by a community of other agents. The answer is yes. As agents regulate their coupling with their environments, they engage with traces of the activity of other agents, which accumulate historically and culturally as patterns of mutually shaped ontogeny. Thus a collective dimension contributes to the path-dependent transformations undergone as the agent makes sense and acts in the shared world (Di Paolo 2016).

This historico-cultural aspect, however, could be seen mainly as a cumulative effect, leaving the fundamental property of agency unchanged: the asymmetrical regulation of the coupling with the world by the self-constituted agent. However, this is not the end of the story. A novel domain is opened up as agents engage in a mutual, concurrent, joint regulation of their couplings with the world and each other. As collective regulation becomes mutual co-regulation, new situations emerge. They include relational phenomena such as coordination and miscoordination (e.g., co-regulation of timing, intonation, and topic in a conversation, or of rhythm and effort when walking together; coordination of moves during a joint task; co-regulation of interpersonal distance; etc.), as well as the possibility of complementary and conflicting acts (salutations, offerings, rituals, etc.). They also include relational patterns that become self-dependent, that is, sustained but underdetermined by the action of the agents engaged in mutual coupling.

These systemic phenomena lead us to the first contribution of the enactive approach: a formal definition of social interaction. Surprisingly, such a definition has been lacking, not only in the psychology literature but also in interaction studies in sociology and conversation analysis, although many of the associated phenomena are clearly addressed by these disciplines. According to the definition,

social interaction is the regulated coupling between at least two autonomous agents, where the regulation is aimed at aspects of the coupling itself so that it constitutes an emergent autonomous organization in the domain of relational dynamics, without destroying in the process the autonomy of the agents involved (though the latter's scope can be augmented or reduced). (De Jaegher and Di Paolo 2007, 493)

We should notice that social interaction is not the mere copresence of two or more autonomous agents (they may just stand there and ignore each other), nor is it just the presence of a mutual coupling between these agents (such as the transfer of body heat or merely noticing each other). The definition demands two strong conditions. Condition (1) requires that there is a co-regulated coupling, which originates a series of dynamical, relational processes that become self-sustaining (autonomous) in the relational domain. This is important because this condition allows us to speak of events and properties as belonging to the interaction or being external to it. Condition (2) requires that the participants are and remain autonomous.

They do not lose the possibility of exerting their powers on the interaction patterns or attempting to terminate it. This does not mean that interactions always unfold according to individual intentions, as this would contradict the first condition, which says that interaction patterns have autonomy.

At this point, we should note already that to speak of a social interaction is to speak about relational patterns and individual participants as equiprimordial. Both conditions are jointly necessary. Too much emphasis on the autonomy of relational patterns, such that individuals are subordinated to them, ignores the second condition of social interaction, which demands the sustained autonomy of individuals. Similarly, too much emphasis on the individual determinants of a social encounter (individual intentions, brain mechanisms for interpreting the actions and intentions of other, etc.) misses the role played by the self-determining aspects of the interactive dynamics.

1.2 Social Understanding and the Individual

A definition of social interaction is a first contribution of the enactive approach toward operationalizing social cognitive phenomena. While an increasingly larger proportion of empirical research in psychology and neuroscience is concerned with social interactions as defined in the previous section, a good part is not, even if the term "interaction" is used. Such is the case of experiments where "interactions" are nonautonomous (e.g., games with preestablished paced turns or a limited set of moves), violating condition (1), or experiments involving nonautonomous others (e.g., images, videos of other persons), violating condition (2). Research of this sort can be valuable from an enactive perspective, but it simply does not involve actual social interactions as defined here. It involves social but noninteractive situations or agent couplings that are nonsocial.

The enactive approach introduces the concept of *sense-making* to describe the key constitutive aspect of all forms of mind, from the simplest to the most complex. Sense-making is what occurs when an adaptive autonomous system (e.g., an organism) regulates its coupling with the world and its own states as a function of the virtual (nonactualized) implications for its continuing form of life (organic, sensorimotor, cognitive, social, etc.) (Varela 1997; Di Paolo 2005, 2009; Thompson 2007). It is an ongoing engagement with the world by an agent that is sensitive to the consequences of this engagement.

Given the definition of social interaction, which involves sense-makers in a particular relational configuration, we describe the sense-making that occurs in these situations as *participatory sense-making*. This is "the coordination of intentional activity in interaction, whereby individual sense-making processes are affected and new domains of social sense-making can be generated that were not available to each individual on her own" (De Jaegher and Di Paolo 2007, 497). It is through mutual (not necessarily symmetrical) participation by cognitive agents in the sense-making activities of others that their understanding depends not just on themselves but on the unfolding of the social engagement. The influence of other participants can take many forms, from "simple" orientations, as when a gesture is used to guide someone who is looking for a missing object, to joint sense-making, or coauthored cognitive activity, such as the collaborative elaboration of a piece of work.

Clearly, participation and social understanding are things that *individuals* do. There are obvious implications from the perspective of the subjectivity of the participants involved in an interactive encounter—implications for their affect and experience during concrete encounters, as well as many other aspects of personality and of social and individual capacities along developmental timescales, in relation to histories of interactions embedded within a particular culture.

One of these implications concerns the personal experience of the alterity of other participants in an interaction, a fundamental issue in the phenomenology of intersubjectivity. Given the autonomy conditions for both interaction patterns and participants, the experience of the other never achieves full transparency or full opacity but rather intermittently moves through regions of understanding and familiarity toward provinces of misunderstanding and bemusement, corresponding to phases of interactive coordination or breakdown respectively (De Jaegher and Di Paolo 2007, 504). Personal experience is, from this perspective, underpinned by a prereflective intercorporeality (Merleau-Ponty [1945] 2012), or mutual incorporation, in which "our body's operational intentionality is partially decentered" (Fuchs and De Jaegher 2009, 476). Here there are "two 'centers of gravity' which both continuously oscillate between activity and receptivity, or 'dominance' and 'submission' in the course of the interaction" (476). This centering and decentering is the pivot of embodied intersubjectivity and implies an ongoing fluctuation between empathy and alterity: "Both partners bring in their dispositions that are based on acquired intercorporeal micro-practices, [and] their retentions and protentions of the process that are partly fulfilled by interactive matches, but also partly disappointed by mismatches" (476).

These intimate relations between intercorporeality and personal experience can reach deep levels of bodily affection. Social interaction processes enter into the core of our self-constitution. In and through social interaction, we can truly affect each other, even each other's self-maintenance and self-affection (De Jaegher 2015). This is illustrated, for instance, in research showing that social interactions provide the conditions for even infants under one year old to behave like clowns (i.e., to do funny things and elicit laughter; Reddy 2001), for the development of self-conscious emotions (Reddy 2008), and for perceptual attitudes (Di Paolo 2016), and that interactions with close others can modulate pain experience (House, Landis, and Umberson 1988; Turk, Kerns, and Rosenberg 1992; Krahé et al. 2013). And since, from an enactive perspective, self-affection is always already an experiential manifestation of the precarious self-constitution of the body, it is also relevant that kind marital relationships can make a spouse's wounds heal faster, whereas hostile relations can slow down their healing (Kiecolt-Glaser et al. 2005; Gouin et al. 2010).

1.3   The Constitutive Roles of Intra- and Inter-personal Processes

The activity of interactive agents is therefore never simply their own, insofar as it arises under the influence of the participation of others as long as interaction is ongoing. But this is not all that happens. Since the definition of social interaction postulates the emergent (though temporary) autonomy of interaction patterns themselves, individual sense-making is influenced by these patterns, which are not under the full command of any participant. Thus we postulate that social interaction itself can play different roles in sense-making in general and social understanding in particular (De Jaegher, Di Paolo, and Gallagher 2010). In some cases, the role of interaction patterns may be contextual; that is, variations in these patterns produce variations in social cognitive phenomena. A stronger possibility is when interaction patterns play an enabling role, that is, facilitate or constrain cognitive phenomena, such that variations in the interaction can result in the absence of such phenomena. These are both cases of a "causal" role played by social interaction in the emergence and explanation of social understanding. But there is also the possibility that social interaction itself plays a constitutive role in social cognition. This is the case when the specific social cognitive activity is at least partially *constituted* by what goes on in the interaction (in addition to what goes on in each individual). Since the constitutive status is conceptual, not causal, one must have a clear and precise description of the phenomenon to determine it. It may be that, under a first approximation, the phenomenon is not sufficiently understood to establish whether something plays a constitutive role or not. And as conceptual categories change, constitutive relations may change too. Empirical evidence cannot by itself serve as proof of constitutive status. However, it can help refine categorical distinctions such that constitutive status may be more clearly revealed.

The possibility of a constitutive status of social interaction for instances of social cognition has generated some discussion (e.g., Herschbach 2012). It may be useful to rehearse one of the cases that, in our view, most clearly exemplifies the claim. Consider the well-known perceptual crossing paradigm (Auvray, Lenay, and Stewart 2009). In it the ecological situation is maximally simplified without eliminating a key factor: the free control of social interaction dynamics by the participants. Two blindfolded participants interact by moving a sensor along a shared virtual line using a computer mouse. Whenever the sensor encounters an object on this line, the participant receives a tap on the finger. The moving objects in this space are controlled by the other participant; one corresponds to the other's own scanning sensor, the other simply shadows the other's sensor at a fixed distance. The situation is the same for both participants. Notice that when a participant's sensor encounters the shadow of the other participant, only the first participant will receive a tactile stimulus. When the two sensors meet, both participants receive a stimulus simultaneously.

In this setup, participants are instructed to click the mouse whenever they judge that they are in contact with the other participant. The findings show that mouse clicks concentrate on each other's sensors and not on the identically moving shadow objects. However, the probability of clicking following stimulation from the other's sensor or from the other's shadow is shown to be approximately the same. This must mean that sensor-sensor encounters occur more frequently than sensor-shadow encounters.

Described in strict computational terms, perceptual crossing is a highly ambiguous, type-2 problem (Clark and Thornton 1997) where stimuli must be actively discriminated spatially and qualitatively using only temporal and proprioceptive cues (all "objects" found in the virtual space produce exactly the same tactile stimulation). In these terms, the task is untypically difficult, since the moving objects that interact with the participant (the other participant's sensor and shadow) move identically and could only be distinguished based on how these objects themselves react to contact. That the difficulty of this computationally tough problem deflates dramatically once we understand the collective dynamics is the theoretically pregnant point of the experiment.

The type-2 regularities present in the perceptual crossing sensory signals that could help distinguish sensors from shadows are statistically invisible in the absence of a systematic sampling strategy. One way to solve the task is to implement a strategy that successfully transforms these type-2 signals into type-1 data, that is, into nonrelational and unambiguous inputs (Clark and Thornton 1997). A type-1 signal by itself contains enough information to determine the next course of action toward the resolution of the task. This route toward solving the task involves a biased sampling of the raw sensory streams. Were this biased sampling to be implemented in the participants' brains, we would not hesitate in acknowledging that the neural processes involved are responsible for the core cognitive workload required to solve the problem. In other words, to solve the perceptual crossing task using this strategy *amounts to* finding the right way of biasing the sampling of sensory inputs so as to transform them from type-2 into type-1.

Now, this sampling bias is precisely what is achieved by the collective dynamics, that is, by the combination of individual strategies. The social interaction process biases the statistical presentation of sensory stimulus toward much more frequent encounters with the other participant's sensor and not the shadow. Mutual scanning of sensors produces mutual sensory feedback and promotes permanence in the shared spatial region, which is more stable than one participant unidirectionally scanning the shadow of the other. This is not done consciously by the participants but accomplished by a relation that emerges between their correlated movements. The cognitive work is neither done externally by a third party nor generated internally within the participants. It is produced by the collective dynamics in which they participate but whose properties do not correspond to individual properties of either agent on its own or to a linear aggregation of these properties. In a clear case of participatory sense-making, the task is transformed from type-2 to type-1—in other words, it is *solved*—by the interaction process. The participants deal with quasi-disambiguated, type-1 stimuli: "If it moves but stays nearby (repeated crossings), then click." If a process (in part) constitutes the solution to a cognitive problem, then it (in part) constitutes an instance of cognition. This is precisely what social interaction does in perceptual crossing.

But even in cases where agents aren't right now interacting, the interaction process is still basic to social understanding, as proposed by the interactive brain hypothesis (Di Paolo and De Jaegher 2012). In the absence of live interaction, the capacities at work in understanding others still rely on interactive skills. People, their bodies, their actions, manifest themselves to other people not as inputs but in the richness of a dynamical coupling full of virtual possibilities, even when the agent is not interacting with them directly. We can describe this as interactive, interpersonal, and even linguistic sensitivities (Cuffari, Di Paolo, and De Jaegher 2015). In other words, social agents are pulled into interactions with others to different degrees, and this pull exists even if actual engagement is absent or not possible. These dispositions have been described as readiness-to-interact (Di Paolo and De Jaegher 2012). This embodied pull (much closer in kind to someone actually pulling our body physically away from its current activity than to something we would have to cogitate about before reacting) is, strictly speaking, something that social agents do to each other even when they don't engage directly (it can be appreciated, e.g., in differential neuromuscular activity when embodied gestures are shown to participants, as opposed to other cues; e.g., Sartori et al. 2009; Ebisch et al. 2014). It is not a result of agents' individually controlled agency, or a sensory input to be processed inferentially, but a direct modulation of bodily self-affection by (even remote) others because our bodies are primarily interactive bodies (De Jaegher 2015).

1.4 Deep Entanglement

To understand the embodied pull of the social interactive domain, we need to thematize the deep entanglement of interactive and individual processes, especially in the case of brain dynamics during social interaction. This entanglement is an aspect of the relation that the enactive approach considers prior to and constitutive of both individual and interactive processes involved in embodied intersubjectivity.

Brain-centered approaches assume that brains are nearly decomposable systems (Simon 1962) with respect to body and environment. Nearly decomposable systems interact with other systems without losing their functionality or significantly altering their internal causal relations. One way in which the brain could be treated as such a system is to treat its couplings with body and world as inputs. There are solid arguments against the disposability of body and world for normal brain function. Some are based on the abundant evidence of the entangled neural, bodily, and environmental dynamics in a wide range of cognitive performance (see Anderson, Richardson, and Chemero 2012). A more conceptual argument is Cosmelli and Thompson's (2011) critique of the brain-in-the-vat thought experiment. They argue that it is inconceivable for a brain to retain its functionality if separated from body and world; in other words, the vat and fake input signals fed into it amount to a surrogate body embedded in the world.

This is borne out empirically. Consider the evidence of the entanglement of brain and interaction dynamics observed in dual-scanning experiments (for a review, see Babiloni and

Astolfi 2014). According to Simon (1969, 204), a nearly decomposable system "has the effect of separating the high-frequency dynamics of a hierarchy—involving the internal structure of the components—from the low frequency dynamics—involving interactions among components." Evidence indicates that this is precisely *not* the case during interbrain synchronization in live interactions. Using dual EEG scanning during an imitation task with interactors visibly moving their hands and allowing spontaneous synchrony and turn-taking, Dumas et al. (2010) have found interbrain phase synchronization in the alpha-mu (8–12 Hz), beta (13–30 Hz), and gamma (31–48 Hz) bands. How is it that an interactive pattern appears to affect the oscillation phase of neural groups occurring in two distinct brains at frequencies more than one order of magnitude *faster* than the interactive movements?

A possible answer is that interaction patterns produce an entanglement between the brains of the participants. Internally, the wave of influence across various temporal and spatial scales may travel from low to high frequencies via cyclical variations in neuronal excitability (see Le Van Quyen 2011). These top-down effects have been associated with different cognitive phenomena, notably with the control of visual attention (Buschman and Miller 2007). From here it is not a big leap to suggest that what explains interbrain synchronization at high frequencies in the experiment by Dumas et al. (2010) and in others (e.g., Astolfi et al. 2010) is a combination of high-to-low frequency integration and low-to-high frequency enslavement, with the difference that, instead of slow neural oscillations, the processes "at the top of the hierarchy" are the emergent rhythms of social interaction.

This interpretation is in line with calls to investigate the braided coordination of neural, behavioral, and social processes (Dumas, Kelso, and Nadel 2014). It also coheres with cumulative evidence of the brain-body as an interaction-dominant system (the opposite of a nearly decomposable one), based on the variability across a wide range of temporal scales in neural processes and behavior (Kelso, Dumas, and Tognoli 2013; Van Orden, Holden, and Turvey 2003). Interaction-dominant systems are characterized by the causal inextricability of the various component processes involved, as well as the unpredictability of the behavior of the whole from knowledge of the isolated parts.

Some evidence of interaction dominance has also been found to involve extraneural factors (e.g., in tool use; Dotov, Nie, and Chemero 2010). Others involve social interaction patterns, which themselves show signatures of interaction-dominant dynamics (e.g., Richardson, Marsh, and Schmidt 2010; Riley et al. 2011; Abney et al. 2014; Fusaroli, Rączaszek-Leonardi, and Tylén 2014).

2 Not One, Not Two

Studying the human mind is a complex endeavor. It would seem that our insistence on the different ways in which extraneural and interpersonal factors can play constitutive roles in particular instances of social understanding only complicates the task beyond hope. This is a hasty reaction. After all, the evidence discussed in the previous section has been gathered with existing scientific methods. And in some cases, like that of perceptual crossing, broadening the range of phenomena under study to include collective dynamics actually simplifies scientific explanation. Choosing the right level for attempting to explain social cognitive phenomena has thus become an additional task for researchers. It used to be a choice dictated by tradition—a neuroscientist would limit herself to looking at brain activity under a range of relevant independent variables, a social psychologist would care about the subject's self-conception and cognitive skills in relation to a situation and to others, and a social scientist might offer structural accounts about expected behavior under given sociocultural norms. Now choosing the relevant explanatory factors necessitates some kind of justification not only in theoretical terms, when this is possible, but also in terms of an awareness of the kind of explanations that we require given the pragmatic context of interest (Garfinkel 1981).

The scientific and philosophical task of understanding embodied intersubjectivity requires an awareness of phenomena outside the boundaries of particular academic traditions. Multidisciplinary efforts, with their difficulties and pitfalls, become unavoidable as we approach a theoretically loaded scientific picture of intersubjective phenomena. However, this must not be read as a recommendation to only approach specific problems by forming multidisciplinary teams. It is fair to think that research within a given discipline will almost surely also be informative about the bigger picture. What is required, however, is an awareness of how a piece of knowledge fits this picture. It is sadly frequent to encounter fashionable, one-size-fits-all solutions, especially in neuroscience, once an important result promises to explain a wide range of phenomena (think mirror neurons). Often such results provide crucial clues to a mystery that has not been well formulated in the first place. The problem facing the scientific and philosophical study of social understanding is to figure out how various sources of knowledge fit together.

2.1 The Trouble with Hybrids

A typical response to the situation we are describing is the emergence of hybrid proposals. Hybrid conceptions of social cognition are driven by the uncritical adoption of a distinction between "online" and "offline" cognition (e.g., Bohl and Van Den Bos 2012; see also De Jaegher and Di Paolo 2013; Gallagher 2015). Online cognition is thought to be involved primarily in direct embodied engagements with the world and with others, while offline cognition concerns "higher" mental functions such as planning and inferences. Some researchers recognize that interactive engagements demand a form of practical embodied coping that is not under the full control of an individual participant. But they suspect that the individual resources that allow this coping involve direct action-based mechanisms that differ from those used in more reflective tasks, such as figuring out the intentions of a remote other or a character in a film. This distinction leads to the proposal of hybrid, two-systems solutions, which are not unlike those already proposed in other contexts as a response to embodied critiques of functionalism (Wheeler 2005). The endgame of hybrid proposals is inevitably a restoration of mental representations.

Proponents of hybrid conceptions worry that enactive accounts of sense-making can merely handle low-level and immediately present or "concrete" phenomena, and that interactive phenomena belong to this class. "The challenge for enactivism will be to show how a richer notion of coupling can be put to work to explain the development from low-level to high-level social cognition" (de Bruin and de Haan 2012, 246). Whereas embodied and dynamical aspects—"resonance," facial mimicry, hand-holding, dancing—can be explained enactively, complex forms of social cognition, such as interpreting mental attitudes, collaborative planning, and so on, cannot reduce to participatory sense-making, say the challengers.

In response, we insist that in the enactive account the distinctions between "low" and "high" levels, and between "online" and "offline" cognition, are the first ones to go. Such widespread, seemingly intuitive distinctions belie the inextricable Cartesian roots hiding in everyday language and the common sense from which they originate. They mischaracterize differences in cognitive complexity and reflexivity as levels of body involvement. I am no less embodied and coupled to the world when I plan my holidays than when I ride a bike; I'm simply doing different things with my body and coupling. To preconceive these differences as high versus low dichotomies is to use different names for the separation between body and mind. Adopting this terminology means implicitly buying into a dualistic perspective. Uncritically assuming this break in the formulation of the question is unlikely to lead to answers that do not perpetuate its inherent dualism. Hybrids of this kind are Cartesian, whether we admit this or not.

Some consider that "higher" forms of social cognition must be "decoupled" (de Bruin and Kästner 2012). These views are built on an erroneous conflation of the operational conditions of cognitive processes with the meaning achieved by a cognizer thanks to those processes. They also suffer from an erroneously reductive understanding of what it means for something to be "right here." In contrast, the point of enaction is to offer an interpretation of "living system" and "niche" that locates meaning, including what may appear to us uncritically as "internal," "detached," "offline," or "abstract" meaning, in the changing relation between the two.

Though we may describe our cognitive abilities as transporting us "beyond" the present moment, it is misguided to attempt to locate the cognitive abilities themselves somewhere beyond the present moment. A fallacy of misplaced concreteness is at play here, which is to associate meaning with coupling. How could we ever mean what we are not coupled to? The question is absurd, like thinking that a house cannot be built with bricks because no brick is as tall as a house. Meaning is the relational activity of sense-making, which holistically involves the autonomous agent's adaptive modulation of its own dynamical tendencies and its coupling with the world (Di Paolo 2005, 2009, 2015; Thompson 2007). Never in any of the descriptions of sense-making in the enactive literature has meaning been equated with coupling. The coupling between agent and environment is only one element in the sense-making process, certainly not the bearer of any meaning by itself (another remnant of representationalist thinking: meaning as content moved about in vehicles). Human sense-making involves a range of sensitivities, including interactive and linguistic sensitivities. When we couple dynamically with other humans, our sense-making reaches a linguistically mediated and layered world of meanings and norms.

In a complex everyday situation, say a group of friends splitting the bill after dinner together, there is no a priori separation between symbolic mental capabilities (e.g., dividing the cost of the wine), subtle sensitivities to socially relevant facts (one of the friends has recently become unemployed), and interactive and affective dimensions of the situation. Sophisticated skills are at play in managing all these different angles in the here and now, and in a situated sense. It is concrete cognition involving some of the most complex human mental capabilities. Further evidence of the breakdown of dualistic distinctions between high and low, or online and offline, cognition is given by Fusaroli, Rączaszek-Leonardi, and Tylén (2014), who discuss the synergistic aspects of real-life conversations that range from the entrainment of physiological variables to the complex coordinations of vocabulary choices and their effect on the rhythms of the dialogic exchange. This inextricability of various levels and timescales is confirmed by evidence of complexity matching in natural conversations (Abney et al. 2014), where synergies appear at several scales from phoneme intonation to syntactic, lexical, and semantic structures. In view of this evidence and for the foregoing theoretical reasons, to adopt an offline/online or low-level/high-level distinction is, at the least, a risky strategy.

2.2 Culture and the Dialectics of Participatory Sense-Making

A different way to approach the problem of how to determine and study the relations between subpersonal, personal, and interpersonal factors is to approach the question in a more principled manner, by looking at the implications of the conceptual categories introduced by enactive theory.

At the core of the enactive conception of mind is the living body as an active network of processes that sustain the organism's identity under precarious conditions. In looking at the relation of life to matter, Hans Jonas (1966, 80) describes it as a "dialectical relation of needful freedom." Because of this, the organism is never passive. The need for active regulation is implied in the primordial tension of life: materials are essential to the living organism, but its identity is dynamic, not tied to the individuation of material constituents but emerging instead as the (risky) ongoing adventure of "riding" material changes "like a crest of a wave" and "as its [the organism's] own feat" (80). The enactive view of life is inherently dynamic and inherently "at risk" because the overcoming of a primordial tension between self-production and self-distinction is an ongoing achievement. Life is always in a dynamic transient, not just empirically but constitutively as the only way of managing the tensions between its own opposing trends.

We find a similar tension in participatory sense-making, and it originates in the two requirements of the definition of social interaction. As we have said, participatory sense-making, in the more general sense, describes the situation in which the sense-making of two or more autonomous agents is mutually modulated as they engage in an interactive encounter. The relational patterns of social interaction also form an autonomous, self-sustaining identity in the space of coordinated and uncoordinated relational "moves." In this basic form, participatory sense-making can happen without sophistications such as the recognition of the other agent as an other. From the perspective of an individual agent, basic participatory sense-making can be experienced as no more than a special sort of engagement with the world, one where the agent's regulation of coupling may be contingently thwarted, extended, challenged, or changed by the interaction dynamics.

The two forms of autonomy in participatory sense-making (individual and interactional) establish a *primordial tension*, one that is managed in increasingly complex ways but never fully disappears. It is important to notice that the tension is *not* between the different participants. The primordial tension is more subtle and pervasive; it is in place even if the other is not present as an other, but manifests as relational patterns that affect my sense-making and are affected by it, such that a social interaction is sustained in time (as in the perceptual crossing experiment). To repeat, the primordial tension, describable only because we have defined social interaction in terms of autonomy, is, from the beginning, not between individuals but between an individual and an interactive (social) order.

An agent acts and makes sense according to her individual embodied norms. These norms relate to the continuity of various forms of autonomous identity or forms of life converging in her body (norms that are biologically, socially, and habitually acquired). However, in an interactive situation, these norms may either accompany or conflict with the autonomous relational dynamics of the encounter. In pursuit of an individual intention, the acts performed by an agent during a social encounter sometimes fuel the interaction process but, through their effect on the social coupling, end up frustrating, in apparent paradox, the originally intended goal (e.g., the classical narrow-corridor example where two people walking in opposite directions get stuck trying to get past each other). The social encounter has an interactive normativity different from individual norms. Independently of how acts and events are evaluated by the participants, they may or may not contribute to the self-sustaining logic that belongs to the social encounter. An individual participant will sometimes perceive a mismatch between what she intends and what actually happens that in general contrasts with noninteractive situations, and this mismatch, this form of heteronomy from the agent's perspective, has its origins in the double normative dimension of participatory sense-making.

How is this tension managed? Both synergy and conflict between interactive and individual normativity have implications for individual acts. These acts suffer an analogous doubling of their nature: they are the acts of an individual agent, but they are also moves in an interactive encounter. In the case of conflict, breakdowns occur, and the space of opportunities for accommodating these breakdowns is where the participatory labor of (re-)creating sense occurs. These are novel sources of frustration that do not occur in solitary existence. In cases of synergy between individual and interactive normativity, acts acquire a "magic power." They achieve more than I intend to. Conversely, I can achieve what I individually intend to with less, through the coordinated completion of the act by the other. Some acts become inherently social; those acts necessarily seek to coordinate interactive and individual normativity.

A single agent cannot manage the tension between acts that simultaneously exist in two different normative domains. On her own, she cannot reliably control the synergy between individual intentions and interactive normativity without terminating the conditions for social interaction (i.e., without overriding the autonomy of other participants). She can only regulate her own coupling to the world in relation to her own individual norms. We have no reason to assume that the credit for sustaining ongoing interactions belongs to individual normativity (the definitions of agency and participatory sense-making do not assume prosociality). Instead, to manage the tension between individual and interactive norms, the regulation of the interactive coupling must involve other participants. Therefore the recovery of interactive breakdowns requires a coregulation of the interactive coupling. This coregulation is directed at managing the mismatches between individual intentions of all participants and the interactive dynamics. This is what we call *social agency*, a specific kind of participatory sense-making whereby the agents not only individually regulate their own couplings and influence other agents but also jointly regulate the mutual coupling (following norms that pertain to the interactive situation).

From this starting point, it is possible to follow the consequences of this dialectical self-deployment of the notion of participatory sense-making. Here we summarize in part the description by Cuffari, Di Paolo, and De Jaegher (2015) of how the primordial tension is transformed in each subsequent attempt at resolving its current manifestation. Along the way, increasingly complex forms of participatory sense-making emerge. Co-regulation leads to the appearance of properly social acts, that is, acts whose conditions of satisfaction cannot be reached by a single individual, such as the act of giving/accepting or of shaking hands. The primordial tension is transformed into the question of how to coordinate the partial contributions to social acts. This can be facilitated by the recursive use of social acts to coordinate other social acts (such as a nod to encourage a particular response in another participant). Social acts become normatively evaluated not only regarding their efficacy as such, but also regarding their power to regulate other social acts.

New manifestations of the primordial tension arise as we enter the community level, where some regulatory acts become strongly normative and therefore meet the requirements of portability among different groups. Ultimately a sociocultural normativity emerges that belongs not just to the interactors here and now but to the larger community in history.

However, the existence of strongly normative acts, such as a loud call to attention, generates a new manifestation of the primordial tension. Because what prevents the abuse of such strong acts? What prevents certain interactors from acquiring a dominant role in all their interactions? Dominated interactions move toward the loss of the autonomy of some participants; so either they cease to be interactions, or a new form of regulation must emerge, one that allows dominant roles to be exchanged in time: a dialogical structure and a mutual recognition between participants as autonomous others. Interactions structured dialogically introduce the phenomenon of the "utterance" (a turn-delimited act) and the conflicting requirements of ease of production and ease of interpretation. The capability of interpreting utterance implies the possibility of self-interpretation and eventually social self-control (the use of dialogical mediation on our own individual agency). Cultural patterns facilitate the tensions of producing and interpreting utterances by precoordinating situation-based interaction styles, or participation genres (salutations, queuing, moving things together, sitting at a waiting room, etc.). Ultimately these categories (social agency, dialogical structure, recognition, self-control, participation genres, etc.) can be used to clarify how we move from the most general forms of participatory sense-making to one of the specialized ones: languaging.

The lesson is in the method. It does not suffice to state obvious facts (cultural factors play a role in the human mind, social norms influence our way of perceiving the world, language strongly impacts conceptual thinking, and so on). Such statements can be true and still leave researchers in the dark as to what is the most useful way of investigating the phenomena. What is necessary is to ascertain whether we already possess the right conceptual categories to describe such broad relations or whether our epistemic instruments are still too blunt.

The dialectics of participatory sense-making are one possible way of generating the necessary concepts. Had we not attempted to do this, we would have little to go on apart from the vague intuition that the fact that we alter our sense-making together with others must somehow relate to sophisticated skills such as our use of language. Once we have done this, then it is possible to consider developmental models that take into account how forms of social agency change by the fact that infants are always already embedded in an encultured, enlanguaged world (Cuffari, Di Paolo, and De Jaegher 2015; Di Paolo 2016).

Conclusion

The point of our discussion is not to insist that everything matters and we must always take into account all the processes involved in intersubjective phenomena if we want to study a particular instance of social action or social understanding. We do not need the ongoing, concurrent involvement of neurophysiologists, psychologists, and sociologists for each aspect of social cognition that we study in the lab or in a natural setting. What we do need is a theory of intersubjectivity that allows us to determine, in each particular instance, the relevant factors that we should pay attention to. The same theory should also allow us to know the significance of what we are *not* including and the reach of our particular investigation. Such decisions are still made intuitively, or by following a tradition. If we work in a neuroscience department, we will be encouraged to look for answers in the activity of the brain; if we are social scientists, we will tend to look for social normative accounts. The knowledge that is produced in this way can be useful, but we are often unaware of its limiting assumptions or of the reasons why it is applicable to certain situations and not others. Perhaps it is time for such decisions to be made in accordance with an overarching theory of intersubjectivity.

References

Abney, D. H., A. Paxton, R. Dale, and C. T. Kello. 2014. Complexity matching in dyadic conversation. *Journal of Experimental Psychology: General* 143:2304.

Anderson, M. L., M. J. Richardson, and A. Chemero. 2012. Eroding the boundaries of cognition: Implications of embodiment. *Topics in Cognitive Science* 4:717–730.

Astolfi, L., J. Toppi, F. de Vico Fallani, G. Vecchiato, S. Salinari, D. Mattia, F. Cincotti, and F. Babiloni. 2010. Neuroelectrical hyperscanning measures simultaneous brain activity in humans. *Brain Topography* 23:243–256.

Auvray, M., C. Lenay, and J. Stewart. 2009. Perceptual interactions in a minimalist virtual environment. *New Ideas in Psychology* 27:32–47.

Babiloni, F., and L. Astolfi. 2014. Social neuroscience and hyperscanning techniques: Past, present, and future. *Neuroscience and Biobehavioral Reviews* 44:76–93.

Barandiaran, X., E. Di Paolo, and M. Rohde. 2009. Defining agency: Individuality, normativity, asymmetry and spatio-temporality in action. *Adaptive Behavior* 17:367–386.

Bohl, V., and W. Van Den Bos. 2012. Towards an integrative account of social cognition: Marrying theory of mind and interactionism to study the interplay of Type 1 and Type 2 processes. *Frontiers in Human Neuroscience* 6:274.

Buschman, T. J., and E. K. Miller. 2007. Top-down versus bottom-up control of attention in the prefrontal and posterior parietal cortices. *Science* 315:1860–1862.

Clark, A., and C. Thornton. 1997. Trading spaces: Computation, representation, and the limits of uninformed learning. *Behavioral and Brain Sciences* 20:57–66.

Cosmelli, D., and E. Thompson. 2011. Brain in a vat or body in a world: Brainbound versus enactive views of experience. *Philosophical Topics* 39:163–180.

Cuffari, E., E. Di Paolo, and H. De Jaegher. 2015. From participatory sense-making to language: There and back again. *Phenomenology and the Cognitive Sciences* 14:1089–1125.

de Bruin, L., and S. de Haan. 2012. Enactivism and social cognition: In search of the whole story. *Journal of Cognitive Semiotics* 4 (1): 225–250.

de Bruin, L. C., and L. Kästner. 2012. Dynamic embodied cognition. *Phenomenology and the Cognitive Sciences* 11:541–563.

De Jaegher, H. 2015. How we affect each other: Michel Henry's "pathos-with" and the enactive approach to intersubjectivity. *Journal of Consciousness Studies* 22:112–132.

De Jaegher, H., and E. Di Paolo. 2007. Participatory sense-making: An enactive approach to social cognition. *Phenomenology and the Cognitive Sciences* 6:485–507.

De Jaegher, H., and E. Di Paolo. 2013. Enactivism is not interactionism. *Frontiers in Human Neuroscience* 6:345.

De Jaegher, H., E. Di Paolo, and S. Gallagher. 2010. Can social interaction constitute social cognition? *Trends in Cognitive Sciences* 14:441–447.

Di Paolo, E. A. 2005. Autopoiesis, adaptivity, teleology, agency. *Phenomenology and the Cognitive Sciences* 4:97–125.

Di Paolo, E. A. 2009. Extended life. *Topoi* 28:9–21.

Di Paolo, E. A. 2015. Interactive time-travel: On the intersubjective retro-modulation of intentions. *Journal of Consciousness Studies* 22:49–74.

Di Paolo, E. A. 2016. Participatory object perception. *Journal of Consciousness Studies* 23(5–6): 228–258.

Di Paolo, E. A., and H. De Jaegher. 2012. The interactive brain hypothesis. *Frontiers in Human Neuroscience* 6:163.

Di Paolo, E. A., M. Rohde, and H. De Jaegher. 2010. Horizons for the enactive mind: Values, social interaction, and play. In *Enaction: Toward a New Paradigm for Cognitive Science*, ed. J. Stewart, O. Gapenne, and E. A. Di Paolo, 33–87. MIT Press.

Di Paolo, E. A., and E. Thompson. 2014. The enactive approach. In *The Routledge Handbook of Embodied Cognition*, ed. L. Shapiro, 68–78. Routledge.

Dotov, D. G., L. Nie, and A. Chemero. 2010. A demonstration of the transition from ready-to-hand to unready-to-hand. *PLoS One* 5 (3): e9433.

Dumas, G., J. A. S. Kelso, and J. Nadel. 2014. Tackling the social cognition paradox through multi-scale approaches. *Frontiers in Psychology* 5:882.

Dumas, G., J. Nadel, R. Soussignan, J. Martinerie, and L. Garnero. 2010. Inter-brain synchronization during social interaction. *PLoS One* 5:e12166.

Ebisch, S. J., F. Ferri, G. L. Romani, and V. Gallese. 2014. Reach out and touch someone: Anticipatory sensorimotor processes of active interpersonal touch. *Journal of Cognitive Neuroscience* 26:2171–2185.

Froese, T., and E. A. Di Paolo. 2011. The enactive approach: Theoretical sketches from cell to society. *Pragmatics and Cognition* 19:1–36.

Fuchs, T., and H. De Jaegher. 2009. Enactive intersubjectivity: Participatory sense-making and mutual incorporation. *Phenomenology and the Cognitive Sciences* 8:465–486.

Fusaroli, R., J. Rączaszek-Leonardi, and K. Tylén. 2014. Dialog as interpersonal synergy. *New Ideas in Psychology* 32:147–157.

Gallagher, S. 2015. The new hybrids: Continuing debates on social perception. *Consciousness and Cognition* 36:452–465.

Garfinkel, A. 1981. *Forms of Explanation*. Yale University Press.

Gouin, J.-P., C. S. Carter, H. Pournajafi-Nazarloo, R. Glaser, W. B. Malarkey, T. J. Loving, J. Stowell, and J. Kiecolt-Glaser. 2010. Marital behavior, oxytocin, vasopressin, and wound healing. *Psychoneuroendocrinology* 35:1082–1090.

Herschbach, M. 2012. On the role of social interaction in social cognition: A mechanistic alternative to enactivism. *Phenomenology and the Cognitive Sciences* 11:476–486.

House, J. S., K. R. Landis, and D. Umberson. 1988. Social relationships and health. *Science* 241:540–545.

Jonas, H. 1966. *The Phenomenon of Life: Toward a Philosophical Biology*. Harper & Row.

Kelso, J. A. S., G. Dumas, and E. Tognoli. 2013. Outline of a general theory of behavior and brain coordination. *Neural Networks* 37:120–131.

Kiecolt-Glaser, J. K., T. J. Loving, J. R. Stowell, W. B. Malarkey, S. Lemeshow, S. L. Dickinson, and R. Glaser. 2005. Hostile marital interactions, proinflammatory cytokine production, and wound healing. *Archives of General Psychiatry* 62:1377–1384.

Krahé, C., A. Springer, J. A. Weinman, and A. Fotopoulou. 2013. The social modulation of pain: Others as predictive signals of salience—a systematic review. *Frontiers in Human Neuroscience* 7:386.

Le Van Quyen, M. 2011. The brainweb of cross-scale interactions. *New Ideas in Psychology* 29:57–63.

Merleau-Ponty, M. [1945] 2012. *Phenomenology of Perception*. Trans. D. A. Landes. Routledge.

Reddy, V. 2001. Infant clowns: The interpersonal creation of humour in infancy. *Enfance* 3:247–256.

Reddy, V. 2008. *How Infants Know Minds*. Harvard University Press.

Richardson, M. J., K. L. Marsh, and R. C. Schmidt. 2010. Challenging egocentric notions of perceiving, acting, and knowing. In *The Mind in Context*, ed. L. F. Barrett, B. Mesquita, and E. Smith, 307–333. Guilford.

Riley, M. A., M. J. Richardson, K. Shockley, and V. C. Ramenzoni. 2011. Interpersonal synergies. *Frontiers in Psychology* 2:38.

Sartori, L., C. Becchio, B. G. Bara, and U. Castiello. 2009. Does the intention to communicate affect action kinematics? *Consciousness and Cognition* 18:766–772.

Simon, H. A. 1962. The architecture of complexity. *Proceedings of the American Philosophical Society* 106:467–482.

Simon, H. A. 1969. *The Sciences of the Artificial.* MIT Press.

Thompson, E. 2007. *Mind in Life: Biology, Phenomenology, and the Sciences of Mind.* Harvard University Press.

Turk, D. C., R. D. Kerns, and R. Rosenberg. 1992. Effects of marital interaction on chronic pain and disability: Examining the downside of social support. *Rehabilitation Psychology* 37:259–274.

Van Orden, G. C., J. G. Holden, and M. T. Turvey. 2003. Self-organization of cognitive performance. *Journal of Experimental Psychology: General* 132:331–350.

Varela, F. J. 1997. Patterns of life: Intertwining identity and cognition. *Brain and Cognition* 34:72–87.

Varela, F. J., E. Thompson, and E. Rosch. 1991. *The Embodied Mind: Cognitive Science and Human Experience.* MIT Press.

Wheeler, M. 2005. *Reconstructing the Cognitive World: The Next Step.* MIT Press.

5 Continuity Skepticism in Doubt: A Radically Enactive Take

Daniel D. Hutto and Glenda Satne

The difference in mind between man and the higher animals, great as it is, certainly is one of degree and not of kind.

—Charles Darwin, *The Descent of Man*, 1871

Introduction: Getting Radical about the Origins of Content

Enactivists of all sorts emphasize the role of active, embodied engagement over representation when it comes to understanding cognition. For radical enactivists about cognition, RECers, this is not just a matter of emphasis: they advance a stronger claim, holding that (1) not all cognition is content involving and, especially, not its root forms (Hutto and Myin 2013).[1] Even so, RECers are not content deniers; they do not embrace global eliminativism about content. RECers hold that if appearances don't deceive us, then (2) some thoughts and linguistic utterances are contentful. Indeed, RECers allow that such thoughts and utterances— judgments about factual matters—are contentful in the full-blooded sense of exhibiting the familiar semantic properties of reference and truth.[2] That (2) is the case not only appears to be borne out by experience but is necessary for explaining certain features of at least some forms of cognition. What's more, RECers are naturalists, albeit of a relaxed sort. They hold that (3) it is possible, in principle, to explain the origins of content-involving cognition in a scientifically respectable, gapless way. RECers aim to do so by appeal, in large part, to the important role played by sociocultural scaffolding (Hutto and Myin 2013; Hutto and Satne 2015).

1. The notion of content that REC rejects is one that assumes the existence of some kind of correctness condition according to which the world is specified as being in a certain way (e.g., "taken," "said," "represented," or "claimed" to be in a certain way) that it might not be in.

2. To accept that some forms of thought and language involve reference and truth conditions is not, of course, to endorse the reductive thesis that all uses of language must be contentful in this sense.

This chapter responds to several accusations that a REC-inspired program for explaining the natural origins of content—the NOC program, for short—is doomed to fail. Section 1 responds to a preliminary general concern that the NOC program is internally incoherent when seen in light of the Hard Problem of Content—a problem identified by RECers themselves. Section 2 considers a different, more softly pitched complaint against the NOC program—namely, that in drawing a sharp distinction between basic, contentless, and content-involving kinds of cognition, REC gives succor to continuity skepticism, the specific complaint being that REC is at odds with evolutionary continuity. Section 3 casts doubt on the idea that REC motivates this kind of continuity skepticism by offering a sketch of how the natural origins of content could be explained in a gapless, REC-friendly way that does not violate evolutionary continuity. Finally, Section 4 considers how REC fares against a different, philosophically motivated variety of continuity skepticism. Although we conclude that REC cannot quell the skeptical worries that a philosophically based continuity skepticism generates, we argue that REC's representationalist rivals fare no better against this brand of skepticism. Thus, in the final analysis, we have good reasons to doubt that a REC-inspired NOC program promotes or is particularly prone to skepticism about continuity.

1   A Fatal Dilemma?

The three REC assumptions outlined in the introduction are incompatible, some claim (Alksnis 2015; Korbak 2015). If these critics of REC are right, its central commitments do not form a coherent set. Their complaint is driven by the impression that RECers are inconsistent in the way they use the Hard Problem of Content, or HPC. RECers invoke the HPC to motivate the adoption of a contentless view of basic minds. But RECers seemingly ignore the HPC's force, selectively, when allowing that some minds—the subset that have benefited from the right kind of scaffolding—have a contentful character, and seek to explain how this can be so. Yet, so the critics of REC insist, if there is a HPC at all, then it must afflict *any and all* attempts to explain how content could be part of the natural order equally. In the view of REC's critics, the HPC is thought to be a universal acid, which, once out of its bottle, cannot be contained in the way RECers hope to stopper it when pursuing the NOC program.

To be consistent, RECers should hold either that (A) content can be understood in a suitably light way such that it makes an appearance wherever we find minds, or that (B) the existence of content should be denied as a naturally occurring phenomenon across the board.[3] For those who see the situation in this all-or-nothing manner, anyone hoping to explain how content might emerge through sociocultural scaffolding, as RECers do, faces not

---

3. In an extreme statement, meant to capture the options available in this apparently forced choice, Korbak maintains that either "basic cognitive systems do genuinely communicate because they are alive, or nothing (not even [Hutto and Myin] themselves) does" (Korbak 2015, 95).

merely a difficult challenge but an intractable dilemma. As Alksnis formulates the problem: "The first horn of the dilemma is a rejection of the compatibility of content with naturalism; the second is the rejection of content in order to preserve naturalism" (Alksnis 2015, 674). According to these assessments, as long as RECers stick with explanatory naturalism, they will find it impossible, in principle, to explain how creatures that begin life with only basic contentless minds could ever come to have minds of a content-involving sort.

Undeniably, RECers make much of insuperable difficulties that the HPC poses for restrictive naturalists—namely, naturalists who limit themselves to *using only a narrow set of resources* when accounting for the place of contentful states of mind in the natural order. That is true enough, but there is no inconsistency in holding that the HPC is hard or even impossible for some explanatory naturalists but not others. The HPC, RECers claim, does not apply universally to all and every variety of explanatory naturalism. It may be impossible to get water from a stone, by any means and under any circumstances, even if it is entirely possible to get water from a sponge when certain conditions are met. By analogy, RECers think that while it is impossible to explain content using only the limited resources of restrictive naturalism, it is entirely possible to explain the origins of content, at least in principle, using the expanded resources afforded by a relaxed naturalism, and to do so in a way that does not presuppose the prior existence of content (Hutto and Satne 2015). A relaxed naturalism is one that avails itself of a wide range of scientifically respectable resources, drawing on the findings of numerous sciences that include not just the hard sciences but also cognitive archaeology, anthropology, developmental psychology, and others.

Most explanatory naturalists are not relaxed naturalists but restrictive ones: they seek to naturalize content by using only the resources of the hard, natural sciences (causation, informational covariance, biological functionality) and nothing more. Such naturalists see no prospect of trying to explain content by appeal to sociocultural factors. There are two main motivations—one general, one specific—for wanting to impose such restrictions, and it is important to disentangle these motivations.

Some strict naturalists are motivated by an uncompromising unification agenda—one that demands reductionist explanations for any phenomenon if it is to qualify as bona fide natural (Rosenberg 2015; Abramova and Villalobos 2015). Extreme naturalists hold that "reality contains only the kinds of things that the hard sciences recognize" (Rosenberg 2014a, 32). They also insist that "natural science requires unification" (Rosenberg 2014b, 41). The combination of these views adds up to zero tolerance for any phenomenon or domain that will not reduce. For example, Rosenberg holds that "science can't accept interpretation as providing knowledge of human affairs if it can't at least in principle be absorbed into, perhaps even reduced to, neuroscience" (2014b, 41).

A serious concern about this austere program is that if naturalists are "too exclusive in what they count as science, naturalism loses its credibility" (Williamson 2014a, 30). An obvious criticism of the extreme naturalism agenda is that it imposes overly strong, ideologically driven constraints on scientific inquiry—constraints that we have little or no reason to suppose will pay off in the end.[4]

There is a second, and prima facie more plausible, rationale for adopting restrictive naturalism, namely, using only the sparse tools of the hard sciences when attempting to naturalize content. This rationale is driven not by general reductive commitments but by the belief that such restrictions are necessary because it is assumed that content *must be* in place before any sociocultural practices appear (or appeared) on the scene. The latter assumption captures the thought that the existence of the relevant kinds of sociocultural practices depends on the logically prior existence of content in ways that make it impossible for sociocultural practices to explain the genesis of content.

Thus if there is content, then restrictive naturalists of this second stripe assume we have no choice but to accept that it originates in states of mind that, necessarily, exist quite independently of, and ontologically prior to, sociocultural practices. This line of argument was made prominent by Fodor (1975) and Searle (1983), and even today influential thinkers take it for granted that "external symbols acquire their meaning from meaningful thoughts—*how could it be otherwise?* ... [External] symbols [cannot] be meaningful independently of the thoughts they have been designed to express. So, if the Hard Problem spells doom for contentful thinking, it ought to spell doom as well for our abilities to understand and use language" (Shapiro 2014, 217–218; italics added).

To assume that content can *only* derive from mental content rules out the possibility of explaining the natural origins of content by appeal to the mastery of sociocultural practices a priori. We are meant to be persuaded by the "how could it be otherwise," "there's no other way" addendum. But surely an alternative explanation is staring us in the face: contentful thoughts only become so when the use of external symbols is mastered. No logical contradiction arises in making such a proposal so long as we assume, along with RE, that contentless basic minds are, at least, possible. If that is allowed, and HPC only spells doom for other restrictive attempts to explain the origins of content without appeal to sociocultural practices, then there are no grounds, pace Shapiro, for ruling out REC-inspired attempts to explain how we come to understand and use contentful language.

To recap, there is no inconsistency in holding, as RECers do, that the HPC troubles only some explanatory naturalists and not others. Nor does Shapiro cite any reason—at least one that does not blatantly beg the question—for denying outright the possibility of naturalistically explaining how contentless minds might become content involving through a process of sociocultural scaffolding.

4. Philosophical criticisms of extreme naturalism are also easy to find. For example, Williamson constructs the following quick but effective argument against extreme naturalism: "If it is true that all truths are discoverable by hard science then it is discoverable by hard science that all truths are discoverable by hard science. But it is not discoverable by hard science that all truths are discoverable by hard science. Therefore the extreme naturalist claim is not true. ... Truth is a logical or semantic property, discoverability an epistemic one, and a hard science a social process" (Williamson 2014b, 37).

2 Evolutionary Discontinuity?

Still, even if we allow that a REC-motivated project of trying to explain the natural origins of content is not internally confused or simply impossible, we might still believe that it is hopeless for a different reason. We might conclude that in drawing a distinction between basic, noncontentful and nonbasic, contentful minds, REC introduces a deep discontinuity in nature (or at least in how we are to understand nature) that is at odds with even a relaxed naturalism. Menary explicates:

> Radicals have a problem bridging the gap between basic cognitive processes and enculturated ones, since they think that meaning, or content, can only be present in a cognitive system when language and cultural scaffolding is present (Hutto and Myin 2013). That, of course, *doesn't sit well with evolutionary continuity*. (Menary 2015, 3n5; italics added)[5]

What exactly is the problem? Why should REC entail evolutionary discontinuity simply by embracing the idea that contentful thinking requires mastery of certain sociocultural practices? A little probing reveals that, according to its critics, the ultimate source of trouble lies with REC's assumption that

> basic minds should not be characterised as representational but that language users take on some of the representational capacities of the language they use; let us call this *the saltationist view*. On the saltationist view representation is only added on as a consequence of using language, narrative, or possibly some other social resources. (Clowes and Mendonça 2015, 17; italics added)

REC is apparently committed to saltationism because it assumes that the arrival of content-involving minds depends on specific kinds of sociocultural practices being in place—such that content is utterly unprecedented in nature. In advancing this view about the origins of content-involving cognition, REC allegedly *"poses a chasm* between representationally enhanced and more basic minds" (Clowes and Mendonça 2015, 18; italics added). And, so the complaint goes, once such a chasm is introduced, we can rule out any possibility of providing a naturalistic explanation of the emergence of content in terms of gradual, continuous change of the sort that evolution favors.

   This kind of worry has been used, dialectically, to supply a reason for embracing a more thoroughgoing representational theory of mind—one that assumes that cognition always involves content (our old friend unrestricted CIC; see Hutto and Myin 2013, 9). Unrestricted CIC apparently offers protection against the nasty consequence of evolutionary discontinuity. A comparison proves instructive. Unlike REC, Sterelny (2015) maintains that although standard-variant teleosemantics are limited in key respects, they nonetheless provide the

5. This comment is made as an aside, in a footnote, but if true, it raises an important challenge for REC. Accusations of discontinuity are, in any case, a running theme in Menary's work. In Menary 2009, he identifies Brentano's formulation of intentionality as a main source of the unwelcome idea that minds are discontinuous with the rest of the world.

right resources for thinking about the content of basic kinds of minds (Millikan 1984, 1993, 2004, 2005; Papineau 1987; McGinn 1989).[6] To assume that some teleosemantics account is essentially correct, if limited, is to assume that minds that occupy the lowest rung of minded-ness are contentful, in a certain rudimentary sort of way. This fits with the CIC assumption that all minds are contentful to some degree. But this still leaves room for CICers to allow that human minds are contentful in uniquely special, impressive ways. Thus, for example, it is open for Sterelny (2015) to recognize the remarkable differences between basic and distinc-tively human forms of cognition while simultaneously holding that these differences should only be deemed to be a matter of degree rather than marking a difference in kind.

The upshot is that if we assume all minds to be contentful, à la CIC, then there is appar-ently no explanatory barrier or difficulty in understanding how "complex agents evolved incrementally from simpler ones" (Sterelny 2015, 552). With the CIC assumption in place, Sterelny is free to hold that once the origins of our scaffolding practices are explained, as well as "how they work, and what their effects are, [then] we are done. There is no extra problem then of explaining intentional content" (562).

In this view, whatever other differences may exist in their respective cognitive profiles, at rock bottom, there is an assumed *"psychological continuity* between human and nonhu-man animals" (Bar-On 2013, 315). What is on offer here is an assurance of gapless evolu-tionary continuity based on the assumption of a fundamental psychological continuity. This psychological continuity putatively exists along the full spectrum of mindedness and is ultimately cashed out in CIC terms. Of course, all of this is perfectly in tune with Dar-win's own take on the issue, and what Penn, Holyoak, and Povinelli (2008, 109) identify as the dominant trend in comparative cognitive psychology, since both regard any relevant cognitive differences between simpler and more complex minds to be a matter of degree and not of kind.

Apparently, going the CIC way with respect to the psychological continuity question has the added advantage of skirting the Scaling Down Objection that Korbak levels against REC

---

6. Teleosemantic theories of content aim to show how "teleology turns into truth conditions" (McGinn 1989, 148). For example, in Millikan's version, very roughly, a device has the teleofunction of representing Xs if it is used, interpreted, or consumed by the system because it has the proper function of representing the presence of Xs. Proper functions are called on to explain how it is that content is fixed by what organisms are supposed to do in their consumptive activity as opposed to what they are merely disposed to do. Sterelny reminds us that "in general, teleosemanticists are representational lib-erals: they are happy to attribute content to the control systems of the simplest of organisms—to bac-teria with their magnetosomes—and to the simple subsystems of more complex agents. For even the simplest organisms systematically register and respond adaptively to some features of their environ-ment: they have control states that vary by design with states of their normal environment and direct behavior that is appropriate, or would be appropriate, were the environment to be in the state to which the control state is supposed to be tuned (for the correlation is imperfect of course)" (Sterelny 2015, 552).

(2015, 92).[7] According to the Scaling Down Objection, "whatever it is that makes human linguistic practices give rise to content should also give rise to content in waggle-dancing honeybees, quorum-sensing in bacteria, and hormones in one's endocrine system" (Korbak 2015, 94). The driving assumption behind the Scaling Down Objection appears to be that, to the extent that content is deemed a naturally occurring psychological phenomenon at all, then, for the sake of consistency, we must find it, at least in some modest form, everywhere in the domain of the psychological. So, again, what makes the unrestricted CIC view attractive, allegedly, is that, unlike REC, it can tell a homogeneous tale about content that "does not need a sophisticated account of spontaneous generation of an elaborate discourse" (Korbak 2015, 93). Thus, following Pattee, Korbak (2015, 93) holds that, on the contrary, "highly evolved languages and measuring devices are only very specialized and largely arbitrary realizations of much simpler and more universal functional principles by which we should define languages and measurements" (Pattee 1985, 26).

A pertinent question to ask about the CIC "content everywhere" proposal is: Must psychological continuity be posited at all to avoid introducing gaps into the evolutionary story? Or more simply: Does evolutionary continuity logically require psychological continuity? Before answering too quickly, we should consider the voiced worry that "the conviction that there *must be* some diachronic emergence story encourages proponents of continuity to overinterpret the mentality and communicative behaviors of existing animal species, and to underplay some of the evidently unique features of human thought and language" (Bar-On 2013, 296; italics added). The risk of overinterpretation worry invites a second question: Is positing CIC-style psychological continuity really the best way to account for the distinctive properties of human minds that mark them out as special? Some decidedly think not; indeed, they take it to be "one of the most important challenges confronting cognitive scientists of all stripes … to explain how the manifest functional discontinuity between extant human and nonhuman minds could have evolved in a biologically plausible manner" (Penn, Holyoak, and Povinelli 2008, 110).

With those concerns in mind, it is natural to ask if there are scientifically tenable alternatives to the psychological continuity demand, or at least the standard unrestricted CIC way of satisfying it, that might account for the important features of human minds that apparently set them apart. As noted, REC assumes that we have good independent reasons for supposing that when it comes to minds, it is a mistake to think of them as possessing content "all the way down." In particular, REC holds, pace Sterelny, that teleosemantics fails to deliver a promised theory of content, and this is precisely because the violation of mere biological norms is never a matter of misrepresenting how things stand with the world. In this, REC agrees with Burge—and many others—in thinking that there is "a root mismatch

---

7. The Scaling Down Objection, as Korbak (2015, 92–93) characterizes it, is an inverted version of the scope objection: the criticism that explanations given in terms of contentless cognition won't scale up (for a discussion, see Hutto and Myin 2013, 45).

between representational error and failure of biological function" (2010, 301; Pietroski 1992; Putnam 1992; Haugeland 1998; Price 2013). It doubts, pace Millikan (2005), that appeals to biological function can solve the rule-following problem.

Price, for example, gives an un-Millikanesque answer to the question of where we find contentful representations in nature. He stresses this requires that it occurs when "something plays a role that is answerable to the kind of 'in-game externality' provided by the norms of the game of [claim making]—the fact that within the game, players bind themselves, in principle, to standards beyond themselves" (Price 2013, 37). This places weight on there being a special sort of practice that opens up the possibility of going wrong and getting things right in ways that do not reduce to purely biological norms or even norms of communal agreement or conformity. The possibility of contentful error requires being a participant in a practice in which the question of truth can arise for what one thinks or says.

According to the REC view, this unique sort of sensitivity to highly particular kinds of norms needed for playing the latter sort of game is a necessary requirement for the existence of contentful ways of thinking and talking. REC holds that the development of such intersubjective practices and sensitivity to the relevant norms comes with the mastery of involving the use of public artifacts. As it happens, this appears only to have occurred with the construction of sociocultural cognitive niches in the human lineage. The establishment and upkeep of sociocultural practices—those that use public representational systems in particular ways for particular ends—are what account for both the initial and the continued emergence of content-involving minds. Only minds that have mastered a certain, specialized kind of sociocultural practice can engage in content-involving cognition. Of course, there are many steps on the way to mastering content-involving practices. But even if some of these steps do require the mastery and use of public symbols, it does not follow that they are somehow content involving, even in a weak sense. For example, consider the alarm calls of vervet monkeys. Despite their sophisticated use of public symbols, these calls nevertheless fail to exhibit the degree of determinacy and context-free characteristics that would qualify them as bona fide referential and claim-making practices (see Hutto 1999, 126). Should creatures with basic minds manage to master more sophisticated practices, they would gain new cognitive properties and become open to new possibilities for engaging with the world and others.

Content-involving minds have features and capacities that other, more basic minds lack. In this sense, they stand apart. This difference can be thought to mark a difference in kind, not just degree, of mindedness. Like Penn, Holyoak, and Povinelli (2008, 110), RECers hold that "Darwin was mistaken: The profound biological continuity between human and nonhuman animals masks an equally profound functional discontinuity between the human and nonhuman mind."

Yet for RECers, the situation is even more complicated than Penn, Holyoak, and Povinelli (2008) make out. This is because REC's distinction between basic, contentless minds and nonbasic, contentful minds does not map neatly onto the nonhuman-animal-versus-human-animal distinction. Plenty of human cognition is basic in the sense that it is contentless; this is true not only of human children who have yet to master the relevant sociocultural practices but also of human adults who have—namely, those who are capable of contentful thought. This follows from the fact that, in the REC picture, even those human minds that become capable of contentful cognition are not *wholly* transformed and thus not fundamentally different from animal minds, pace McDowell (1994).[8]

Instead REC holds that the basic character of cognition—including content-involving human cognition—is always and everywhere interactive and dynamic in character. Content-involving cognition need not be content based; it need not be based in cognitive processes that involve the manipulation of contentful tokens. In an absolutely key respect, in the REC view, human minds—even when content involving—are alike those of all cognitive creatures in terms of their deep, nonrepresentational nature. This is a non-CIC way of understanding the psychological continuity that exists between contentless and contentful forms of cognition. The difference is that REC conceives of the commonalities present in all forms of cognition in terms of interactive engagement (with the world, with others) as opposed to understanding such commonalities in terms of representational content. Thus REC assumes that all cognitive creatures share a basement-level similarity, and—*if we focus on those basic features of cognition*—human beings are not cognitively unique.

Nevertheless, despite this important acknowledgment, REC holds that there is a distinction in kind not just between nonhuman animals and humans but also within the human sphere, which is marked by the fact that *only some* minds are capable of content-involving cognition.

REC assumes that content-involving cognition has special properties that are not found elsewhere in nature, and minds capable of contentful thought differ in kind, in this key respect, from more basic minds. Very well. But does this mean that REC embraces or entails a kind of continuity skepticism? Does REC's insistence that some minds have special, contentful properties that set them apart entail that "we must recognize a sharp discontinuity in the natural history of our species" (Bar-On 2013, 294)?[9]

In assessing this charge, it is important to disentangle two forms of continuity skepticism that are frequently run together. For example, as formulated by Bar-On (2013), continuity

8. For a detailed discussion of how the REC take on the transformative powers of scaffolded minds compares to and differs from that of McDowell's, see Hutto 2006b.

9. Here we focus exclusively on worries that REC promotes what Bar-On (2013) identifies as a radical, deep-chasm, diachronic version of continuity skepticism. This is because the synchronic variant of such skepticism, which focuses on the gaps that exist between currently existing cognitive creatures, is entirely compatible with there having been a full range of intermediaries, with no missing links, over the course of natural history.

skepticism—of the worrisome diachronic variety we are considering here—boils down to the following claim:

There can be *no philosophically cogent or empirically respectable* account of how human minds could emerge in a natural world populated with just nonhuman creatures of the sort we see around us. Few would deny that, biologically speaking, we "came from" the beasts. But the diachronic deep-chasm claim says that we must accept an unbridgeable gap in the natural history leading to the emergence of human minds—or, at the very least, in our ability to tell and make sense of such a history. (Bar-On 2013, 294; italics added)

Let us bracket, for the moment, the question of philosophical cogency, as it needs separate treatment. Rather, let us concentrate on this issue, of primary concern for naturalists: Does going the REC way in understanding the genesis of contentful minds imply an empirically disrespectable discontinuity in our understanding of the fundamental fabric of the world— one that might incline us to think there is a sort of unbridgeable gap or schism in nature? Returning to Menary's (2015) charge, it seems that REC is at odds with evolutionary continuity if and only if the following is true: That REC's claim that content-involving cognition differs from basic, contentless cognition in kind and not just degree, in key respects, entails that any explanation of how content-involving minds might have arisen introduces an unbridgeable or at least a scientifically inexplicable gap in natural history from the evolutionary point of view. Does it?

3   A Sketch of a Possible Story

REC's preferred diachronic explanation of the natural origins of content is kinky. It is kinky because it doesn't play out along a single dimension; it isn't a simple tale of the mere elaboration of existing forms. It is not an account of mere elaborations because it does not see all cognitive properties as just more complex versions of what has come before. It is a multistorey story, one that centrally involves special sociocultural platforms and constructions. REC's tale is one of cognitive niches and how they scaffold and introduce novel cognitive features and capacities. REC assumes that sociocultural practices introduce something genuinely new and qualitatively distinct into the cognitive mix. Through their acquaintance with culture, some cognitive creatures come to be able to think about the world in wholly new ways. Through mastering what is for them novel practices, they become capable of new forms of thinking of a unique kind.

Does assigning a pivotal role to sociocultural niches in the construction of kinky cognitive capacities put REC at odds with a gapless evolutionary account of the origins of content? That would only be the case if there were some explanatory step between (a) contentless minds and (b) content-involving thought for which REC could not account. If we look closely, from the point of view of understanding our natural history, it is difficult to see exactly where the missing step is supposed to be.

In the abstract, the REC recipe for explaining NOC is fairly simple. A first requirement is that some purely biologically based forms of basic cognition are in place and shared across the species. These are a ground-floor requirement and can serve as the first main platform for understanding how and under what conditions contentful forms of cognition could have arrived on the scene.

The positive account on offer draws on, but significantly adapts, some core resources of teleosemantics—the most promising naturalistic theory of content to date—by putting them to a different theoretical use. The teleosemantic apparatus is used to give an account of contentless attitudes exhibiting basic intentional directedness—a.k.a. intentional attitudes—as opposed to providing a robust semantic theory of content. This allows us to understand basic cognition in terms of active, informationally driven, world-directed engagements, where a creature's current tendencies for active engagement are shaped by its ontogenetic and phylogenetic history. Basic minds target, but do not contentfully represent, specific objects and states of affairs. Fundamentally, cognition is a matter of sensitively and selectively responding to information, but it does not involve picking up and processing informational content or forming representational contents.

Target-focused but contentless intentional attitudes exhibit only ur-intentionality. Ur-intentionality can be explicated by making theoretical adjustments to teleosemantics, thus supplying a fundamental explanatory tool for those working in the enactivist framework. It enables enactivists and others to make a clean and radical break with intellectualist ways of thinking about the basis of cognition and opens up the logical possibility of explaining the emergence of contentful states of mind as dependent on mastery of sociocultural practices.

To borrow from Dennett (1995), we can take our biologically inherited form of cognition —our first natures—as a starting point for the development of more sophisticated forms of cognition. That is, we can think of evolution as putting in place platforms that act as launch pads, not leashes. Beyond this, for the sociocultural emergence of content, we need to assume that our ancestors were capable of social processes of learning from other members of the species, and that they established cultural practices and institutions that stabilized over time.

Crucial to NOC's story is that biological capacities gifted by evolution could have given rise to social learning. So far, so good: surely, there is nothing mysterious or gappy on the table yet. The capacities in question can be understood in biological terms as mechanisms through which basic minds are *set up to be set up* by other minds and *to be set off* by certain things. As a baseline, all such setting up requires is that basic minds can target and respond to certain things in ways that are relevant for learning how to do so from their fellows. According to REC's proposed modification to teleosemantics—its teleosemiotic view—individuals can respond to the environment, and to each other, in ways that allow for emulation, imitation, and regulation of what is targeted and how one responds to what is targeted in ways that are characteristic of social learning.[10] We have no reason to suppose that the cognition driving such social engagements and interactions must be grounded in awareness or rules based on representations of any other kind. Rather, all that we need to assume is that normally developing participants in such practices are already set up, nonaccidentally, to target and tune into the expressively rich intentional attitudes of others.

None of the basic cognitive activity just described requires any purposive rule compliance by participants. Positing mechanisms of social conformity would suffice to explain how creatures with only basic minds could come to be set up to be set up (by others) and to set up (others). If such an account is tenable, it could explain how the practice of social learning might get off the ground in an evolutionarily respectable way without bringing any contentful attitudes into the story.

Critics have argued that the REC story, as we have sketched it here, won't fly. But this is not because it is evolutionarily unsound (indeed, quite the contrary); it is because REC's naturalistically respectable resources are too crude to tell the story properly. The REC story is gappy, not because it introduces kinks or evolutionary gaps but because it can't fill in all the relevant details. This gappiness, so its critics claim, is because REC's vision of contentless basic cognition is just too meager and weak to serve as a tenable foundation for the relevant action.

This negative assessment is based on the repeated accusation that REC's content-free account of basic minds reduces to, or has no more resources than, a crude stimulus-response form of behaviorism (O'Brien and Opie 2015; Kiverstein and Rietveld 2015).[11] Should the "mere stimulus-response behaviorism" characterization turn out to be true of REC, then its

10. Tomasello (1999), Rakoczy, Warneken, and Tomasello (2008, 2009), and Csibra and Gergely (2009) provide accounts of capacities for learning and teaching as species-relative, biologically inherited capacities.

11. O'Brien and Opie are perfectly clear in making this charge, and in promoting a CIC alternative on the back of it: "There is a fundamental problem with the idea of contentless intentionality: it's been tried before, and it doesn't work. Back then the scheme was known as 'behaviorism,' rather than 'targeted directedness,' but the two ideas are of a piece. Behaviorists sought to explain animal behavior, including all the complexities of human problem solving and language, in terms of the history of stimulus-response events to which organisms (of each kind) are typically exposed. The bankruptcy of this approach consists in the fact that moment-by-moment stimuli are simply too impoverished to account for the richness, variety, and specificity of the behaviors that animals exhibit. *It just isn't possible* to explain the ability of evolved creatures to selectively engage with features of the environment—in other words, engage in targeted behavior—*without supposing they employ internal states that in some way represent those features.* ... [The NOC program, and by implication REC] misunderstands the broader explanatory project [which is to explain] ... intelligence rather than just intentionality" (2015, 724; italics added). Kiverstein and Rietveld are a tad more cagey in accusing REC of endorsing nothing more than stimulus-response behaviorism, but they too think its vision of ur-intentionality "implies a problematic view of animal behavior as either hard-wired or learned dispositions to respond to fixed and stable

radically enactive account of basic cognition would really be nothing more than a radically enactive account of behavior—REC would be REB, after all. Exposed as REB, REC would clearly lack the resources for supplying a biologically credible story about how creatures, with only contentless states of mind, could have engaged in flexible kinds of basic and social cognition of the sort needed for triangulating in primitive, nonlinguistic ways with each other.

If the "REC is really REB" analysis proves correct, there is no prospect of REC's giving the sort of explanation, as sketched earlier, about how children could have become players in the relevant sociocultural games without assuming the existence of some kind of prior content-ful mentality. REC would have no real chance of accounting for even the most basic forms of social cognition. Making this latter worry explicit, Lavelle argues that REC, explicated only in terms of intentional attitudes directed at natural signs (à la Hutto 2008), encounters spe-cial problems when it comes to explaining "the flexibility of pre-linguistic social interac-tions" (Lavelle 2012, 469). If Lavelle's diagnosis is right, it gives us reason to believe that to engage with others in ways needed to form even the most primitive intersubjective triangles requires participants to have states of mind with content and to be able to ascribe, by some means, contentful states of mind to others. Naturally, these observations, if true, would sup-port the claim that "mindreading is required to learn language in the first place" (Carruthers 2011, 321).[12]

Should the "REC equates to REB" charge hold up, then the possibility of giving a credible REC account of the emergence of content along NOC lines, let alone a gapless one, can be pretty safely ruled out. The big question is: does it? Consider Lavelle's (2012) claim that REC only gives an account of basic cognition in terms of natural signs. Although an account of how basic minds respond to natural signs in a hardwired way is part of the REC story, it is not the whole of it.

RECers nowhere claim that contentless ways of responding to worldly offerings—including the attitudes of others—must be simple, fixed, or slavishly automatic. This is clearly not true of the most basic infantile ways we have of responding to and engaging with others. And even in most cases in which Mother Nature initially fixes intentional targets, it is perfectly consistent with REC that how we respond to such targets can be shaped in dynamic, spontaneous, and context-sensitive ways (see Hutto 2006a, 2006b).

Exactly how such enactive engagements will unfold in any particular situation will depend on a number of hard-to-predict factors about what is targeted and how it is targeted. The

environmental cues. The selective responsiveness of animal behavior which we have taken to be consti-tutive of skills is thereby conceived of as the animal being equipped with mechanisms that are set off only by specific environmental triggers" (2015, 708).

12. Mind reading is here defined as a process that involves the attribution of mental contents in ways that involve the manipulation of contentful mental representations—whether theoretical inferences, simulations, or both.

pattern of current responses depends on individuals' past interactions with similar situations and the unique features of the current context. Such factors make a difference not only to the manifest intentional attitudes of participants but also to how they respond to the manifest intentional attitudes of others. Yet there is no need for the participants in such engagements (or some subpersonal cognitive system within them) to be aware of, or have knowledge of, or represent the steps of these dynamical processes to respond flexibly to situations.

That REC does not reduce to REB can clearly be seen in the way it allows that basic minds can have quite flexible and expressive modes of operation. REC is perfectly congenial with, and embraces, Bar-On's (2013) account of expressive attitudes given in terms of animal signs that are not mere natural signs.[13]

By Bar-On's lights, expressive attitudes of the kind that feature in prelinguistic triangulation are not to be understood in terms of natural signs precisely because responses to natural signs alone would not account for the way animal mental capacities are deployed in interesting and flexible ways. Animal signs cannot be understood as mere triggered reactions to the reliable indication of proximal stimuli. Rather, they are subtle and adjustable responses to sophisticated patterns of expressive behavior; to understand them requires getting a grip on "complex networks of animal communication" (Bar-On 2013, 300). The richly expressive behaviors in question include, among their number, "yelps, growls, teeth-barings, tail-waggings, fear barks, and grimaces, lip smacks, ground slaps, food-begging gestures, 'play faces' and play bows, copulation grimaces and screams, pant hoots, alarm, distress, and food calls, grooming grunts, ... so on" (Bar-On 2013, 317).

Sensitive responses to such expressive behaviors can be understood fully in terms of contentless intentional attitudes being directed at other contentless but expressive intentional attitudes. As Bar-On makes clear:

Acts of expressive communication often involve an overt gaze direction, head tilt, or distinctive bodily orientation guiding the receiver's attention not only to the expressive agent's affective state but also to the object of that state—the source or target of the relevant state. [For example] ... a dog's cowering demeanor upon encountering another will show to a suitably endowed recipient the dog's fear (kind of state), how afraid it is (quality/degree of state), of whom it is afraid (the state's intentional object), and how it is disposed to act—for example, slink away from the threat (the state's dispositional "profile"). (Bar-On 2013, 320)

The important thing to note about expressive signs is, as Medina highlights, that they should not be understood

on the Gricean model of conventional signs, that is, as involving or requiring fully formed communicative intentions and internal representations. Expressive behavior is not self-reflective intentional-inferential communication among rational agents who are representing each other's minds and their

13. In defending REC on precisely this score, Medina relates that for Bar-On (2013), "expressive behavior is far richer than mere 'brute signaling' or causally produced indication" (Medina 2013, 324).

contents. The production and the uptake of expressive behavior place much weaker representational demands on their producers and responders than self-reflective intentional-inferential communication does. ... [Even so] expressive acts have a significant degree of spontaneity that distinguishes them from automatic physiological reactions. (Medina 2013, 326)[14]

These observations are revealing in two crucial respects. First, they show that in offering an account of contentless minds and ur-intentionality, REC does not reduce to REB. It should be clear now that the "nothing but REB" charge is a complete red herring: it seriously underestimates and mischaracterizes REC's resources. Second, it should also be clear that it is entirely possible to offer alternatives to mind-reading explanations of basic social cognition and primitive triangulation that do not bring mental contents into the picture at all, pace familiar CIC advertisements.

In sum, once REC is properly characterized, we have no reason to think it cannot get a firm toehold in explaining the origins of noncontentful forms of intersubjective triangulation. With these pieces of the puzzle in place, it should also be clear that there is nothing in REC's assumptions about basic minds that ought to make its NOC story about the emergence of content evolutionarily implausible. The forms of social interaction it requires are surely possible without presupposing any contentful attitudes.[15]

That takes care of the foundation. But how, when, and where does content enter into the picture? Content only arises when special sorts of sociocultural norms are in place. The norms in question depend on the development, maintenance, and stabilization of practices involving the use of public artifacts through which the biologically inherited cognitive capacities can be scaffolded in very particular ways.

The practices in question are claim-making practices—and they are special because they require participants, not only to respond to things but to do so by *representing them as being thus and so* independently of what might be said about them. The claims in question must satisfy what McDowell identifies as the "familiar intuitive notion of objectivity," which is bound up with a "conception of how things could correctly be said to be anyway—whatever, if anything, we in fact go on to say about the matter" (1998, 222).

14. Although in this extract Medina (2013) speaks of "weaker representational demands," it is clear that he endorses much stronger REC claims about the noncontentful and nonintensional character of the cognition in question. In the same way, Bar-On (2013, 357) also speaks of expressive signals exhibiting aboutness—but scrutiny reveals that her remarks are best understood in terms of directed intentionality only, and not content-involving intentionality (see Hutto and Satne 2015).

15. This is not an exhaustive rebuttal of representationalist approaches to social cognition. Indeed, we have not fully developed or explicated our alternative nonrepresentationalist approach to basic social cognition in this chapter. We have so far only provided a reminder that the default tendency to posit representations is not logically necessary. Representationalism about the basis of social cognition is not the only game in town. These short replies and observations only scratch the surface; for fuller arguments that discuss recent empirical findings about infant cognition, see Satne 2014 and Hutto 2015.

Only with the appearance of such claim making does it become possible to make the special kinds of semantic error unique to contentful thought and speech. Getting things wrong in a truly representational sense is not just a matter of being literally misguided in the way purely biological entities and creatures can be. It involves being subject to the censure of others—not just in the sense of being in or out of line with what is acceptable or not for some community, but also in the sense of being able to get things right or wrong in a game in which it is at least possible to be right according to how things are anyway.

Only those in a position to play this sort of game can be said to have content-involving thoughts and speech. And those capable of playing such games have an enhanced cognitive repertoire that is of a quite different order and kind than is deployed in only basic cognitive operations. Despite these special features, the emergence of content-involving cognition can be explained in naturalistic terms, without residue, by appeal to the way—with the right support from others—purely biologically basic cognitive capacities get used under the right conditions.

REC assumes that the normative practices required for claim making arose with the advent of special kinds of sociocultural niches.[16] In assuming this, REC's NOC story follows Clark in supposing that content-generating practices are part of a "cognition-enhancing animal-built structure ... a kind of self-constructed cognitive niche" (2006a, 370; see also Clark 2006b). From a naturalistic point of view, unless this assumption about cognitive niches is at odds with evolutionary continuity, it is difficult to see what in this sketch of a REC-inspired NOC story entails a discontinuity in nature. If the trick to understanding the emergence of content is to understand the emergence of a special sort of normative sociocultural practice in the use of public symbols, then unless there is something deeply mysterious about social conformity and cultural evolution, nothing in the proffered explanation introduces any inexplicable gap into nature.

4 Philosophical Discontinuity?

Perhaps the real worry with REC and continuity is not about its scientific but its philosophical credentials. After all, in raising concerns about continuity accounts, McDowell was clear that "we must sharply distinguish natural-scientific intelligibility from the kind of intelligibility something acquires when we situate it in the logical space of reasons" (1994, xix).

The same goes for Davidsonian motivations for believing in continuity skepticism. Davidson doubted that there could be "a sequence of emerging features of the mental ... described

16. Tomasello has long defended the idea that "the amazing suite of cognitive skills and products displayed by modern humans is the result of some sort of species-unique mode or modes of cultural transmission" (1999, 4). But see Satne (forthcoming) for a REC-friendly critique of Tomasello's CICish cognitivism.

in the usual mentalistic vocabulary" (1999, 11). In his most famous passage on this topic, Davidson writes:

In both the evolution of thought in the history of mankind, and the evolution of thought in an individual, there is a stage at which there is no thought followed by a subsequent stage at which there is thought. To describe the emergence of thought would be to describe the process which leads from the first to the second of these stages. What we lack is a satisfactory vocabulary for describing the intermediate steps. (1999, 11)

For these philosophers, this implies that, as Bar-On makes clear, "any *scientific* account of the emergence of our mental states and the sort of communication they underwrite is bound to be philosophically inadequate" (2013, 303; italics added). The problem is apparently that there can be no "legitimate *philosophical characterization* of such a progression" (303; italics added).

Bar-On attempts to address this philosophical challenge head-on by appealing to expressive attitudes. As she explains:

I have proposed that our commonsense descriptions of expressive behavior may be a fit starting place for *the conceptual task of fusing the scientific image and the naive commonsense image* regarding relevant continuities between us and the beasts. For, although these are mentalistic descriptions, which do not carve behavior in purely causal terms, they do not presuppose the full battery of concepts that inform our descriptions of each other. (2013, 329; italics added)

Her ambition in addressing this challenge, as she describes it, is to provide commonsense descriptions of the expressive behavior that "can guide us towards a natural intermediate stage in a diachronic path connecting the completely unminded parts of the animal world with the fully minded, linguistically infused parts that we humans now occupy" (Bar-On 2013, 330).

However, we agree with Sultanescu that closing this sort of imaginative gap is a fool's errand. The trouble is that "the intermediate steps between primitive intentionality and contentful intentionality cannot in fact fully be accounted for" (Sultanescu 2015, 639). Yet however much we might narrow the gap by appeal to REC resources, we have no way to close it completely. Even if REC's expressive intentional attitudes are allowed into the story, the question can then be recast as how exactly the gap between such attitudes and "contentful goings-on [is] supposed to be bridged" (646).

Yet as long as we distinguish the explanatory and imaginative limitations we face (distinguishing between scientific and philosophical accounts of continuity), it is possible to show how the NOC proposal explains structurally what needs explaining (how content could have arisen in the natural world without gaps) without our trying to do the impossible: imagine a cognitive missing link, an intermediate state of mind that sits somewhere between expressive, intentional attitudes and contentful attitudes. There is no shame in failing to do the impossible. And if one thinks that not giving a straight answer to the philosophically motivated continuity skeptic is a cop-out for RECers, then we would be interested to know how our CIC rivals have tried to answer this conundrum. To our knowledge, they have never even tried. They too draw a line between minds and nonminds, wherever exactly that is assumed to be, but they do not attempt to explain how we might imagine what comes between, and they feel no compulsion to do so.[17] If we are right in thinking philosophical continuity skepticism is an insoluble problem for everyone, not just RECers, then perhaps it belongs on the long list of philosophical problems that need dissolving, not solving.

Conclusion

In the final analysis, we doubt that REC and its NOC program face any special problems with respect to scientific or philosophically motivated forms of continuity skepticism. Moreover, we have tried to sketch how to pursue the NOC program, and to show that no compelling reason prevents us from trying to do so. Of course, making room for the logical possibility of a NOC story in REC terms and giving a basic sketch of it is one thing, and telling it credibly, in a detailed way that shows its superiority to its rivals, is quite another. But that's a task for another occasion.

References

Abramova, K., and M. Villalobos. 2015. The apparent (ur-)intentionality of living beings and the game of content. *Philosophia* 43:651–668. doi:10.1007/s11406-015-9620-8.

Alksnis, N. 2015. A dilemma or a challenge? Assessing the all-star team in a wider context. *Philosophia* 43:669–685. doi:10.1007/s11406-015-9618-2.

Bar-On, D. 2013. Expressive communication and continuity skepticism. *Journal of Philosophy* 110 (6): 293–330.

Burge, T. 2010. *Origins of Objectivity*. Oxford University Press.

Carruthers, P. 2011. *The Opacity of Mind: An Integrative Theory of Self-Knowledge*. Oxford University Press.

Clark, A. 2006a. Language, embodiment and the cognitive niche. *Trends in Cognitive Sciences* 10 (8): 370–374. doi:10.1016/j.tics.2006.06.012.

Clark, A. 2006b. Material symbols. *Philosophical Psychology* 19 (3): 291–307.

Clowes, R. W., and D. Mendonça. 2015. Representation redux: Is there still a useful role for representation to play in the context of embodied, dynamicist and situated theories of mind? *New Ideas in Psychology* 40 (A): 26–47. doi:10.1016/j.newideapsych.2015.03.002.

Csibra, G., and G. Gergely. 2009. Natural pedagogy. *Trends in Cognitive Sciences* 13 (4): 148–153.

17. As Fodor tells us, "Wherever precisely the line is to be drawn, and however thick it may be, it is vastly plausible that we fall on one side and the paramecium fall on the other" (1986, 12).

Davidson, D. 1999. The emergence of thought. *Erkenntnis* 51 (1): 7–17.

Dennett, D. C. 1995. *Darwin's Dangerous Idea*. Simon & Schuster.

Fodor, J. A. 1975. *The Language of Thought*. Harvard University Press.

Fodor, J. A. 1986. Why paramecia don't have mental representations. *Midwest Studies in Philosophy* 10 (1): 3–23.

Haugeland, J. 1998. Truth and rule-following. In *Having Thought: Essays in the Metaphysics of Mind*, 305–361. Harvard University Press.

Hutto, D. D. 1999. *The Presence of Mind*. John Benjamins.

Hutto, D. D. 2006a. Unprincipled engagements: Emotional experience, expression and response. In *Radical Enactivism: Intentionality, Phenomenology, and Narrative*, ed. R. Menary, 13–38. John Benjamins.

Hutto, D. D. 2006b. Both Bradley and biology: Reply to Rudd. In *Radical Enactivism: Intentionality, Phenomenology, and Narrative*, ed. R. Menary, 81–105. John Benjamins.

Hutto, D. D. 2008. *Folk Psychological Narratives: The Sociocultural Basis of Understanding Reasons*. MIT Press.

Hutto, D. D. 2015. Basic social cognition without mindreading: Minding minds without attributing contents. *Synthese*. doi:10.1007/s11229-015-0831-0.

Hutto, D. D., and E. Myin. 2013. *Radicalizing Enactivism: Basic Minds without Content*. MIT Press.

Hutto, D. D., and G. Satne. 2015. The natural origins of content. *Philosophia* 43:521–536. doi:10.1007/s11406-015-9644-0.

Kiverstein, J., and E. Rietveld. 2015. The primacy of skilled intentionality: On Hutto and Satne's "The natural origins of content." *Philosophia* 43:701–721. doi:10.1007/s11406-015-9645-z.

Korbak, T. 2015. Scaffolded minds and the evolution of content in signaling pathways. *Studies in Logic, Grammar and Rhetoric* 41 (54): 89–103. doi:10.1515/slgr-2015-0022.

Lavelle, J. S. 2012. Two challenges to Hutto's enactive account of pre-linguistic social cognition. *Philosophia* 40:459–472.

McDowell, J. 1994. *Mind and World*. Harvard University Press.

McDowell, J. 1998. *Mind, Value and Reality*. Harvard University Press.

Medina, J. 2013. An enactivist approach to the imagination: Embodied enactments and fictional emotions. *American Philosophical Quarterly* 50 (3): 317–335.

Menary, R. 2009. Intentionality, cognitive integration and the continuity thesis. *Topoi* 28 (1): 31–43.

Menary, R. 2015. Mathematical cognition: A case of enculturation. In *Open MIND*, ed. T. Metzinger and J. M. Windt, 25 (T). MIND Group. doi:10.15502/9783958570818.

McGinn, C. 1989. *Mental Content*. Blackwell.

Millikan, R. G. 1984. *Language, Thought, and Other Biological Categories*. MIT Press.

Millikan, R. G. 1993. *White Queen Psychology and Other Essays for Alice*. MIT Press.

Millikan, R. G. 2004. *Varieties of Meaning: The 2002 Jean Nicod Lectures*. MIT Press.

Millikan, R. G. 2005. *Language: A Biological Model*. Oxford University Press.

O'Brien, G., and J. Opie. 2015. Intentionality lite or analog content? A Response to Hutto and Satne. *Philosophia* 43:723–729. doi:10.1007/s11406-015-9623-5.

Papineau, D. 1987. *Reality and Representation*. Oxford University Press.

Pattee, H. 1985. Universal principles of measurement and language functions in evolving systems. In *Complexity of Language and Life: Mathematical Approaches*, ed. J. Casati and A. Karlqvist, 268–281. Springer.

Penn, D. C., K. J. Holyoak, and D. J. Povinelli. 2008. Darwin's mistake: Explaining the discontinuity between human and nonhuman minds. *Behavioral and Brain Sciences* 31:109–178.

Pietroski, P. 1992. Intentionality and teleological error. *Pacific Philosophical Quarterly* 73:267–282.

Price, H. 2013. *Expressivism, Pragmatism, and Representationalism*. Cambridge University Press.

Putnam, H. 1992. *Renewing Philosophy*. Harvard University Press.

Rakoczy, H., F. Warneken, and M. Tomasello. 2008. The sources of normativity: Young children's awareness of the normative structure of games. *Developmental Psychology* 44 (3): 875–881.

Rakoczy, H., F. Warneken, and M. Tomasello. 2009. Young children's selective learning of rule games from reliable and unreliable models. *Cognitive Development* 24:61–69.

Rosenberg, A. 2014a. Why I am a naturalist. In *Philosophical Methodology: The Armchair or the Laboratory?* ed. M. C. Haug, 32–35. Routledge.

Rosenberg, A. 2014b. Can naturalism save the humanities? In *Philosophical Methodology: The Armchair or the Laboratory?* ed. M. C. Haug, 39–42. Routledge.

Rosenberg, A. 2015. The genealogy of content or the future of an illusion. *Philosophia* 43:537–547. doi:10.1007/s11406-015-9624-4.

Satne, G. 2014. Interaction and self-correction. *Frontiers in Psychology* 5:798. doi:10.3389/fpsyg.2014.00798.

Satne, G. Forthcoming. A two-step theory of the evolution of human thinking: Joint and (various) collective forms of intentionality. *Journal of Social Ontology*.

Searle, J. R. 1983. *Intentionality: An Essay in the Philosophy of Mind*. Cambridge University Press.

Shapiro, L. 2014. Radicalizing enactivism: Basic minds without content. *Mind* 123 (489): 213–220.

Sterelny, K. 2015. Content, control and display: The natural origins of content. *Philosophia* 43:549–564. doi:10.1007/s11406-015-9628-0.

Sultanescu, O. 2015. Bridging the gap: A reply to Hutto and Satne. *Philosophia* 43:639–649. doi:10.1007/s11406-015-9625-3.

Tomasello, M. 1999. *The Cultural Origins of Human Cognition.* Harvard University Press.

Williamson, T. 2014a. What is naturalism? In *Philosophical Methodology: The Armchair or the Laboratory?* ed. M. C. Haug, 29–31. Routledge.

Williamson, T. 2014b. The clarity of naturalism? In *Philosophical Methodology: The Armchair or the Laboratory?* ed. M. C. Haug, 36–38. Routledge.

II Intersubjectivity, Selfhood, and Persons

6   The Primacy of the "We"?

Ingar Brinck, Vasudevi Reddy, and Dan Zahavi

The capacity to engage in collective intentionality is a key aspect of human sociality. Social coordination might not be distinctive of humans—various nonhuman animals engage in forms of cooperative behavior (e.g., hunting together)—but humans seem to possess a specific capacity for intentionality that enables them to constitute forms of social reality far exceeding anything that can be achieved even by nonhuman primates. Consider, for example, how a piece of paper or a string of code in certain contexts, such as financial transactions or insurance policies, is constituted as money by the sheer intentional act of "collective acceptance."

During the past few decades, collective intentionality has been discussed under various labels in a number of empirical disciplines including social, cognitive, and developmental psychology, economics, sociology, political science, anthropology, ethology, and the social neurosciences. Much of the empirical work in the area has relied on and drawn inspiration from the theoretical work of a few influential philosophers such as Searle, Bratman, Gilbert, Pettit, and Tuomela. Building on these standard proposals, a rich philosophical discussion has emerged, further exploring the nature of shared or collective intentions, collective emotions, group and corporate agency, the constitution of social and institutional facts, and the status of group and corporate personhood.

Despite all this work, however, many foundational issues remain controversial and unresolved. In particular, it is by no means clear exactly how to characterize the nature, structure, and diversity of the *we* to which intentions, beliefs, emotions, and actions are often attributed. Is the *we* or *we-perspective* independent of, and perhaps even prior to, individual subjectivity, or is it a developmental achievement that has a first- and second-person-singular perspective as its necessary precondition? Is it something that should be ascribed to a single owner, or does it perhaps have plural ownership? Is the *we* a single thing, or is there a plurality of types of *we*?

1 A Prereflective "Sense of Us"

Let us start our investigation by first considering a recent proposal by Hans Bernhard Schmid. In various publications, Schmid has argued that the *we* is conceptually and developmentally foundational. It does not originate in any kind of agreement or commitment or communication or joint action. It is not founded on any form of social cognition and does not presuppose any experience or givenness of another subject, let alone any kind of reciprocal relation between *I* and *you* or self and other. Rather, the *we*, the "sense of us" or "plural self-awareness," precedes the distinction between yours and mine, is prior to any form of intersubjectivity or mutual recognition (Schmid 2005, 138, 145, 149, 296), and is itself the irreducible basis for joint action and communication.[1]

One consideration that might support Schmid's reluctance to let the *we* arise out of any *I-you* relation is the following: Consider a couple enjoying a movie together. Their focus of attention is precisely on the movie and not on each other. What is salient here is not the relation between the two of them, but the extent to which they jointly share a perspective on a common object. As Schmid argues, in the case of shared experiences, there is a sense in which my experience isn't really mine, and yours isn't yours, but ours. That is, shared experiences are experiences whose subjective aspect is not singular ("for me") but plural ("for us") (Schmid 2014, 9), or to put it differently, the what-it-is-like of experiential sharing is necessarily a what-it-is-like-*for-us*.

In "Plural Self-Awareness," Schmid acknowledges the transitory character and status of the *we*: "Two people team up spontaneously and thereby think and act from a common perspective, based on a 'sense of "us"'; barely a minute later, they part ways never to meet

---

1. Ideas found in some parts of phenomenology are reflected in Schmid's proposal. On the one hand, Schmid often draws on Scheler and in particular references a place in *The Nature of Sympathy* where Scheler denies that shared emotions can be analyzed as a mere conjunction of matching individual experiences and reciprocal knowledge and instead employs the notion of a *feeling-in-common* (Scheler 2008, 12–13). Schmid has interpreted this passage in support of what we might call the *token identity account* of emotional sharing (Schmid 2009, 69) and writes: "There is a sense in which it is literally true that when a group of people has an emotion, there is one feeling episode, one phenomenal experience in which many agents participate. Group emotions are shared feelings. Shared feelings involve some 'phenomenological fusion.' They are 'shared' in the strong straightforward sense in which there is one token affective state in which many individuals take part" (Schmid 2014, 9). On the other hand, Schmid also stresses the extent to which his position resonates with ideas found in Heidegger. There is, for instance, an obvious similarity between Schmid's view and the view expressed in the following quote: "The With-one-another [*Miteinander*] cannot be explained through the I-Thou relation, but rather conversely: this I-Thou relation presupposes for its inner possibility that *Dasein* functioning as I and also as Thou is determined as with-one-another; indeed even more: even the self-comprehension of an I and the concept of I-ness arise only on the basis of the with-one-another, not from the I-Thou relation" (Heidegger 2001, 145–146).

again, and so whatever 'plural self' might have existed between them is gone" (Schmid 2014, 22). This might be quite right, but it also gives rise to the question of how such a process comes about. How is it that the two persons can come to share a perspective? What are the cognitive and affective preconditions for a shared *we*-perspective? Does it come about through communication, joint attention, joint declaration, group identification, perspective taking, and so on? Schmid, however, denies the legitimacy of this line of questioning, since he takes it to involve a commitment to a form of unacceptable reductionism, one that involves either a petitio principii or an infinite regress (Schmid 2014, 10–11). Group identification, for instance, is always after the fact, according to Schmid. It merely confirms a felt sense of "us-ness" that is already in place. To claim that groups come about through some process of declaration also ignores the fact that the declarative act is something the participating individuals have to perform *jointly*. Shared intentionality is consequently presupposed (2014, 10). The same is true with any act of communication (including even preverbal dyadic attention): such acts cannot establish shared meaning, since to be at all communicative, they must be jointly accepted as having meaning. To put it differently, communication is an irreducible joint action and therefore presupposes *we*-intentions. It is *we* who are communicating together, and since communication presupposes a preexisting "sense of us," the former cannot explain or establish or secure the latter (Schmid 2014, 11).

It is important for Schmid to emphasize that the *we* must be understood as minds-in-relation, rather than as some kind of undifferentiated unity. The *we* involves a plurality and is not some kind of larger-scale *I* (Schmid 2009, 156). But he also considers the *we* as a fundamental explanans and rejects any attempts to explain it further. He even rejects the idea that plural self-awareness (and having a "sense of us") might presuppose and build on singular self-awareness. Pointing to developmental research, for instance, findings pertaining to social referencing, Schmid suggests that small children do not seem to draw a clear line between their own goals and the goals of others, nor do they seem to be aware of their own beliefs as theirs, in a singular rather than in a plural way. In fact, they first seem to come to be aware of themselves as members of a group, and only subsequently do they become aware of themselves as individuals. To that extent, plural self-awareness and group membership precede and ground singular self-awareness (Schmid 2014, 23).

In the rest of the paper, we outline an alternative view that insists on the distinction between plural subjecthood and common ground and highlights the importance of reciprocity and second-person engagements for social agency. The claim that agents interact in a common, meaningful world is compatible with the view that individual encounters create new forms of sharing. To some extent, the issues under discussion are also developmental: to ask whether and to what degree individual subjectivity precedes, is co-occurring with, or follows from shared subjectivity is also to ask which comes first in ontogenetic development. That is why we address the philosophical and empirical questions in tandem: we look first at the origins of self–other differentiation and consciousness of self; second, at infant behavior with other persons immediately after birth; and third, at evidence of infant ability to act in acknowledgment of, or with reference to, a jointness of experience or activity with others.

2   Self–Nonself Differentiation

The idea that infants are born in an initial adualism, without a distinction between self and other or self and world, is an old one and could, if true, well be used to support the argument that individual self-experience is a latecomer. In the psychoanalytic literature, it was often assumed that the infant is initially incapable of distinguishing itself from the caregiver, not only in the obvious sense that it is unable to *conceptualize* the difference between self and other, but also in the sense that the infant exists in a "state of undifferentiation, of fusion with mother, in which the 'I' is not yet differentiated from the 'not-I' and in which inside and outside are only gradually coming to be sensed as different" (Mahler, Pine, and Bergmann 1975, 44). Piaget also made a related claim, characterizing the initial adualist state as one where there

does not yet exist any consciousness of the self; that is, any boundary between the internal or experienced world and the world of external realities. ... Insofar as the self remains undifferentiated, and thus unconscious of itself, all affectivity is centred on the child's own body and action, since only with the dissociation of the self from the other or non-self does decentration, whether affective or cognitive, become possible. (Piaget and Inhelder 1969, 22)

In this view, the infant is unable to distinguish events enabled by herself and events originating in an external world. Building on such a developmental basis, it is easy to see why some would interpret the pervasive undifferentiation as equivalent to a jointness, a general *we*-ness of consciousness, which is the starting point for development and for a gradually separating consciousness of self and other. However, there are various reasons for challenging both the empirical claims and their conceptual interpretation.

At an empirical level, considerable evidence now suggests that the assumption of self–nonself undifferentiation at birth is simply wrong. We will mention three types of evidence here: infant discrimination at birth between internally and externally originating sensory stimulation, fetal distinctions between own and other bodies as targets for action, and early forms of social interactions.

Healthy neonates have an innate rooting response. When the corner of the infant's mouth is touched, the infant turns her head and opens her mouth toward the stimulation. By recording the frequency of rooting in response to either external tactile stimulation or tactile self-stimulation, researchers discovered that newborns (twenty-four hours old) showed rooting responses almost three times more frequently in response to external stimuli. Even newborns can consequently discriminate double touch stimulation combined with proprioception from single touch of exogenous origin. That is, they can pick up the intermodal invariants that specify self- versus non-self-stimulation (Rochat and Hespos

1997). Further, even at birth, newborns are coherent organisms able to detect the concordance between internal (feeling touch) and external (seeing touch) stimulation (Filippetti et al. 2013).

Infants are not merely able to distinguish self from nonself but also, from very early on, able to relate to social partners. Consider, for instance, findings discussed by Brinck (2008) and Csibra (2010). Csibra argues that infants are able to recognize that they are being addressed by someone else's communicative intentions before they are able to specify what those intentions refer to; that is, infants are sensitive to the presence of communicative intentions before they are able to access the content of the information communicated (Csibra 2010, 143). He takes this ability to be one of the crucial sources for the development of communicative skills and then highlights three stimuli that are discriminable even by newborns, which specify that the infant is the addressee of the communicative act: (1) eye gaze, (2) "motherese," and (3) turn-taking contingency (Csibra 2010, 144). Already neonates show a preference for looking at faces that make eye contact with them (rather than at faces with closed eyes or averted eyes) (Farroni et al. 2002; Farroni et al. 2004). Just like eye contact, the prosody of motherese indicates to the infant that she is the addressee of the utterances, and it elicits preferential orientation to, and positive affect toward, the source of such stimuli (Csibra 2010, 148). The dramatic difference in neural responses that comes from being a recipient rather than an observer has been shown with adults (Kampe, Frith, and Frith 2003; Schilbach et al. 2008; Schilbach et al. 2006) and is affected by not only current but also previous engagement (Kourtis, Knoblich, and Sebanz 2013). Similar neural discrimination of gaze toward the self can be seen in four- and five-month-old infants (Grossman et al. 2007; Parise et al. 2011)

The ability to discriminate (and act differently toward) self and other may even precede birth. In a suggestive study, Castiello and colleagues (2010) found that fourteen- and eighteen-week-old twin fetuses controlled their arm movements in a far-from-random manner, using lower velocities when directed toward the twin than toward their own body or the uterine wall, and when directed toward their own faces, going slower toward their eye region than toward the mouth region. Already in the third trimester, fetuses have developed a large repertoire of self- and externally directed bodily movements (sucking, kicking, turning, making a fist, etc.) adapted to the environment of the uterus, with some coordination (e.g., hand–mouth movement) and with the velocity adapted to the target (e.g., touch of mouth or eye). They regulate behavior relative to external stimuli and selectively respond to the mother's touch and vocalization (Marx and Nagy 2015; Zoia et al. 2007; Zoia et al. 2013). These findings seem to suggest that self–nonself differentiation and the sense of agency are more fundamental phenomena than often supposed.

Some developmental psychologists have argued that an objective sense of self requires a self-concept, an "idea of me," that emerges toward the end of the second year of life and can be measured by means of the so-called mirror self-recognition task (Lewis 1995; Gallup 1977). Others have argued that acting in the world would be impossible if there were self-world confusion, and that infants have an early sense of themselves as differentiated, environmentally situated, and agentive entities not only long before they are able to pass any mirror self-recognition tasks (Neisser 1993; Stern 1985) but also before they are aware of themselves as members of a group. It is this last view that we adopt here. Only at a later stage of development do children start to show sensitivity to group affiliation and its necessary counterpart: social ostracism. Buttelmann et al. (2013) report that fourteen-month-olds selectively imitate in-group members, although they do not show selectivity in a preference task. Around three to four years, children start to show in-group biases and come to share the view and preferences of the group with which they identify. Classic instances of strong group conformity (Asch 1956) have also been replicated in three- to four-year-old children, who tend to reverse their own objective perceptual judgments to fit a peer group's majority opinion (Corriveau and Harris 2010; Haun and Tomasello 2011). By four years, children also manifest out-group gender and racial stereotypes and other implicit group attitude biases toward others (Cvencek, Greenwald, and Meltzoff 2011).

The proposal regarding a nondifferentiation between self and other or self and world not only can be found in research on early infancy but also resurfaces in claims concerning the inability of young children to differentiate between their own goals and beliefs and those of others (cf. Schmid 2014). Such claims are equally problematic. Infants of five months (and even three months with specific training) can distinguish the targets of others' actions, even when the infants themselves are merely observers of others' actions and thus could be argued not to share the same goals (Woodward 1998; Sommerville, Woodward, and Needham 2005). Even at two months, infants respond to others' intentional actions toward them (e.g., an adult reaching to pick the infant up) with anticipatory bodily adjustments, suggesting that the other's specific goal of reaching toward the infant is sufficiently perceived to enable the infant's specific goal—to *be* picked up (Reddy, Markova, and Wallot 2013). From six months, infants distinguish the other-directed goals of animated "persons" on a screen, distinguishing and preferring "helpers" from "hinderers" (Hamlin, Wynn, and Bloom 2007). The infants' own goals in these experimental observations were not only sympathetic toward the helpers but also independently negative toward the hinderers, showing no trouble distinguishing between at least two opposing goals. Probably most significant in this regard is infant teasing—evident from around nine months—where infants directly obstruct, provoke, and violate others' goals and expectations, albeit playfully (Reddy 1991, 2007, 2008). Around the end of the first year, nonverbal false-belief tasks measuring spontaneous looking behavior instead of answers to questions have shown that infants distinguish between their own beliefs (e.g., about where an object really is) and the false beliefs of others in the scenario (Onishi and Baillargeon 2005; Surian, Caldi, and Sperber 2007; Luo 2011). The idea that a confusion exists between own and others' goals and beliefs is consequently problematic even when applied to one-year-olds, let alone young children. Indeed, the suggestion that infants should fail to appreciate the divergence between their own and others' points of view is also contradicted by the existence of joint attention interactions. The whole point of protodeclaratives is to bring someone else's focus of attention in line with one's own (cf. Roessler 2005).

3   Social Relatedness, Common Ground, and Mutual Engagement

As we have seen, Schmid (2014) argues that plural self-awareness, the awareness of us sharing intentions, emotions, and experiences, precedes both self-experience and other-experience, and that plural self-awareness cannot come about through interaction or communication, since interaction and communication are precisely joint actions and therefore presuppose *we*-intentions. This argument seems to conflate three separate issues: social relatedness, common ground, and *we*-intentionality. Face-to-face-based communication (including preverbal dyadic attention) does indeed presuppose a form of social connectedness, as well as a shared contextual grounding in space and time. It does not occur between disembodied and disembedded agents. The participants' engagements with each other and the shared context of interaction provide sufficient resources for getting communication going. But to claim that this already amounts to *we*-intentionality is to miss out on important distinctions. Communicative events can be harmonious or antagonistic or unequally initiated or entirely novel. In some harmonious conversations, our perspectives "slip into each other," forming, as Merleau-Ponty puts it, "a single fabric" ([1945] 2012, 370). In conflictual conversations, by contrast, the presence of a *we*-perspective is far more questionable.

It might be true that even a conflict needs coordination, since the interaction would otherwise soon break down (Fusaroli, Rączaszek-Leonardi, and Tylén 2013). But although the coordination in question, the shared context and automatic attunement, might provide a common ground, that does not per se amount to *we*-intentionality. To insist that I constitute a plural self, a *we*, with whomever I am socially related to, regardless of the character of the social relationship (be it commanding, hostile, abusive, or dismissive), is to miss out on the distinctiveness and peculiarity of *we*-ness. It involves a special kind of social interaction and addressing and isn't simply a synonym for any kind of social relatedness whatsoever. The abusive bully might employ various forms of perspective taking when harassing his victim, but that doesn't give rise to a shared *we*-perspective or any what-it-is-like-for-us-ness (either on the side of the bully or on the side of the victim).

Another sort of common ground that also fails to amount to any kind of recognizable *we*-ness is the one established through evolutionary history: for instance, neonatal preferences for warmth, skin contact, sound in a certain range, faces, and so on. These preferences are meaningful to caregivers in the sense of having bodies and tendencies that fit them. The newborn might trust the solidity of the parent's supporting arms and thus share adaptation to gravity and ground; the newborn even shares a potentially similar set of motor and sensory capacities with the parent. From this distant viewing perspective, the communication of newborns thus already contains shared meaning of some kind. But an entirely different kind of sharing is present if one looks at communication close up and considers a case of mutual addressing, say a newborn looking with interest at the parent's face and the parent saying hello. We need to differentiate the common ground, shared capacities, and adapted bodies and systems from what happens when subjects address each other, even in simple cases of looking and being looked at. It is the latter phenomena we first have to study if we wish to understand emergent *we*-intentionality, shared perspectives, and the feeling of togetherness.

What exactly happens in the mutuality of addressing? A classical analysis can be found in Husserl, who argues that the particular being-for-another in immediate face-to-face communication takes the form of an *I-you* relation. In the *I-you* relation, the subject is not merely standing next to the other; rather, I motivate her, just as she motivates me, and through this reciprocal interaction, a unity is established (Husserl 1973a, 171). In elucidating the structure of *we*-intentionality, Husserl also highlights the importance of being aware of oneself in the accusative, as attended to or addressed by the other (Husserl 1973a, 211). As he puts it in a late manuscript from 1932: "I am not simply for me, and the other is not given opposite me as an other, rather the other is my you, and speaking, listening, responding, we already constitute a we, one that is unified and communalized in a particular manner" (Husserl 1973b, 476). Similar ideas can be found in other classical phenomenologists such as Stein, Walther, and Schutz, who all insist that the most fundamental form of *we*-relationship is the one that is established in and through dyadic interactions (cf. Zahavi 2015; Zahavi and Salice, forthcoming; Zahavi and Satne 2015; León and Zahavi 2016).

More recently, Peter Hobson has defended a comparable approach and in particular stressed the importance of early affective engagement. That is what provides young infants with interpersonal experiences encompassing an interplay between similarity and difference, connectedness and differentiation (Hobson 2007, 270). From the earliest age, children engage in dynamic coregulation with others that amounts to an open-ended system of negotiation, where this includes the dynamic process of constant affect monitoring and emotional alignment with others, that is, a mutual adjustment between self and others' experience. Throughout this process, the perspectives coregulate, even overlap, but do not fuse with one another; they remain differentiated sufficiently to allow for coordination. Inversely, such coordination appears to emerge not from an original state of fusion or symbiosis with the other, as psychoanalysts tend to suggest, but rather from an original self-world differentiation combined with an early ability to discriminate people as distinct sentient and animated entities in the world (cf. Zahavi and Rochat 2015). Hobson further suggests that infant-adult engagement constitutes the cradle of thought including thoughts about *I*, *you*, *we*, and others as situated in a common world. It permits the infant to be mentally moved by the caregiver to occupy another stance in relation to the world. This repeated moving through somebody else, Hobson claims, is necessary for the infant to understand that the world can be experienced in different ways and that meanings can consequently be negotiated and shared (Hobson 2002).

If shortly after birth, infants start to share experiences in face-to-face, open-ended proto-conversation with others, things change again by seven to nine months, when infants break away from mere face-to-face reciprocal exchanges to engage in referential sharing with others *about* things in the world outside the dyadic exchange. This transition is behaviorally indexed with the emergence of social referencing and triadic joint attention, whereby triangular reciprocal exchanges emerge between child and others in reference to objects or events in the environment (Striano and Rochat 2000; Tomasello 1995). However, findings suggest that already by two months infants react to ostensive gestures and can incorporate objects into protoconversation, having eye contact, moving the gaze to the object and back to the adult, and vocalizing (Rodríguez et al. 2015). By triangulation of attention, objects become jointly captured and shared and start feeding into the exchange dynamically (Rossmanith et al. 2014). These early indicators of a "secondary" intersubjectivity (Trevarthen and Hubley 1978) add to the first exchanges of two- to six-month-olds.

We suggest that *we*-ness is a changing developmental achievement gradually emerging in repeated episodes of engagement between infant and adult, with emotional expressivity playing an essential role in the communication and creation of meaning. The interaction is bidirectional, multimodal, continuous, and dynamic, shaped by local situational constraints and the subjects' skills and knowledge. Infant and adult coordinate their behavior in real time, jointly influencing the direction of the process as a whole with the character of the relation changing over time (Pitsch et al. 2014). Engagement and openness to the other allow the agents to cocreate the interaction, thereby developing (new) contextualized shared intentions and meanings as the interaction unfolds. The participants of joint activities typically address each other as independent agents but do not act or behave independently of each other. Engagement promotes *we*-ness, as in sharing experiences and thoughts together, while performing similar actions and taking part in joint activities such as exploring novel objects or movements, for example, walking with your feet far apart or taking turns to roll a ball across the floor. Fantasia, De Jaegher, and Fasulo (2014) point to an important qualitative aspect of stable yet progressive engagement, the sense of the interaction as unfolding as it were by itself, "taking over" and acquiring a momentum of its own, marked by the precision of timing and synchronicity of actions and the rhythm of the ongoing process considered as a whole.

Certain aspects of mutual engagement may suggest a normative dimension to the *we*, the interaction being structured around shared values such as pleasure, functionality, and care. Thus Rochat and Passos-Ferreira (2009) assert that infant-adult imitation involves a reciprocal willingness to learn and teach. Structured, multimodal routines in infant-caregiver play support young infants' participation in joint sense making and have cooperative qualities (Fantasia, Fasulo et al. 2014). In a study on diaper change in three-months-olds, Rączaszek-Leonardi and Nomikou (2015) observe that mothers treat their babies with respect, as responsible and independent partners with similar communicative skills, although the interaction is asymmetric and requires compensatory actions. Mothers adjust their behavior to the infant, using props such as objects or songs to direct attention and facilitate new behavior. While younger infants seek the instant joy of interaction, the direction and continuation of the activity soon acquire a value to them and move to center stage, allowing infants to take the lead. Face-to-face intentional communication in adults can show a similar emergent normative dimension (Hodges and Fowler 2010), with speakers minding each other, compensating for each other's weaknesses to maintain the conversation and avoid breakdown beyond repair (Dale et al. 2014).

These findings confer a central role to normativity in mutual engagement. One might argue that some forms of normativity have their basis in the reciprocity of second-person engagement and cannot be fully understood from the first- or third-person perspectives. A less-bold claim is that norms and values are at the core of engagement, since episodes of engagement cannot occur in their absence. To avoid breakdown, there have to be routines and mutual expectations or at least a passing agreement concerning how to address each other and how to go on. Establishing such routine or agreement thus presupposes certain general norms or values, while the engagement itself will create shared, local norms tuned to the subjects and the present situation (Brinck 2014). The normative dimension of the "how" arises within engagement and conditions the duration and persistence of any joint activity: the more caring the engagement, the more resilient it is.

4   Varieties of the "We"

Compare the case of two people looking and smiling at each other with the case of two people carrying a sofa up the stairs together. Clearly the situations are distinct, comprising different types of activities. However, on a closer inspection, looking beyond the surface, they also seem to have something in common: they presuppose bodily and emotional engagement and communication, timing and synchrony, and complementary actions. Yet both examples involve short-lived relations between ad hoc pairs of individuals in the here and now. But this surely is only one type of *we*. Other forms of *we* are not in the same way tied to reciprocal bodily interaction but are far more anonymous. A subject can experience herself as a member of a community (say, a religious community), can identify with other members of the same community, and can have group experiences even if she is alone and temporally and spatially removed from the others. Far from being merely dyadic, the *we* can also occur as something more akin to an enduring culture (often involving conventions, norms, and institutions), where *we* do things in a certain way. This kind of group identity can obviously persist even when negative interpersonal emotions are present, and sometimes will amount to a form of communal being-together that is in no way deliberately chosen or voluntarily adopted, but rather a result of a shared heritage. Sometimes, however, it is chosen, and the members can be devoted to their group, sharing rituals and carrying symbols that display their belonging. Examples to consider might include the fan club of a football team with many thousands of members or a local golf club whose members are certainly also involved in the group but might be so to an emotionally less intense degree than the football fans.

Research on collectivity and joint action has shown a tendency to conceive of the *we* and the individuals that form part of it as separate entities that do not essentially influence each other, regardless of how they interact. Individuals are the building blocks, the *we* is the composed unity, and there is nothing more to it. However, this conception neglects a crucial characteristic of human social cognition: how we perceive others and others perceive us, whether as partners in crime, mere associates, or handymen paid to do the dirty stuff, fundamentally influences both the quality and the quantity of the interaction and communication (Becchio, Sartori, and Castiello 2010; Becchio et al. 2013). We perceive and act differently when performing actions in parallel for a common good or when acting individually but simultaneously in pursuit of the same goal. Furthermore, many cases of large-scale collective action include different phases and forms of interaction. One phase may involve engagement, another might be more impersonal and disengaged, and a third one might require deep involvement by reflection and analysis. Thus the *we* is not a simple notion, and it is not merely from the viewpoint of development that we need to consider its multiple forms.

The importance of verbal language and symbolization for developing the forms of *we* must not be neglected. Consider Isa (twenty-three months old), who talks of herself as forming a group with others, now Isa and Daddy; then Isa, Aunt Yvette, and Granny; then again Isa and Bowie, the dog; or Isa, Sam, and Albin at nursery school. These constellations enable alternative activities and form microscopic protocultures, organized around their own rituals and norms and are bound to distinct places. Thus Cousin Liam may provide the springboard to playing football, something that *we* do, a role that will soon be taken over by the toddler football classes, and subsequently a more general and conventional *we*-formation might develop from participating in the focused activities down at the football club. From early on, infants take part in different activities and with different people, stepping in and out of groups that persist even when the infant is not there. Once language starts to develop, labeling the groups helps to memorize them and discriminate one group from another, and also to keep track of them over time, adding certain structure to everyday life. In a longer perspective, this will lead to a flexible conception of the *we* that will comprise a number of alternative instrumental and normative, emotionally charged ways of interacting with others depending on the setting.

We also need to recognize the difference between identifying and being united with a certain group and knowing this reflectively. In fact, rather than being reflectively affirmed, a *we*-perspective is most often simply lived through prereflectively. But this can change, for instance, through the intervention of a third party. A central idea from social psychology is that one's tendency to experience oneself as part of a *we* increases when confronted with outgroup members and contrasting groups. According to one theoretical approach, it is where intergroup differences tend to be perceived as larger than intragroup differences that we tend to categorize ourselves as a *we* instead of an *I*, and to see the included other(s) as similar rather than different (Turner et al. 1994; Brewer and Gardner 1996). Something important also happens when one's own group is recognized as such not simply by its own members but also by members belonging to another group. Such "external" recognition might further solidify one's own group formation and provide it with a new and more objective status. Similarly, this same development—what we might call the experience of the *we* as an "us," a tacit and dynamic objectification—might occur ontogenetically in triadic interactions where a dyad is addressed by a third partner (cf. Fivaz-Depeursinge, Favez, and Frascarolo 2004).

Conclusion

We need to recognize the existence of several forms of *we*, but we also need to ask how they relate to the dyadic form we primarily have been focusing on here. The answer is straightforward: one acquires familiarity with tradition, one becomes enculturated, through one's upbringing. And we cannot make sense of such a process unless we start with the beginning, namely, emotional engagement between specific others.

Mutual engagement involves emotional openness. It is multimodal and characterized by timing and synchrony, reciprocity and complementarity, diversity and similarity, perturbation and repair. Many of these aspects are also operative on the large-scale, more or less impersonal *we* that the traditions and institutions of modern society support. Mutual engagement may not be directly involved in all forms of *we*, but it is fundamental and not merely in a developmental sense. Sometimes things do not unfold as expected, and sometimes unforeseen problems take the form of a crisis—at work, at the level of national or international politics, or elsewhere. The structures that normally support social life collapse. However, it rarely happens that the problems cannot be handled. People fall back on face-to-face negotiation and take it from there, and they soon manage to work out alternative ways of managing the situation. Mutual engagement has prepared us for this kind of event and will do in almost all situations. Although its role remains invisible in many forms of social action, it constitutes the backbone of human social life. Sociocultural and institutional forms of *we*-ness presuppose and cannot be fully explained without reference to the *I* and the *you*.

Summing up, let us now distinguish three options: First, the *we* is conceptually and developmentally prior to the *I* and the *you*. Second, the *I*, the *you*, and the *we* are equiprimordial. Third, the *I* and the *you* are conceptually and developmentally prior to the *we*. We have in the preceding rejected the first option. If we are to speak meaningfully of a *we*, of a first-person plural, we need to preserve plurality and differentiation. To conceive of the *we* as an undifferentiated oneness is to misunderstand the very notion. But what about the second option? Whereas a *we*-experience, a shared perspective, a sense of togetherness, involves an interplay of identification and differentiation, integration and distinctness, and thereby presupposes a preservation of the self–other distinction, it is by no means evident that each and every self-experience and other-experience necessarily requires a concomitant *we*-experience. I can be aware of myself (for instance, as a subject of experience or embodied agent) without being reflectively or prereflectively aware of myself as part of a *we*, and I can be aware of another without that awareness necessarily giving rise to a shared *we*-perspective.

The self emerges, develops, and differentiates itself in relation. But so does the *we*. For the neonate to perceive an action directed to her by a focused dynamic face, and for her to be interested in the action and to respond with some facial action herself, there does not need to be a fully developed *I*, nor does there need to be a full-blown recognition of a *you*. The infant's very response to the actions directed to her promotes more responses, and slowly a recognition develops, both of the phenomenon of being addressed and of being addressed by this face and body. In support of the third option, we argue that it is within this process of engagement (between a fragile *I* dealing with a dawning recognition of a *you*) that the *we* gradually emerges and develops. As the baby's capacities grow and multiply and as the caregiver increases the difficulty of the interaction, new behaviors, new vocalizations, new objects and functions, new turns and meanings, are added. Step by step symbolization is introduced to the infant, and the space of actions acquires semiotic properties, transforming the senses of *I*, *you*, and *we*. Even if self-experience and other-experience precede *we*-experience, all three continue to develop in tandem, mutually specifying and transforming each other throughout development.

References

Asch, S. E. 1956. Studies of independence and conformity: A minority of one against a unanimous majority. *Psychological Monographs* 70 (9): 1–70.

Becchio, C., M. Del Giudice, O. Dal Monte, L. Latini-Corazzini, and L. Pia. 2013. In your place: Neuropsychological evidence for altercentric remapping in embodied perspective taking. *Social Cognitive and Affective Neuroscience* 8 (2): 165–170.

Becchio, C., L. Sartori, and U. Castiello. 2010. Toward you: The social side of actions. *Current Directions in Psychological Science* 19 (3): 183–188.

Brewer, M. B., and W. Gardner. 1996. Who is this "we"? Levels of collective identity and self representations. *Journal of Personality and Social Psychology* 71 (1): 83–93.

Brinck, I. 2008. The role of intersubjectivity for the development of intentional communication. In *The Shared Mind: Perspectives on Intersubjectivity*, ed. J. Zlatev, T. Racine, C. Sinha, and E. Itkonen, 115–140. John Benjamins.

Brinck, I. 2014. Developing an understanding of social norms and games: Emotional engagement, nonverbal agreement, and conversation. *Theory and Psychology* 24 (6): 737–754.

Buttelmann, D., N. Zmyj, M. Daum, and M. Carpenter. 2013. Selective imitation of in-group over out-group members in 14-month-old infants. *Child Development* 84 (2): 422–428.

Castiello, U., C. Becchio, S. Zoia, C. Nelini, L. Sartori, L. Blason, G. D'Ottavio, M. Bulgheroni, and V. Gallese. 2010. Wired to be social: The ontogeny of human interaction. *PLoS One* 5:e13199.

Corriveau, K. H., and P. L. Harris. 2010. Preschoolers (sometimes) defer to the majority in making simple perceptual judgments. *Developmental Psychology* 46 (2): 437–445.

Csibra, G. 2010. Recognizing communicative intentions in infancy. *Mind and Language* 25 (2): 141–168.

Cvencek, D., A. G. Greenwald, and A. N. Meltzoff. 2011. Measuring implicit attitudes of 4-year-olds: The Preschool Implicit Association Test. *Journal of Experimental Child Psychology* 109:187–200.

Dale, R., R. Fusaroli, N. Duran, and D. C. Richardson. 2014. The self-organization of human interaction. *Psychology of Learning and Motivation* 59:43–96.

Fantasia, V., H. De Jaegher, and A. Fasulo. 2014. We can work it out: An enactive look at cooperation. *Frontiers in Psychology* 5:874.

Fantasia, V., A. Fasulo, A. Costall, and B. López. 2014. Changing the game: Exploring infants' participation in early play routines. *Frontiers in Psychology* 5:522.

Farroni, T., G. Csibra, F. Simion, and M. Johnson. 2002. Eye contact detection in humans from birth. *Proceedings of the National Academy of Sciences of the United States of America* 99 (14): 9602–9605.

Farroni, T., S. Massaccesi, D. Pividori, F. Simion, and M. Johnson. 2004. Gaze following in newborns. *Infancy* 5 (1): 39–60.

Filippetti, M. L., M. H. Johnson, S. Lloyd-Fox, D. Dragovic, and T. Farroni. 2013. Body perception in newborns. *Current Biology* 23 (23): 2413–2416.

Fivaz-Depeursinge, E., N. Favez, and F. Frascarolo. 2004. Threesome intersubjectivity in infancy: A contribution to the development of self-awareness. In *The Structure and Development of Self-Consciousness: Interdisciplinary Perspectives*, ed. D. Zahavi, T. Grünbaum, and J. Parnas, 21–34. John Benjamins.

Fusaroli, R., J. Rączaszek-Leonardi, and K. Tylén. 2013. Dialog as interpersonal synergy. *New Ideas in Psychology* 32:147–157.

Gallup, G. G. 1977. Self-recognition in primates: A comparative approach to the bidirectional properties of consciousness. *American Psychologist* 32:329–338.

Grossman, T., M. Johnson, T. Farroni, and G. Csibra. 2007. Social perception in the infant brain: Gamma oscillatory activity in response to eye gaze. *Social Cognitive and Affective Neuroscience* 2 (4): 284–291.

Hamlin, J. K., K. Wynn, and P. Bloom. 2007. Social evaluation by preverbal infants. *Nature* 450:557–559.

Haun, D., and M. Tomasello. 2011. Conformity to peer pressure in preschool children. *Child Development* 82 (6): 1759–1767.

Heidegger, M. 2001. *Einleitung in die Philosophie, Gesamtausgabe 27*. Vittorio Klostermann.

Hobson, R. P. 2002. *The Cradle of Thought: Exploring the Origins of Thinking*. Macmillan.

Hobson, R. P. 2007. Communicative depth: Soundings from developmental psychopathology. *Infant Behavior and Development* 30:267–277.

Hodges, B. H., and C. A. Fowler. 2010. New affordances for language: Distributed, dynamical, and dialogical resources. *Ecological Psychology* 22:239–254.

Husserl, E. 1973a. *Zur Phänomenologie der Intersubjektivität II: Texte aus dem Nachlass; Zweiter Teil, 1921–28*. Ed. I. Kern. *Husserliana 14*. Martinus Nijhoff.

Husserl, E. 1973b. *Zur Phänomenologie der Intersubjektivität III: Texte aus dem Nachlass; Dritter Teil, 1929–35*. Ed. I. Kern. *Husserliana 15*. Martinus Nijhoff.

Kampe, K., C. Frith, and U. Frith. 2003. "Hey John": Signals conveying communicative intention toward the self activate brain regions associated with "mentalizing," regardless of modality. *Journal of Neuroscience* 23 (12): 5258–5263.

Kourtis, D., G. Knoblich, and N. Sebanz. 2013. History of interaction and task distribution modulate action simulation. *Neuropsychologia* 51 (7): 1240–1247.

León, F., and D. Zahavi. 2016. Phenomenology of experiential sharing: The contribution of Schutz and Walther. In *The Phenomenological Approach to Social Reality: History, Concepts, Problems*, ed. A. Salice and H. B. Schmid, 219–234. Springer.

Lewis, M. 1995. Aspects of the self: From systems to ideas. In *The Self in Infancy: Theory and Research*, ed. P. Rochat, 95–116. Elsevier.

Luo, Y. 2011. Do 10-month-old infants understand others' false beliefs? *Cognition* 121 (3): 289–298.

Mahler, M. S., F. Pine, and A. Bergmann. 1975. *The Psychological Birth of the Human Infant*. Basic Books.

Marx, V., and E. Nagy. 2015. Fetal behavioural responses to maternal voice and touch. *PLoS One* 10 (6): e0129118.

Merleau-Ponty, M. [1945] 2012. *Phenomenology of Perception*. Trans. D. A. Landes. Routledge.

Neisser, U. 1993. The self perceived. In *The Perceived Self: Ecological and Interpersonal Sources of Self-Knowledge*, ed. U. Neisser, 3–21. Cambridge University Press.

Onishi, L., and R. Baillargeon. 2005. Do 15-month-old infants understand false beliefs? *Science* 308 (5719): 255–258.

Parise, E., A. Handl, L. Palumbo, and A. Friederici. 2011. Influence of eye gaze on spoken word processing: An ERP study with infants. *Child Development* 82 (3): 842–853.

Piaget, J., and B. Inhelder. 1969. *The Psychology of the Child*. New York: Basic Books.

Pitsch, K., A. Vollmer, K. Rohlfing, J. Fritsch, and B. Wrede. 2014. Tutoring in adult-child interaction: On the loop of the tutor's action modification and the recipient's gaze. *Interaction Studies: Social Behaviour and Communication in Biological and Artificial Systems* 15 (1): 55–98.

Rączaszek-Leonardi, J., and I. Nomikou. 2015. Beyond mechanistic interaction: Value-based constraints on meaning in language. *Frontiers in Psychology* 6:1579.

Reddy, V. 1991. Playing with others' expectations: Teasing and mucking about in the first year. In *Natural Theories of Mind*, ed. A. Whiten, 143–158. Blackwell.

Reddy, V. 2007. Getting back to the rough ground: Deception and "social living." *Philosophical Transactions of the Royal Society B: Biological Sciences* 362:621–637.

Reddy, V. 2008. *How Infants Know Minds*. Harvard University Press.

Reddy, V., G. Markova, and S. Wallot. 2013. Anticipatory adjustments to being picked up in infancy. *PLoS One* 8 (6): e65289.

Rochat, P., and S. J. Hespos. 1997. Differential rooting response by neonates: Evidence of an early sense of self. *Early Development and Parenting* 6 (3–4): 105–112.

Rochat, P., and C. Passos-Ferreira. 2009. From imitation to reciprocation and mutual recognition. In *Mirror Neuron Systems: The Role of Mirroring Processes in Social Cognition*, ed. J. A. Pineda, 191–212. Humana Press, Springer Science.

Rodríguez, C., A. Moreno-Núñez, M. Basilio, and N. Sosa. 2015. Ostensive gestures come first: Their role in the beginning of shared reference. *Cognitive Development* 36:142–149.

Roessler, J. 2005. Joint attention and the problem of other minds. In *Joint Attention: Communication and Other Minds*, ed. N. Eilan, C. Hoerl, T. McCormack, and J. Roessler, 230–259. Oxford University Press.

Rossmanith, N., A. Costall, A. F. Reichelt, B. Lopez, and V. Reddy. 2014. Jointly structuring triadic spaces of meaning and action: Book sharing from 3 months on. *Frontiers in Psychology* 5:1390.

Scheler, M. 2008. *The Nature of Sympathy*. Transaction Publishers.

Schilbach, L., S. Eickhoff, A. Mojzisch, and K. Vogeley. 2008. What's in a smile? Neural correlates of facial embodiment during social interaction. *Social Neuroscience* 3 (1): 37–50.

Schilbach, L., A. Wohlschlaeger, N. Kraemer, A. Newen, J. Shah, G. Fink, and K. Vogeley. 2006. Being with virtual others: Neural correlates of social interaction. *Neuropsychologia* 44:718–730.

Schmid, H. B. 2005. *Wir-Intentionalität: Kritik des ontologischen Individualismus und Rekonstruktion der Gemeinschaft*. Karl Alber.

Schmid, H. B. 2009. *Plural Action: Essays in Philosophy and Social Science*. Springer.

Schmid, H. B. 2014. Plural self-awareness. *Phenomenology and the Cognitive Sciences* 13:7–24.

Sommerville, J. A., A. L. Woodward, and A. Needham. 2005. Action experience alters 3-month-old infants' perception of others' actions. *Cognition* 96 (1): B1–B11.

Stern, D. N. 1985. *The Interpersonal World of the Infant*. Basic Books.

Striano, T., and P. Rochat. 2000. Emergence of selective social referencing. *Infancy* 1 (2): 253–264.

Surian, L., S. Caldi, and D. Sperber. 2007. Attribution of beliefs by 13-month-old infants. *Psychological Science* 13 (7): 580–586.

Tomasello, M. 1995. Joint attention as social cognition. In *Joint Attention: Its Origins and Role in Development*, ed. C. J. Moore and P. Dunham, 103–130. Erlbaum.

Trevarthen, C., and P. Hubley. 1978. Secondary intersubjectivity: Confidence, confiding, and acts of meaning in the first year. In *Action, Gesture, and Symbol: The Emergence of Language*, ed. A. Lock, 183–229. Academic Press.

Turner, J. C., P. J. Oakes, S. A. Haslam, and C. A. McGarty. 1994. Self and collective: Cognition and social context. *Personality and Social Psychology Bulletin* 20 (5): 454–463.

Woodward, A. L. 1998. Infants selectively encode the goal object of an actor's reach. *Cognition* 69 (1): 1–34.

Zahavi, D. 2015. You, me, and we: The sharing of emotional experiences. *Journal of Consciousness Studies* 22 (12): 84–101.

Zahavi, D., and P. Rochat. 2015. Empathy ≠ sharing: Perspectives from phenomenology and developmental psychology. *Consciousness and Cognition* 36:543–553.

Zahavi, D., and A. Salice. Forthcoming. Phenomenology of the we: Stein, Walther, Gurwitsch. In *The Routledge Handbook of Philosophy of the Social Mind*, ed. J. Kiverstein. Routledge.

Zahavi, D., and G. Satne. 2015. Varieties of shared intentionality: Tomasello and classical phenomenology. In *Beyond the Analytic-Continental Divide: Pluralist Philosophy in the Twenty-first Century*, ed. J. Bell, A. Cutrofello, and P. Livingston, 305–325. Routledge.

Zoia, S., L. Blason, G. D'Ottavio, M. Biancotto, M. Bulgheroni, and U. Castiello. 2013. The development of upper limb movements: From fetal to post-natal life. *PLoS One* 8 (12): e80876.

Zoia, S., L. Blason, G. D'Ottavio, M. Bulgheroni, E. Pezzetta, A. Scabar, and U. Castiello. 2007. Evidence of early development of action planning in the human foetus: A kinematic study. *Experimental Brain Research* 176:217–226.

7 Selfhood, Schizophrenia, and the Interpersonal Regulation of Experience

Matthew Ratcliffe

This paper addresses the view, currently and historically popular in phenomenological psychopathology, that schizophrenia involves disturbance of a person's most basic sense of self, the *minimal self*. The concept of "minimal self" is to be understood in wholly phenomenological terms. Zahavi (2014) offers what is perhaps the most detailed characterization to date. All our experiences, he maintains, have a "first-personal character"; their structure incorporates a sense of mineness, of their originating in a singular locus of experience. So the minimal self is neither an object of experience/thought nor an experience of subjectivity that is separate from one's various experiences. Rather, it pertains to "the distinct manner, or *how*, of experiencing" (Zahavi 2014, 22). Those who subscribe to this view do not insist that minimal self is the only kind of self. As Zahavi acknowledges, "self" may legitimately refer to a range of different phenomena, all of which need to be carefully distinguished from one another. But the minimal self is the most fundamental of these, a condition for the integrity of experience that all other kinds of self-experience presuppose.

Some have further proposed that this basic sense of self is altered (but not entirely lost) in schizophrenia. There is a "disturbance of minimal- or core-self experience," an "ipseity" disturbance (Sass 2014, 5). What is eroded is a sense of "mineness" or "first-person perspective," of a kind that is more usually "automatic" (Parnas et al. 2005, 240). It is in this context that seemingly more localized symptoms such as delusions and hallucinations emerge and are to be understood. A characteristic type of self-disturbance, it is maintained, distinguishes the phenomenology of schizophrenia from that of other psychiatric conditions. The latter either do not involve changes in the minimal self or involve changes that are qualitatively different and less profound than those arising in schizophrenia (Raballo, Sæbye, and Parnas 2009).

I think it is plausible to maintain that human experience includes something along the lines of the minimal self and also that diagnoses of schizophrenia are sometimes associated with changes to this aspect of experience. Nevertheless, I argue that the minimal self needs to be reconceptualized in *interpersonal* terms. Zahavi states that minimal selfhood does not depend on social interaction for its development and/or its sustenance: it is not "a product of social interaction or the result of a higher cognitive accomplishment" but "a basic and indispensable experiential feature," one that is not "constitutively dependent upon social interaction" (Zahavi 2014, 63, 95). When applied to schizophrenia, this view can lend itself to a somewhat individualistic approach. Insofar as schizophrenia originates in disturbance of a presocial self, the social world is perhaps not the place to look for causes. Instead one could seek to identify preexisting "self-disorders" that "antedate the onset of psychosis" and render one susceptible to it (Raballo and Parnas 2011, 1018). It has been further suggested that these disorders have genetic causes. Hence their "primary relevance" is to "etiological research into the genetic architecture of schizophrenia" (Raballo, Sæbye, and Parnas 2009, 348).

Then again, it would also be quite consistent to maintain that (a) schizophrenia involves disruption of a presocial, minimal self, and (b) schizophrenia has social and interpersonal causes. Something that did not depend on other people for its development or maintenance could still be disrupted by them. By analogy, having a brain does not depend on a history of social interaction, but brains are not immune to injuries inflicted by other people. However, I argue in what follows that the most basic sense of self is indeed developmentally dependent on interactions with other people. Furthermore, it is interpersonally sustained and continues to depend on other people even in adulthood. I begin by showing why even the most minimal experience of selfhood must include not just the sense that one is in an intentional state (where intentionality is construed in phenomenological terms) but also the sense that one is in an intentional state of one or another kind, such as perceiving or remembering.[1] Minimal self, we might say, encompasses an appreciation of the *modalities of intentionality*. I then offer some remarks on the extent to which experience and thought are interpersonally regulated. These build up to the claims that (a) the modalities of intentionality depend on a certain way of experiencing and relating to other people in general, which involves a primitive, affective, nonconceptual form of *trust*; (b) trust of this kind develops within, and is then sustained by, the interpersonal environment; (c) its development can be derailed at various stages by traumatic events involving other people, and it can also be disturbed—to differing degrees—by traumatic events in adulthood; and (d) such derailments or disturbances are implicated in many cases of schizophrenia, although probably not all. It follows that the minimal self cannot be distinguished from an interpersonally constituted self, and self-disturbances in schizophrenia are equally *relational* disturbances. However, the relevant sense of self remains prereflective. Furthermore, it is not an object of experience. Rather, it is integral to the experience of being in a given intentional state and inextricable from the structure of intentionality. So we can continue to distinguish it from various other

1. I conceive of intentionality in phenomenological terms throughout, as an *experience* of something or other, such as remembering that *p* or perceiving that *q*. For convenience, I sometimes use the term "intentional state" to refer to an experience of type *y*, which has *p* as its content. However, one should not read too much into the term "state," which I use in a noncommittal way.

notions of self, of the kind that might appeal to narrative, higher-order thought or to rich conceptions of personhood.

1 The Modalities of Intentionality

Minimal self involves a sense of first-person perspective that is inseparable from one's being in an intentional state of one or another kind. However, what is not so clear is whether it further includes a sense of the kind of intentional state one is in. For any type of experience, such as perceiving, remembering, or imagining, we can ask whether minimal self involves a sense of (a) being a locus of experience, or (b) being the locus of a type x experience, rather than a type y or z experience. The answer, I suggest, has to be (b). When I remember that p, I have a prereflective, immediate appreciation not just that I am experiencing p but that I am encountering p in a certain way, as remembered rather than as perceived or imagined. Granted, I might think that I am remembering that p when I am in fact imagining that p. But I am not concerned with the epistemic issue of whether and how an experience of remembering p gives me grounds for thinking that I really am remembering p (an issue that applies equally to other kinds of intentional state). My concern is with the relevant phenomenology: whether or not I am mistaken when I take myself to be remembering something, I have a sense *of* remembering. Integral to being in an intentional state is the experience of being in one or another kind of intentional state. It is this aspect of experience that I am concerned with here. One might argue that *the sense of having a type x experience* is distinct from *the having of a type x experience*: I could have an experience of imagining while at the same time having the sense that it is an experience of perceiving. However, I reject that view. My sense that I am encountering something "here, now," as opposed to something past, imagined, or anticipated, influences my ongoing behavior in ways that shape how the experience unfolds. If I did not experience the entity as "here, now," I would not anticipate possibilities such as walking around it and seeing it from other angles, picking it up, or showing it to others. It is plausible, I think, to maintain that such possibilities are integral to an experience of perceiving (Ratcliffe 2015, chap. 2). Without them, one would not anticipate or feel drawn to act in ways that are consistent with perceiving. Hence the experience would not unfold in ways that are characteristic of specifically perceptual experiences, a point that generalizes to other kinds of intentional state. It follows that having an experience of type x is inseparable from the sense that one is having an experience of type x, at least if we conceive of experience in a dynamic rather than static way.

Why should such distinctions be inseparable from even the most minimal sense of self? If one's experience did not respect the distinction between perceiving and remembering, one would lack a sense of temporal location. And if one could not distinguish imagining from perceiving, experienced boundaries between self and environment would break down, to the extent that one would lack any sense of spatial location. My sense of perceiving a, b, and *c* while imagining *d*, *e*, and *f* constrains my potential actions. I cannot make the unseen sides of *a*, *b*, and *c* perceptually available without acting in specific ways, which I may or may not be capable of. This does not apply to *d*, *e*, and *f*. Without the distinction between perceiving and imagining, such constraints would be lacking. Everything would appear experientially and practically accessible in the same ways, thus amounting to a lack of self-location, of having a particular, contingent, changeable standpoint on the world. Without some sense of spatiotemporal location, it is difficult to see how the experience of being a singular, coherent locus of experience could be sustained. Hence I suggest that minimal self experience must discriminate between types of intentional state. This interpretation is consistent with how the concept is applied to the phenomenology of schizophrenia. Consider the following:

> Note that in a normal experience, e.g. in a perceptual act, the perceptual act is immediately and prereflectively aware of itself; it is an instance of ipseity. In other words, when I perceive or I think something, I do not become aware of the fact of my perceiving or thinking by some reflective/introspective examination of my current mental activity and comparing it with other possible modalities of intentionality (e.g. fantasizing). Any experience, any intentional act, is normally articulated as ipseity, i.e. it is automatically prereflectively aware of itself. The difficulties in this domain point to a profound disorder of ipseity. (Parnas et al. 2005, 242)

This passage does not explicitly endorse the less-minimal conception of minimal self. Nevertheless, it states that we do *not* take ourselves to perceive rather than fantasize by reflectively comparing two kinds of experience. Hence, although the subsequent claim that an intentional act is "prereflectively aware of itself" need not be interpreted as its being "prereflectively aware of the type of intentional act that it is," the stronger interpretation is implied by preceding context. Sass (2014, 6) similarly links minimal self to the modal structure of intentionality: "Disturbances of spatiotemporal structuring of the world, and of such crucial experiential distinctions as perceived-vs-remembered-vs-imagined, are grounded in abnormalities of the embodied, vital, experiencing self." His view, as I understand it, is not that disturbance of selfhood *causes* erosion of these distinctions or that disturbances of selfhood imply other kinds of phenomenological disturbance but not vice versa. Rather, the relationship is one of mutual implication: minimal selfhood, the coherence of world experience, and a sense of relating to the world are all aspects of a unitary phenomenological structure, one that includes the modalities of intentionality.

When employed to interpret first-person accounts of schizophrenia, the self-disturbance view has some plausibility. Consider the following: "The real 'me' is not here any more. I am disconnected, disintegrated, diminished. Everything I experience is through a dense fog, created by my own mind, yet it also resides outside my mind" (Kean 2009, 1034). Here, altered and diminished experience of self is associated with erosion of the boundaries between self and nonself, a blurring of the phenomenological distinction between what is self-generated and what is not.[2] Of course, a breakdown of the modalities of intentionality would also imply changes in the structure of intersubjectivity. If one could not distinguish self from nonself in the usual way, one could not—by implication—distinguish self from others either. More specific distinctions, such as the distinction between one's own thoughts or experiences and those of others, would be similarly affected. Now, one might adopt the view that intersubjective disturbances are secondary: disruption of interpersonal experience is implied by erosion of a more fundamental, pre-intersubjective sense of self. Indeed, even those who emphasize the interpersonal, relational phenomenology of psychiatric illness sometimes write as though this were so. For example, Fuchs states that a "disturbance of the pre-reflective, embodied self must necessarily impair the patient's social relationships" (Fuchs 2015, 199) and adds that self-disturbance can be exacerbated by subsequent social problems. But he does not go so far as to say that the relationship between disturbances of minimal self and altered interpersonal experience is one of mutual implication. I propose that the integrity of intentionality, the sense of being in one kind of intentional state rather than another, depends on a certain *way* of experiencing and relating to others, both developmentally and constitutively. Hence the most minimal or primitive sense of self that we possess is also an interpersonal self. It follows that certain events, of a kind that disrupt intersubjective development and/or interpersonal experience in adulthood, can also impact on the structure of intentionality.

It might seem that there is an obvious objection to my position: if the modalities of intentionality arise during social development, young infants and all nonsocial organisms are implausibly denied the status "minimal self." Infants may well be born with various rudimentary intersubjective abilities, but to the extent that the structure of intentionality is contingent on developmental processes, it is lacking at the outset. That is not my view at all, though. One option would be to maintain that infants are born with the capacity for intentional states of type x and y, and that, as they develop, they come to distinguish other intentional state types as well. Hence a richer sense of minimal self develops, and disturbances of intersubjectivity involve regression to something more primitive. However, I suggest that we instead construe self-development as a transformative process. Suppose infants are born with a capacity for intentional state type y, which does not correspond to *any* of the intentional state types possessed by typical adults. So it is not that one first experiences y and later comes to experience z as well; y is a developmental precursor to a modal structure that does not retain y in unadulterated form. Furthermore, the development of y into z does not revert back to y when it goes awry; one is left with neither. More generally, development does not have a rewind button—capacities that enter into developmental processes

2. An example often quoted in support of the minimal self account of schizophrenia is Elyn Saks's memoir *The Centre Cannot Hold*: "Consciousness gradually loses its coherence. One's center gives way. The center cannot hold. The 'me' becomes a haze, and the solid center from which one experiences reality breaks up like a bad radio signal" (2007, 12).

become hostages to fortune, and there is no going back. All sorts of mundane examples serve to illustrate this point: starved of light from birth, one is not left with the capacities of infant eyes; deprived of the opportunity to walk, one is not left with the musculature of infant legs. In the case of minimal self, it is equally coherent to maintain that infants possess a minimal self of type A, a capacity for certain types of distinguishable and coherently integrated intentional state, but that adults possess a minimal self of type B, involving a repertoire of intentional states qualitatively different from A. When development is derailed, the resulting lack of phenomenological coherence is quite unlike both the typical infant and adult cases. Hence there is a need to distinguish the *possible* kinds of minimal self from the *actual* kind(s) possessed by adult humans. Even if we accept that selfhood is possible without intersubjectivity, there are no grounds for maintaining that this same kind of self-experience is preserved in unadulterated form throughout social development, as an underlying kernel.

2 Interpersonal Experience Regulation

Interactions with other people affect our experience and thought in a range of often subtle ways. How our surroundings appear is partly a reflection of whether we are with others, what they are doing, and how we relate to them. This is not to suggest that whether or not one can see a cup or a car, a red thing or a green thing, is contingent on interpersonal context. Rather, what appears relevant to us and the kind of significance it has for us is partly attributable to our relations with others. Van den Berg makes the following observations:

We all know people in whose company we would prefer not to go shopping, not to visit a museum, not to look at a landscape, because we would like to keep these things undamaged. Just as we all know people in whose company it is pleasant to take a walk because the objects encountered come to no harm. These people we call friends, good companions, loved ones. (van den Berg 1972, 65)

Interaction with a specific individual can drain one's surrounding world of its significance or, alternatively, enrich it with salient possibilities that would otherwise be lacking. This observation is not specific to experience of one's current surroundings. It applies equally to interpreting and finding significance in past, present, and anticipated events, to anything that *matters* to us:

We all know people with whom it is best not to share anything that matters to us. If we have experienced something exciting, and if we tell it to those people, it will seem almost dull. If we have a secret, we will keep it safe from those people, safe inside us, untold. That way it won't shrivel up and lose all the meaning it has for us. But if you are lucky, you know one person with whom it is the other way around. If you tell that person something exciting, it will become more exciting. A great story will expand, you will find yourself telling it in more detail, finding the richness of all the elements, more than when you only thought about it alone. Whatever matters to you, you save it until you can tell it to that person. (Gendlin [1978] 2003, 115)

Of course, one might challenge the reliability of casual phenomenological reflection. However, such claims are also supported by empirical studies on the effects of other people's gaze and expression. Bayliss et al. (2007, 644) found that others' emotional reactions to target objects influence one's appraisal of them: objects looked at with a happy expression by someone else are subsequently liked more than those looked at with disgust. Becchio, Bertone, and Castiello (2008, 254) thus maintain that "observing another person gazing at an object enriches that object of motor, affective and status properties that go beyond its chemical or physical structure." Furthermore, it seems that the experienced relevance and significance of perceived entities reflect not only one's own capacities but also what could be achieved in cooperation with others who are present, what "we" could do together (Sebanz, Bekkering, and Knoblich 2006). It should be added that whether and how something appears significant to us is symptomatic of wider projects and concerns as well, the integrity of which depends on various kinds of relation with specific individuals, and with people in general. Consider the potential effects of a marital breakup on how various things matter to a person: "This is no longer the bar where we go together to enjoy free time"; "This is no longer *our* home"; and so forth. But the significance of an experience or activity need not depend on anyone in particular. In writing this paper, I presuppose that there are people who are interested in the topic, some of whom may read it, but I need not have any specific individuals in mind. Importantly, longer-term commitments and projects give the significance of our surroundings greater coherence and temporal consistency. Insofar as our concerns are stable, things will matter to us in consistent and enduring ways. So there is a need to distinguish between the regulatory effects of particular interpersonal encounters and the wider interpersonal, social, and cultural contexts in which those encounters are embedded.

One might retort that none of this pertains to our immediate experiences of things. Rather, one experiences p, then registers someone else's presence, direction of attention, and emotional reaction, and then makes a value judgment that is separate from how p is experienced. Likewise, one experiences p and then makes a value judgment in the light of the projects and concerns one already has. However, there are reasons for preferring the view that we *experience* our surroundings as we do in virtue of how we have related to others in the past, how we currently relate to them, and how we anticipate relating to them, both specific individuals and people in general. For the most part, we *do* experience the significance of a thing as inherent in it: a pen appears relevant to us in the context of writing; an ice cream looks enticing on a hot day. We do not usually have to make a separate, explicit value judgment that accompanies our experience of an object or situation. It would be implausible to maintain that we have to resort to this in all those cases where others are somehow implicated. This view is corroborated by the empirical studies that I have mentioned, which take themselves to be exploring a low-level phenomenon, involving an involuntary effect that arises even when we are not encouraged to attend to or reflect on what others are doing.[3]

The view that we regulate each other's experience, thought, and activity in a range of ways, often without the mediation of explicit propositional thought, is also consistent with various recent enactivist approaches to intersubjectivity, according to which coordination and understanding between two or more parties depend on patterns of interaction between them that cannot be reduced to sets of individual cognitive achievements (e.g., De Jaegher and Di Paolo 2007). For instance, Froese and Fuchs (2012, 205) describe a kind of "inter-bodily resonance between individuals" that can "give rise to self-sustaining interaction patterns." It is important to distinguish two aspects of interaction-experience. Interaction with another person can foster the ability to interpret her mental life, but in addition to this, it shapes how the surrounding world is experienced:

Making sense of the world together (in a social process) is not the same thing as making sense of another person within our interactive relationship. ... The presence of others calls forth a basic and implicit interaction that shapes the way we regard the world around us. (Gallagher 2009, 298, 303)

Take the example of reading a book to one's child. In his absence, the book might appear tedious, devoid of significance. Yet as he smiles, frowns, expresses curiosity, and gestures to something, and as one responds in complementary ways, a shared experience of the book develops. One continues to distinguish "my experience of the book" from "his experience," but there is also "our experience" of it, a sense of its significance "for us." Interaction with one's child also reshapes "my experience of the book"; it is not currently boring even "for me." To offer a further example, as one walks through a city with one's child, the salience of one's surroundings is partly a reflection of the child's needs, vulnerabilities, and concerns, and how these relate to one's own actual and potential activities. A road appears hazardous, a steep escalator treacherous, and a box of Star Wars Lego in a shop window salient. One does not have to think, "He likes Lego," or somehow come to like Lego oneself, to experience it as immediately relevant in the context of a shared situation.

There are also stronger claims in the enactivist literature, concerning the dependence of subjectivity itself on interpersonal interaction. For example, McGann and De Jaegher (2009, 430) write that "a subject is not fully constituted outside of the interaction, independently of it. A subject, instead, is partly constituted in and through the interaction." A number of complementary claims have been made regarding intersubjective development, to the effect that infant-parent interactions are integral to the developmental process through which a sense of self arises. As Reddy (2008, 148-149) puts it, the self is "a dialogic entity,

3. One might object that what I am describing is distinct from "sensory perception," according to one or another definition of it. But my concern is simply with how the world appears to us, as distinct from what we might explicitly infer from appearances, regardless of whether or not all immediate experience of our surroundings is to be construed in terms of "sensory perception." When I refer to "perception" here, I am thinking of kinds of experience that present their object as "here, now."

existing only in relation and therefore knowable only as a relation. Other-consciousness, therefore, is inseparable from self-consciousness, and perhaps both should be called self-other-consciousness." Zeedyk (2006, 321) similarly maintains that "subjectivity arises out of intimate engagement with others," that emotional interaction with caregivers enables the development of capacities including "self-awareness" and "even consciousness." However, as Zahavi (2014, 27–28) rightly responds, such claims are often unclear about the kind of self at stake. For example, when suggesting that selfhood arises through interpersonal processes, Maclaren (2008, 65) refers to capacities for "agency, self-possession, and self-governance." These arguably relate to a richer conception of self, one that involves *reflective* regulation or "governance" of conduct.

Furthermore, even if we accept that interpersonal processes can enrich, diminish, or transform the nature of *what* one perceives, remembers, or imagines, it remains the case that one thinks it, remembers it, or imagines it. In other words, the modal structure of intentionality, the sense of being in state x rather than y, is unaffected. Nevertheless, it is arguable that, for at least *some* experiential contents, interpersonal processes do contribute to the sense of which intentional state one is in or was in. We often turn to others in order to interpret and validate our experiences. Suppose one sees something utterly strange and surprising, such as a person walking past on the street, wearing a Darth Vader costume, pulling a zebra on a lead, singing "House of the Rising Sun," and kicking passersby. One's first reaction might be to wonder, "What is happening?" or even "Is this really happening?" and to look for similar reactions of surprise and bewilderment from others. There may also be a felt need to tell people about it afterward, something that involves interpreting and reinterpreting experienced events in cooperation with them. In the absence of interpersonal processes, at the time and afterward, one might start to wonder, "Did I really see that?" "Did I misperceive?" "Did I imagine it?" "Am I really remembering something?" "Did that actually happen?" Even if that is so, such indeterminacies and doubts are ordinarily restricted to localized and unusual events. But now let us suppose that the *possibility* of confirming, sharing, interpreting, and reinterpreting one's experience through relations with others were altogether absent from one's world; suppose one ceased to encounter others as offering these *kinds* of possibility. In such a case, I argue, the effects would be much more widespread and could serve to erode one's more general grasp of the modalities of intentionality, of the boundaries between intentional state types.[4]

---

4. Although I focus on cases of psychiatric illness here, the effects of prolonged solitary confinement also provide some evidence for the view. Instead of the usual reciprocity between people, solitary confinement involves a unilateral relationship of control and surveillance. Along with this, the prisoner is denied all those forms of interpersonal contact that more usually serve to shape and maintain the integrity of experience. There is a difference between being forcibly denied x and ceasing to recognize the possibility of x. Nevertheless the former could lead to the latter, to a progressive change in one's sense of the kinds of interpersonal relation that the world offers. Lisa Guenther thus maintains that solitary confinement threatens "the most basic sense of identity" (Guenther 2013, xi). Other people, she says,

3   Intentionality and Interpersonal Relations

In his *General Psychopathology*, Karl Jaspers describes how delusions and hallucinations in schizophrenia crystallize out of an all-enveloping delusional "mood" or "atmosphere" (1963, 93–107). One might think that this is best construed in individualistic terms: an endogenous phenomenological disturbance, of a kind that is difficult to characterize, induces an unpleasant feeling of indeterminacy and tension, which is resolved through the formation of delusions with more determinate contents. However, Jaspers also offers the following remarks, which suggest the need for greater emphasis on the interpersonal and social:

> Normal convictions are formed in a context of social living and common knowledge. Immediate experience of reality survives only if it can fit into the frame of what is socially valid or can be critically tested. ... The source for incorrigibility therefore is not to be found in any single phenomenon by itself but in the human situation as a whole, which nobody would surrender lightly. If socially accepted reality totters, people become adrift. ... Reality becomes reduced to an immediate and shifting present. (Jaspers 1963, 104)

We can discern at least three interrelated themes here. First of all, beliefs are ordinarily formed in the context of relations with others, who provide testimony, instruction, confirmation, clarification, and correction. Thus, in the absence of such relations, belief-formation processes would be quite different. Second, where beliefs are not formed through interpersonal processes that align them with consensual knowledge, their contents will differ. Third, and most importantly, the attitude of belief itself presupposes a grasp of the distinction between *x* being the case and *x* not being the case. Someone who lacked the usual sense of that distinction could not *believe* in quite the same way as those who take it as given. The same applies to intentionality more generally: a grasp of the distinction between something's being the case and not being the case is implicated in the structure of all intentional state types, along with more specific distinctions such as "the case now" / "not the case now," "here now" / "somewhere else now," "could be the case" / "could not be the case," and so forth. And taking something to be the case or not the case ordinarily involves taking it to be part of a publicly accessible consensus world, or otherwise. If none of one's beliefs developed in the context of that world or were ultimately embedded in it, one's sense of the contrast between being and not being the case would be altered or diminished. This is why Jaspers (1963, 104) remarks that "reality" for the patient "does not always carry the same meaning as that of normal reality." The boundaries between intentional state types would be

---

more usually "support our capacity to make sense of the world, to distinguish between reality and illusion, and even to tell where our own bodily existence begins and ends" (146). In solitary confinement, one's sense of the distinction between one's body and one's surroundings can break down, and one can have difficulty distinguishing waking from dreaming. However, it is difficult to tease apart the effects of interpersonal privation from more general sensory deprivation.

differently configured. As Jaspers says, reality becomes "an immediate and shifting present," given that the temporal consistency of belief depends in part on its being anchored in an enduring, structured, public world. Furthermore, states of the kind "belief" are more usually susceptible to certain distinctive kinds of interpersonal regulatory process. They are accountable to others in a way that our imaginings are not. Isolated from the social world, one's convictions would become more ephemeral, transient, idiosyncratic, and thus closer in structure to one's imaginings.

These points apply equally to perceptual experience. The kinds of significant possibility attached to entities in the surrounding environment are, to a substantial degree, consistent and enduring, a consistency that depends partly on relations with others. What we find salient and how it is significant to us reflect various coherent and enduring projects, commitments, and concerns, all or almost all of which implicate other people in one or another way. In addition, one's sense of what things offer is regulated by anticipated and actual interactions with others. In a world where others appeared unpredictable or only in the guise of threat, one's experience of everything else would be riddled with uncertainty and doubt: "the fact that other, qualified people have built this device gives me no reason to think it will work"; "this object could be used by someone to harm me"; "others might do any number of things with this." Different kinds of intentional states have coherent but qualitatively different temporal profiles. As Straus remarks, "Waking experience has its own peculiar order and precision. Every moment is directed to the following one in a meaningful anticipation. ... Only in physiological wakefulness do we have the power of anticipation, and in the continuum of anticipation we grasp our wakefulness" (Straus 1958, 164). In other words, wakeful perception has a different and more consistent, constrained pattern of anticipation and fulfillment to that of dreaming, a contrast that applies equally to imagining and, to a lesser extent, to remembering. If extricated from the regulatory roles of a consensus world and from certain kinds of structured interaction with other people, the anticipation-fulfillment structure would be profoundly disrupted. This would result in changes to the temporal profile of perceptual experience, of a kind that would bring it closer to the more anarchic temporal profiles of other intentional state types.

It is clear how phenomenological changes of this nature might render one more vulnerable to psychosis. If a person is alienated from an interpersonal context that more usually regulates the formation, maintenance, revision, and content of her beliefs, she will be more likely to form eccentric beliefs that are recalcitrant to change in the light of others' interventions. Furthermore, if the integrity of intentionality is altered, such that her grasp of the distinction between "is" and "is not" is compromised, she will be less able to distinguish what is and was the case from what is merely imagined. Hence delusions and hallucinations, in this scenario at least, are not simply anomalous beliefs or perceptions. They have a type of intentionality that differs from mundane experiences of believing, remembering, imagining, or perceiving. This is consistent with the observation that schizophrenic delusions can involve a form of double-bookkeeping, where the person seems to both believe that p and not believe that *p*, to hear that *q* and not hear that *q*: "Many schizophrenic patients seem to experience their delusions and hallucinations as having a special quality or feel that sets these apart from their 'real' beliefs and perceptions, or from reality as experienced by the 'normal' person" (Sass 1994, 3).[5] Such ambiguity is to be expected, if delusions and hallucinations involve a kind of intentionality that differs from other forms of believing and perceiving.

4 Trauma and Trust

The view that the structure of intentionality depends both developmentally and constitutively on the interpersonal is consistent with the association between interpersonally induced trauma and psychosis. That view also serves to illuminate the nature of the relationship. Traumatic events, I suggest, can alter the overall structure of interpersonal experience, so as to deprive one of interpersonal relations that the integrity of experience depends on. In addressing the relationship between trauma and psychosis, it is important to distinguish the effects of trauma during development, which interferes with the formation of intersubjective capacities, from trauma in adulthood, which disrupts already established ways of relating to others. Various combinations of the two should also be considered. So there are many different potential trajectories and outcomes. Furthermore, these depend not just on the immediate effects of traumatic events but also on the subsequent responses of specific individuals and people in general, responses that are influenced—to varying degrees—by wider social and cultural structures.[6]

Of course, traumatic experiences are not invariably associated with psychosis or, more specifically, with schizophrenia diagnoses. Nevertheless, there is a substantial literature on trauma and psychosis, which points to strong links, especially in cases of childhood trauma. According to Read et al. (2005), around 85 percent of adults with schizophrenia diagnoses have suffered some form of childhood abuse, with sexual abuse in 50 percent of cases, figures that are far higher than the population base rate. Abuse and trauma in adulthood also increase vulnerability to psychosis, and certain symptoms are most reliably associated with a combination of childhood and adulthood abuse. Abuse is specifically linked to hallucinations, and the contents of hallucinations often resemble—to varying degrees and in different ways—the nature of traumatic events (see also Read et al. 2003). One might question the extent to which self-reports of abuse are reliable. However, Read et al. (2005, 334) note that

5. There is much more to be said about *why* certain things are experienced in these ways and others not, and precisely *how* various seemingly localized experiences and convictions arise out of global changes in the structure of intentionality and intersubjectivity. However, for current purposes, I seek only to maintain that there *is* a plausible link here.

6. For example, symptoms of "war trauma" can be mitigated or exacerbated by the attitudes, narratives, and interpretive resources of a person's community (Hunt 2010).

where corroboration is possible, reliability is high. Similar findings are reported by Kilcommons and Morrison (2005), who conducted a study involving thirty-two people with diagnosed psychosis, 94 percent of whom reported at least one traumatic lifetime event, and 53 percent of whom also met the criteria for post-traumatic stress disorder. They found correlations between severity of trauma and severity of traumatic events, between physical abuse and positive symptoms, and between sexual abuse and hallucinations. Many others have noted similarly strong and specific links. Varese et al. conducted a meta-analysis of forty-one studies (all judged to be methodologically sound), which together point to the conclusion that "childhood adversity is substantially associated with an increased risk of psychosis" (Varese et al. 2012, 669).

Now, correlation does not imply causation. Nevertheless, it is at least suggestive of it. How, though, do episodic and/or sustained traumatic experiences affect how a person subsequently relates to others in general and—with this—the structure of intentionality? I suggest that diminution or loss of what we might call a primitive form of "trust" has a central role to play. Trust, in the relevant sense, is something that first arises in infancy. It is a bodily, affective set of interpersonal expectations, which develop through patterned interactions with caregivers and later come to regulate encounters with people more generally.[7] The developmental process and the effects of deviation from it are described by Fonagy and Allison (2014). They maintain that secure attachments in early life foster a sense of trust in others that later generalizes, disposing one to accept credible communications, while also instilling confidence in one's own judgments and abilities (something that relies on feedback from others). Those who are unable to relate to others in this way are left in a "state of interminable searching for validation of experience" (Fonagy and Allison 2014, 374). In addition, they have a pervasive sense of being unsafe or threatened, which leaves them in a state of "epistemic hypervigilance," constantly on the lookout for potential dangers (374). The person needs something from others and may continue to seek it from them, but she cannot obtain it, as she is unable to participate in the required kinds of relationship. Implied by an experience of others in general as potentially or actually hostile is a breakdown of what Csibra and Gergely call "natural pedagogy," the transmission of generalizable cultural knowledge during development, through a process that depends on the assumed "communicative cooperation and epistemic benevolence of the communicative partner" (Csibra and Gergely 2009, 148). Fonagy and Allison (2014, 374) add that attachment insecurity in adults is associated with epistemic biases that include intolerance of ambiguity, inflexible and dogmatic thinking, and a tendency to make judgments based on insufficient information. It is easy to see why a general lack of trust in other people would be associated with such biases. If others are not regarded as credible sources of information, one will not be disposed to enter into interactions with them of the kind that might challenge one's opinions, draw attention to

7. See Colombetti and Krueger 2015 for a complementary discussion of interpersonal trust and affect regulation.

alternative possibilities, and thus foster recognition of ambiguities. As for dogmatic thinking, insofar as one is cut off from interpersonal processes that more usually shape not only the content but also the nature of belief, one's thinking will be impervious to the influence of others. One will also arrive at "judgments" without consulting all the relevant information, because recognizing information as relevant involves respecting the testimonies of others, something one cannot do if the interpersonal world as a whole fails to offer the possibility of trusting relations.

Even in the adult case, this kind of trust is not to be construed primarily in terms of forming however many judgments along the lines of "B is trustworthy" or "I trust B to do *p*." What is at stake is a more encompassing way of encountering people in general, a capacity to trust that is presupposed by the possibility of adopting a trusting attitude in any given case (for a discussion, see also Ratcliffe, Ruddell, and Smith 2014). It amounts to an overall *style* of anticipation that shapes all experiences of, relations with, and anticipated interactions with other people. My claim is not that we more usually trust everyone but that we have the capacity to trust. Moreover, we are disposed to trust; it is usually the default attitude toward other people in general. Whenever we enter into a relationship with someone, of a kind that might enrich our experience, influence our beliefs, and shape our expectations, we inevitably render ourselves vulnerable to being affected in other ways, to having our world somehow diminished. Hence being open to the other person in the required way involves anticipating that one will not be harmed. This is usually implicit in interpersonal interactions and respected by them. Løgstrup describes the interplay between vulnerability and habitual trust as follows:

Trust is not of our own making; it is given. Our life is so constituted that it cannot be lived except as one person lays him or herself open to another person and puts her or himself into that person's hands either by showing or claiming trust. ... By our very attitude to one another we help to shape one another's world. By our attitude to the other person we help to determine the scope and hue of his or her world; we make it large or small, bright or drab, rich or dull, threatening or secure. We help to shape his or her world not by theories and views but by our very attitude toward him or her. Here lies the unarticulated and one might say anonymous demand that we take care of the life which trust has placed in our hands. (Løgstrup [1956] 1997, 18)

The capacity—or at least the experienced capacity—for this kind of relation can be disrupted by traumatic events, especially when one is deliberately harmed by others. First-person accounts of trauma in adulthood consistently describe the loss of something so wide-ranging and deep-rooted that it is seldom reflected on and easy to overlook altogether until it is lost. For instance, Brison (2002, xii) describes how "trauma shatters one's most fundamental assumptions about the world." A habitual confidence and sense of continuity are replaced by a pervasive sense of unpredictability, uncertainty, and threat: "When the inconceivable happens, one starts to doubt even the most mundane, realistic perceptions" (8–9). Traumatic events can thus affect the overall *style* in which events and actions are anticipated. As

Stolorow (2007, 16) writes, one is exposed to "the inescapable contingency of existence on a universe that is random and unpredictable and in which no safety or continuity of being can be assured." This shift in what one anticipates from other people and from the world in general is referred to in a range of ways, but the most conspicuous and consistent theme is loss of trust. Herman describes a "shattering" of "basic trust," a loss of one's "sense of safety in the world" (Herman 1997, 51). Bernstein (2011) and Janoff-Bulman (1992) also describe the effects of trauma in terms of losing trust, while Jones (2004) opts for the term "basal security."

Phenomenological changes that sometimes precede the onset of psychosis are plausibly interpreted in this way. For instance, Conrad (2012, 177) attempts to convey the early, "trema" phase of schizophrenia through the analogy of walking through a dark forest:

Nothing is "taken for granted" anymore. Nothing is experienced as "natural." In the darkness, precisely where one cannot see, there lurks something, behind the trees—one does not ask what it is that lurks. It remains undefined. It is the lurking itself. In that area between what is visible and what is "behind" the visible (e.g., the particular tree), what we call the background, where what we cannot grasp becomes uncanny. The background, from which the things we do grasp stand out, loses its neutrality.

It can be added that what "lurks" is more specifically a sense of *interpersonal* menace. This permeates experience more generally, given that how one's surroundings appear salient and significant depends on what one anticipates from others people. Although the relationship between trauma, trust in other people, and the integrity of world experience is not such a prominent theme in phenomenological psychopathology, it is explicit in the work of Wolfgang Blankenburg. According to Blankenburg ([1969] 2001, [1971] 2012), localized symptoms of schizophrenia, such as delusions and hallucinations, arise in the context of a more general disturbance of "common sense," in a background of habitual confidence that more usually shapes all experience, thought, and activity. This common sense is inextricable from how we relate to others. It is "primarily related to an intersubjective world [*mitweltbezogen*]," and certain patterns of interpersonal development fail to nurture a "basic trust" on which common sense is founded (Blankenburg [1969] 2001, 307, 310). Again, the kind of trust in question is not simply trust in persons A, B, and C to do p, q, and r. A global "loss of natural self-evidence" (*Verlust der natürlichen Selbstverständlichkeit*), of the kind that characterizes schizophrenia, is bound up with loss of the "capacity for trust" (*Vertrauenkönnen*).[8] This loss of natural self-evidence can also be plausibly related to the "hyperreflexivity" that Sass takes as central to self-disturbance in schizophrenia, a kind of involuntary attentiveness to aspects of experience and thought that are more usually unproblematic and inconspicuous: "A condition in which phenomena that would normally be inhabited, and in this sense experienced as part of the self, come instead to be objects of focal or objectifying awareness" (Sass 2003, 153). Insofar as there is an experience of doubt or uncertainty over the nature of p and what

8. See also Fuchs 2015 for a good discussion of Blankenburg on affective trust.

to expect from it, p will stand out in a way that it does not when encountered in a habitually confident way. The point applies not only to experience of one's surroundings. Confident, habitual action is regulated by accepted backgrounds of shared norms and practices, which prescribe—with varying degrees of determinacy—what is to be done in a given situation. The confidence we have in our beliefs is also partly a reflection of their being affirmed by others and integrated into a consensus world. And, I have also suggested, the unproblematic sense of being in intentional state x rather than y similarly depends on a certain *style* of interpersonal experience. Hence, without trust, ordinarily inconspicuous features of the environment may appear salient, questionable, and problematic, as may one's own activities and thoughts. Furthermore, the sense that one is in one kind of intentional state and not another may be diminished, rendering one's experiencing conspicuous in unfamiliar and troubling ways.

To further illuminate the nature of trust and show how the structure of intentionality depends on it, it is instructive to turn to the later writings of Edmund Husserl (e.g., [1948] 1973), who offers a more developed account of how the modalities of intentionality presuppose a sense of confidence or certainty. When I perceive an entity, I may be uncertain as to what exactly its unseen side looks like or how it might feel to the touch. Nonetheless I have at least an indeterminate sense of what to expect, and my experiences generally unfold in line with this sense. For the most part, experience involves the confident (albeit variably determinate) anticipation of what will happen next, what will be revealed to us when we act in a certain way or as we watch an event occur. For example, as I watch a drinking glass fall to the floor, the anticipation of seeing and hearing it break takes the form of certainty. I need not form an explicit judgment to that effect, but even so, I anticipate that events will unfold in a given way. This is illustrated by experiences of surprise. If the glass were to land without a sound, the situation would appear immediately incongruous. More generally, Husserl maintains, experience incorporates a structured framework of confident anticipation and fulfillment. There are of course experiences of uncertainty too: "perhaps p is not what I take it to be"; "perhaps it is not there at all." There are also experienced doubts with more determinate contents: "perhaps it is not p that I see but q." However, Husserl maintains that localized experiences of uncertainty and doubt only arise—and indeed are only intelligible—within a wider context of certainty. It is only in relation to confident, practical immersion in the world that one can encounter something as potentially or actually anomalous. Anomalies stand out against a more generally harmonious backdrop of unproblematic anticipation and fulfillment. Without this wider confidence and cohesiveness, there would be nothing to rupture. The point applies equally to the explicit sense that p is the case. We only take something to be the case in that way after uncertainty or doubt has arisen:

The simplest certainty of belief is the primal form and ... all other phenomena, such as negation, consciousness of possibility, restoration of certainty by affirmation or denial, result only from the modalization of this primal form and are not juxtaposed, since they are not on the same level. (Husserl [1948] 1973, 100–101)

Husserl also maintains that, as with experiences of potential and actual anomalies, explicit judgments concerning what is, is not, might be, or might not be the case depend for their intelligibility on this primitive sense of certainty and its localized disruption. Furthermore, as all the modalities of intentionality depend in some way on the distinction between what is and what is not, the overall integrity of intentionality presupposes a type of anticipation-fulfillment structure. If we accept something along these lines, we can further see why loss of "common sense" is implied by certain kinds of disruption to intersubjectivity. What we anticipate from our impersonal surroundings is contingent on what we anticipate from others: the stability of the world is inextricable from established norms and shared projects; the kind of confidence we have in our own beliefs involves their rootedness in a consensus world; and the significance that attaches to things—what we anticipate from them—is shaped by relations with others, by specific individuals and by others in general. A change in how one experiences and relates to other people thus implies a more-enveloping change in the structure of experience. The ordinarily presupposed sense of confidence or certainty, in which the modalities of intentionality are rooted, implicates other people through and through.

5   Schizophrenia and the Interpersonal World

I have argued that (i) minimal self cannot be extricated from the modalities of intentionality; (ii) the modalities of intentionality depend on certain kinds of interpersonal relation for their development and sustenance; (iii) traumatic events involving other people can disrupt these relations; and (iv) disturbance of selfhood in schizophrenia can be understood in this way. I conclude by briefly considering what the implications of this position are for our understanding of schizophrenia. To start with, I suggest that schizophrenia should be conceived of as a relational rather than individual phenomenon (for this view, see also Fuchs 2015).[9] Indeed, I have at least some sympathy with views expressed by the "voice hearing" movement. In brief, it is claimed that anomalous experiences with distressing contents arise due to problematic interpersonal experiences and relations, often involving childhood trauma and abuse. They are not "symptoms" of "illness" but meaningful phenomena to be interpreted in the light of past experience and treated accordingly. Instead of seeking simply to extinguish them, the aim should be to make sense of them, help people come to terms with life events, and thus reduce distress (e.g., Romme et al. 2009). Hence "psychosis" is to be reconceptualized as an "emotional crisis" (Romme and Escher 2012).

However, there is a frequent lack of clarity in this literature over whether (a) the diagnosis "schizophrenia" is overextended, or (b) the diagnostic category "schizophrenia" is misguided

9. This is also consistent with therapies that involve nurturing a trusting relationship, allowing the client to "relinquish the rigidity that characterizes individuals with enduring personality pathology" and thus the "relearning of flexibility" (Fonagy and Allison 2014, 372).

and should be jettisoned altogether. For instance, Longden, Madill, and Waterman (2012) maintain that phenomena such as verbal hallucinations should be construed as dissociative rather than psychotic, and cite numerous studies pointing to the link with trauma, especially childhood trauma. So it would seem that the issue is one of whether something should be categorized as *x* rather than *y*. At the same time, they question the legitimacy of a distinction between *x* and *y* in suggesting that "trauma-induced dissociation" and "psychosis" are different ways of interpreting a common phenomenon, the latter inappropriately. So it is unclear what the implications are supposed to be for the schizophrenia construct.

At the same time, the association between trauma and psychosis remains compatible with the view that schizophrenia diagnoses are sometimes associated with disturbances of the minimal self (insofar as the relevant sense of self remains both prereflective and integral to the structure of intentionality), so long as a distinction between minimal and interpersonal self is no longer insisted on. What that association might suggest, however, is that these self-disturbances are not *specific* to schizophrenia. After all, trauma is associated with a range of different psychiatric conditions (e.g., Mueser et al. 2002). Indeed, Read et al. (2005, 341) go so far as to suggest that whether post-traumatic stress disorder or schizophrenia is diagnosed depends, in some cases, on whether or not symptoms are interpreted in the light of past interpersonal experiences. Where one clinician sees a "hallucination," another may see a "flashback." On the other hand, claims regarding the phenomenological distinctiveness of schizophrenia should not be prematurely dismissed, a move that Parnas (e.g., 2013) warns us against. One characteristic that might distinguish schizophrenic self-disturbances—or, to put it more cautiously, a distinctive subset of self-experiences accommodated by current applications of the diagnostic category "schizophrenia"—is a degree of salience dysregulation and temporal fragmentation that goes beyond what I have so far described. Consider delusional "mood" or "atmosphere," as described by Jaspers (1963, 98), which involves one's surroundings appearing salient in ways that are unfamiliar and strange. A sofa that should present itself as something to sit on and offer comfort may appear oddly menacing, or significant in some other incongruous way. A wide-ranging mismatch of entities and types of significance arguably amounts to a more specific and perhaps more profound kind of disturbance. Even so, it is plausible to suggest that something like this could be brought about by an interpersonal process, as a *potential* outcome rather than something *necessitated* by loss of trust and resultant changes in the structure of intentionality. For example, Fuchs describes a loss of familiarity and significance from the world, a kind of experience that is essentially intersubjective in structure and can be socially caused. He then adds the following:

With this diminished intentionality, the second major change in schizophrenic perception arises: the objects in the perceptual field, having lost their "objectivity" and meaning, may gain an overwhelming wealth of physiognomic expressions, and new, peculiar meanings may arise. Single aspects or details of the perceptual field, now no longer framed and kept in distance by active intention, may become prominent, leap at the perceiving subject, catch him or penetrate into him. Especially the gaze of others, the quintessence of expression, obtains a captivating and piercing power. The breakdown of the intentional, active perception thus releases an archaic communication of the lived body with its environment. Instead of the common and intersubjective significance of things or situations (for example, "this is a table set for the meal"), there arise idiosyncratic fragments of meaning, always alluding to the patient and his body. ... The intersubjective constitution of reality is replaced by the idiosyncratic experience of the lived body. (Fuchs 2008, 283)

If this is right, then loss of trust could estrange one from interpersonal regulatory processes and erode the structure of intentionality, in such a way as to make one vulnerable to further disruption of saliences and significances that are more usually held together in a coherent, stable way by relations with other people. However, I have suggested that, even without this degree or kind of salience disruption, a change can occur in the structure of intentionality implicating the most primitive sense of self. Furthermore, this change would plausibly render one susceptible to psychosis, even without a widespread pairing of entities with incongruous salience types.

A further complication is the likelihood that something along the lines of "delusional atmosphere" can also arise due to endogenous or at least impersonal processes. Kilcommons and Morrison (2005, 352) hypothesize that there are two distinct pathways to psychosis: an endogenous route, involving a predominance of negative symptoms, and a trauma-driven route, where positive symptoms are more conspicuous. Myin-Germeys and van Os (2007) similarly note that schizophrenia is heterogeneous in its symptoms and causes. They suggest an "affective pathway" to psychosis, involving childhood trauma as a significant risk factor. Of course, various in-between scenarios are also possible. For instance, it could be that certain endogenous symptoms—perhaps along with the stigma of potential or actual diagnosis—generate social anxiety, of a kind that further disrupts relations with others and erodes interpersonal trust (Birchwood et al. 2006). Ways in which others respond (which are also socially and culturally mediated) could in turn exacerbate the erosion of trust, leading to growing isolation from the social world and from interpersonal regulatory processes. For example, involuntary detention on mental health grounds can itself be traumatic.[10] Morrison, Frame, and Larkin (2003) thus suggest that causal relationships between trauma and psychosis can go both ways, and that vicious cycles sometimes develop.[11]

So things could turn out to be messy, but this does not detract from the general need to interpret symptoms in more relational terms. With this in mind, it is instructive to revisit a classic text that is often quoted in support of the view that schizophrenia involves a

10. See, e.g., Rooney et al. 1996 for a study of differing short- and long-term reactions to voluntary and involuntary hospital admission. See also Beveridge 1998.

11. Such views are not necessarily inconsistent with phenomenological accounts of self-disturbance in schizophrenia. For instance, Sass acknowledges the heterogeneity of the relevant phenomenology and also suggests that it is most likely "a final common pathway with diverse etiological origins" (Sass 2014, 5).

distinctive form of self-disturbance: *Autobiography of a Schizophrenic Girl*. The author, "Renee," describes "a disturbing sense of unreality," which began when she was five years old (Sechehaye 1970, 21), followed by increasingly profound changes in her experience of self, world, other people, and the relationships between them. However, what is less often remarked on is the familial context in which the disturbances are said to have first arisen: "It was during this period that I learned my father had a mistress and that he made my mother cry. This revelation bowled me over because I had heard my mother say that if my father left her, she would kill herself" (22). Throughout the remainder of the text, there are numerous references to intensifying feelings of anxiety and interpersonal isolation, which could be interpreted in terms of an unraveling process that begins with the erosion of basic trust. This is not to challenge phenomenological interpretations of the text but to point out that there is also an interpersonal backstory to be told, involving unsettling events, interpersonal estrangement, loss of trust and associated social anxiety, departure from consensus reality, and the eventual onset of anomalous experiences with more specific contents.

All of this is compatible with the view that schizophrenia involves disruption of the minimal self, and also the view that a certain kind of self-disruption is specific to a subset of schizophrenia diagnoses, so long as it is admitted that even this level of self-experience is intersubjective in nature. There are insufficient grounds for postulating a further, "even more minimal" and presocial sense of self in adult humans, unless we assume a dubious conception of development where earlier capacities are preserved in developmental outcomes. Some self-disturbances are more profound than others, but differences between them are not to be conceived of in terms of interpersonal and pre-interpersonal selves.

Acknowledgments

Thanks to Louis Sass, Christoph Durt, and an audience at the University of Vienna for helpful comments on an earlier version of this chapter.

References

Bayliss, A. P., A. Frischen, M. J. Fenske, and S. P. Tipper. 2007. Affective evaluations of objects are influenced by observed gaze direction and emotional expression. *Cognition* 104:644–653.

Becchio, C., C. Bertone, and U. Castiello. 2008. How the gaze of others influences object processing. *Trends in Cognitive Sciences* 12:254–258.

Bernstein, J. M. 2011. Trust: On the real but almost always unnoticed, ever-changing foundation of ethical life. *Metaphilosophy* 42:395–416.

Beveridge, A. 1998. Psychology of compulsory detention. *Psychiatric Bulletin* 22:115–117.

Birchwood, M., P. Trower, K. Brunet, P. Gilbert, Z. Iqbal, and C. Jackson. 2006. Social anxiety and the shame of psychosis: A study in first episode psychosis. *Behaviour Research and Therapy* 45:1025–1037.

Blankenburg, W. [1969] 2001. First steps towards a psychopathology of "common sense." Trans. A. L. Mishara. *Philosophy, Psychiatry, and Psychology* 8:303–315.

Blankenburg, W. [1971] 2012. *Der Verlust der natürlichen Selbstverständlichkeit: Ein Beitrag zur Psychopathologie symptomarmer Schizophrenien*. Parodos.

Brison, S. J. 2002. *Aftermath: Violence and the Remaking of a Self*. Princeton University Press.

Colombetti, G., and J. Krueger. 2015. Scaffoldings of the affective mind. *Philosophical Psychology* 28:1157–1176.

Conrad, K. 2012. Beginning schizophrenia: Attempt for a gestalt-analysis of delusion. Trans. A. L. Mishara. In *The Maudsley Reader in Phenomenological Psychiatry*, ed. M. R. Broome, R. Harland, G. S. Owen, and A. Stringaris, 176–192. Cambridge University Press.

Csibra, G., and G. Gergely. 2009. Natural pedagogy. *Trends in Cognitive Sciences* 13:148–153.

De Jaegher, H., and E. Di Paolo. 2007. Participatory sense-making: An enactive approach to social cognition. *Phenomenology and the Cognitive Sciences* 6:485–507.

Fonagy, P., and E. Allison. 2014. The role of mentalizing and epistemic trust in the therapeutic relationship. *Psychotherapy* 51:372–380.

Froese, T., and T. Fuchs. 2012. The extended body: A case study in the neurophenomenology of social interaction. *Phenomenology and the Cognitive Sciences* 11:205–235.

Fuchs, T. 2008. Beyond descriptive phenomenology. In *Philosophical Issues in Psychiatry: Explanation, Phenomenology, and Nosology*, ed. K. S. Kendler and J. Parnas, 278–285. The John Hopkins University Press.

Fuchs, T. 2015. Pathologies of intersubjectivity in autism and schizophrenia. *Journal of Consciousness Studies* 22 (1–2): 191–214.

Gallagher, S. 2009. Two problems of intersubjectivity. *Journal of Consciousness Studies* 16 (6–8): 289–308.

Gendlin, E. T. [1978] 2003. *Focusing: How to Gain Direct Access to Your Body's Knowledge*. Rider Books.

Guenther, L. 2013. *Solitary Confinement: Social Death and Its Afterlives*. University of Minnesota Press.

Herman, J. 1997. *Trauma and Recovery*. 2nd ed. Basic Books.

Hunt, N. C. 2010. *Memory, War and Trauma*. Cambridge University Press.

Husserl, E. [1948] 1973. *Experience and Judgment*. Trans. J. S. Churchill and K. Ameriks. Routledge.

Janoff-Bulman, R. 1992. *Shattered Assumptions: Towards a New Psychology of Trauma*. Free Press.

Jaspers, K. 1963. *General Psychopathology*. Trans. J. Hoenig and M. W. Hamilton. Manchester University Press.

Jones, K. 2004. Trust and terror. In *Moral Psychology: Feminist Ethics and Social Theory*, ed. P. DesAutels and M. O. Walker, 3–18. Rowman & Littlefield.

Kean, C. 2009. Silencing the self: Schizophrenia as a self-disturbance. *Schizophrenia Bulletin* 35:1034–1036.

Kilcommons, A. M., and A. P. Morrison. 2005. Relationships between trauma and psychosis: An exploration of cognitive and dissociative factors. *Acta Psychiatrica Scandinavica* 112:351–359.

Løgstrup, K. E. [1956] 1997. *The Ethical Demand*. University of Notre Dame Press.

Longden, E., A. Madill, and M. G. Waterman. 2012. Dissociation, trauma, and the role of lived experience: Toward a new conceptualization of voice hearing. *Psychological Bulletin* 138:28–76.

Maclaren, K. 2008. Embodied perceptions of others as a condition of selfhood? Empirical and phenomenological considerations. *Journal of Consciousness Studies* 15 (8): 63–93.

McGann, M., and H. De Jaegher. 2009. Self–other contingencies: Enacting social perception. *Phenomenology and the Cognitive Sciences* 8:417–437.

Morrison, A. P., L. Frame, and W. Larkin. 2003. Relationships between trauma and psychosis: A review and integration. *British Journal of Clinical Psychology* 42:331–353.

Mueser, K. T., S. D. Rosenberg, L. A. Goodman, and S. L. Trumbetta. 2002. Trauma, PTSD, and the course of severe mental illness: An interactive model. *Schizophrenia Research* 53:123–143.

Myin-Germeys, I., and J. van Os. 2007. Stress-reactivity in psychosis: Evidence for an affective pathway to psychosis. *Clinical Psychology Review* 27:409–424.

Parnas, J. 2013. On psychosis: Karl Jaspers and beyond. In *One Century of Karl Jaspers' General Psychopathology*, ed. G. Stanghellini and T. Fuchs, 208–228. Oxford University Press.

Parnas, J., P. Møller, T. Kircher, J. Thalbitzer, L. Jansson, P. Handest, and D. Zahavi. 2005. EASE: Examination of anomalous self-experience. *Psychopathology* 38:236–258.

Raballo, A., and J. Parnas. 2011. The silent side of the spectrum: Schizotypy and the schizotaxic self. *Schizophrenia Bulletin* 37:1017–1026.

Raballo, A., D. Sæbye, and J. Parnas. 2009. Looking at the schizophrenia spectrum through the prism of self-disordered: An empirical study. *Schizophrenia Bulletin* 37:344–351.

Ratcliffe, M. 2015. *Experiences of Depression: A Study in Phenomenology*. Oxford University Press.

Ratcliffe, M., M. Ruddell, and B. Smith. 2014. What is a sense of foreshortened future? A phenomenological study of trauma, trust and time. *Frontiers in Psychology* 5 (Article 1026): 1–11.

Read, J., K. Agar, N. Argyle, and V. Aderhold. 2003. Sexual and physical abuse during childhood and adulthood as predictors of hallucinations, delusions and thought disorder. *Psychology and Psychotherapy: Theory, Research and Practice* 76:1–22.

Read, J., J. van Os, A. P. Morrison, and C. A. Ross. 2005. Childhood trauma, psychosis and schizophrenia: A literature review with theoretical and clinical implications. *Acta Psychiatrica Scandinavica* 112:330–350.

Reddy, V. 2008. *How Infants Know Minds*. Harvard University Press.

Romme, M., and S. Escher, eds. 2012. *Psychosis as a Personal Crisis: An Experience-Based Approach.* Routledge.

Romme, M., S. Escher, J. Dillon, D. Corstens, and M. Morris, eds. 2009. *Living with Voices: 50 Stories of Recovery.* PCCS Books.

Rooney, S., K. C. Murphy, F. Mulvaney, E. O'Callaghan, and C. Larkin. 1996. A comparison of voluntary and involuntary patients admitted to hospital. *Irish Journal of Psychological Medicine* 13:132–137.

Saks, E. R. 2007. *The Centre Cannot Hold: A Memoir of Schizophrenia.* Virago.

Sass, L. A. 1994. *The Paradoxes of Delusion: Wittgenstein, Schreber, and the Schizophrenic Mind.* Cornell University Press.

Sass, L. A. 2003. "Negative symptoms," schizophrenia, and the self. *International Journal of Psychology and Psychological Therapy* 3:153–180.

Sass, L. A. 2014. Self-disturbance and schizophrenia: Structure, specificity, pathogenesis. *Schizophrenia Research* 152:5–11.

Sebanz, N., H. Bekkering, and G. Knoblich. 2006. Joint action: Bodies and minds moving together. *Trends in Cognitive Sciences* 10:70–76.

Sechehaye, M., ed. 1970. *Autobiography of a Schizophrenic Girl.* Signet.

Stolorow, R. D. 2007. *Trauma and Human Existence: Autobiographical, Psychoanalytic, and Philosophical Reflections.* Analytic Press.

Straus, E. W. 1958. Aesthesiology of hallucinations. In *Existence,* ed. R. May, E. Angel, and H. F. Ellenberger, 139–169. Simon & Schuster.

van den Berg, J. H. 1972. *A Different Existence: Principles of Phenomenological Psychopathology.* Duquesne University Press.

Varese, F., F. Smeets, M. Drukker, R. Lieverse, T. Lataster, W. Viechtbauer, J. Read, J. van Os, and R. P. Bentall. 2012. Childhood adversities increase the risk of psychosis: A meta-analysis of patient-control, prospective- and cross-sectional cohort studies. *Schizophrenia Bulletin* 38:661–671.

Zahavi, D. 2014. *Self and Other: Exploring Subjectivity, Empathy, and Shame.* Oxford University Press.

Zeedyk, M. S. 2006. From intersubjectivity to subjectivity: The transformative roles of emotional intimacy and imitation. *Infant and Child Development* 15:321–344.

8   The Touched Self: Psychological and Philosophical Perspectives on Proximal Intersubjectivity and the Self

Anna Ciaunica and Aikaterini Fotopoulou

Introduction: The Mentalization of the Body and Minimal Selfhood

Whenever I perceive something or feel an emotion, these perceptions and feelings are somehow given to me as *mine*. The idea that our everyday experiences are characterized by a prereflective sense of self, referred to as the "minimal" self, has been highlighted by a long-standing phenomenological tradition (Husserl, Sartre, Merleau-Ponty), as well as more recent authors (Gallagher 2000; Metzinger 2003; Zahavi 2005; Hohwy 2007; Blanke and Metzinger 2009; Blanke 2012). There is wide agreement about the importance of examining the bodily foundations of such prereflective forms of self-awareness, in the sense that one needs to view the mind as a support system that facilitates the functioning of the body and not the other way round. Crucially, bodily self-awareness is not an awareness of the body in passive isolation from the physical and social world. Indeed, both classic phenomenologists such as Husserl and Merleau-Ponty and, more recently, researchers working within the embodied and enactive cognition paradigm (Varela, Thompson, and Rosch 1991) insist on the idea that prereflective self-awareness ought to be understood primarily by taking into account the larger brain-body-environment dynamics (De Jaegher and Di Paolo 2007; Menary 2007). This emphasis has also been adopted by influential recent models of brain function in theoretical neuroscience (Friston 2010; see also Clark 2013), as we briefly outline later. The development of the mind, and selfhood more specifically, can therefore be viewed as the consequence of embodiment within its environment.

The question of what, if anything, makes the "self" a unifying phenomenon has attracted a considerable number of empirical studies and theoretical accounts. A detailed review of the literature dedicated to clarifying the notion of the self lies beyond the scope of this chapter. Rather, for our limited purposes here, it is important to note that despite disagreements on crucial questions about whether there is a self, to what degree it is prereflective, and what exactly constitutes this prereflective sense of self, most of the contemporary accounts share the important assumption that minimal selfhood is not to be conceived as a *static* internal snapshot of some mysterious substance called the "self." Instead, minimal selfhood is conceived as an ongoing process of tracking and controlling bodily properties as a whole (Blanke and Metzinger 2009). If this is so, then one of the main challenges for both theoretical and empirical accounts of minimal forms of self-awareness consists in characterizing the nature of the *relational* components of selfhood by taking into account the role of worldly engagements in shaping its different facets. Indeed, while there is wide agreement over the idea that prereflective self-awareness is a *dynamic* and more *primitive* form of awareness, it is still unclear whether the "ongoing" dynamic and "primitive" aspects refer exclusively to a self-centered continuity or encompass self–other relatedness, as well. For example, Zahavi (2014, 2015a, 2015b) has recently drawn a careful analysis of minimal forms of selfhood and self–other relatedness by distinguishing between (1) "experiential minimalism" (EM) and (2) "social constructivism" (SC), which can be seen as two opposing poles of the debate (Zahavi 2015a). On the one hand, experiential minimalism claims that our experiential life is characterized by a prereflective sense of self or mineness that can and should be understood without any contrasting *others*. On the other hand, according to social constructivism, the minimal self is not innate but a later socioculturally determined acquisition, emerging in the process of social exchanges and mutual interactions.[1]

Against this background, we aim to argue in favor of a reconceptualization of minimal selfhood that transcends such debates and instead traces the relational origins of the self to fundamental principles and regularities of the human embodied condition, which includes social, embodied interactions and practices. Specifically, our position is motivated by the following five theoretical and empirical observations: (1) The progressive integration and organization of sensory and motor signals constitute the foundations of the minimal self, a process that we have elsewhere named "mentalization" of the body (Fotopoulou 2015). (2) Minimal selfhood is best understood by a conceptualization that takes into account all sensory and motor modalities, along with their distinct properties and rules of integration, instead of relying mostly on a "detached" visuospatial model of perceptual experience, and by extension a model of "detached" social understanding. (3) Crucially, as some of these sensorimotor modalities are specialized to respond to experiences both "within" and "on" the physical boundaries of the body (e.g., the skin), an experiencing subject is not primarily understood as being "here" and facing a perceptual object or subject "there," that is, in a separate physical location. (4) Instead, interactions with other people are motivated and constrained by the same principles that govern the "mentalization" of sensorimotor signals in the singular individual, and hence the mentalization of one's body includes *any* body in physical proximity and interaction. (5) Finally, given the premature birth and social dependency of humans in early infancy, there is a "homeostatically necessary," genetically

---

1. For example, W. Prinz (2012) argues that the self is essentially a social and cultural construct, and he emphasizes the socially constructed character of phenomenal consciousness. H.-B. Schmid holds the view that what is shared precedes the self–other distinction (Schmid 2005, 145, 149, quoted in Zahavi 2015a, 157; Schmid 2014).

prescribed, and culturally enriched plethora of such embodied "proximal" experiences and interactions. Collectively, such experiences of proximal intersubjectivity "sculpt" the mentalization process and hence the constitution of the self, including the progressive sophistication of mental distinctions between "subject–object," "here–there" and "good–bad." We unpack these points by focusing predominately on the domain of touch as a paradigmatic example of proximal intersubjectivity.

Before proceeding, a few brief conceptual clarifications are needed. What is prereflective self-awareness? Philosophers usually start with the intuitive idea of reflecting on our inner experiences: for example, I can introspect what I am experiencing right now while drinking my jasmine tea. I can also recognize myself in a mirror and reflect aloud, "This face is mine." Hence one convenient way to define the notion of *prereflective* self-awareness is by contrasting it with *reflective* self-awareness. The latter occurs, for instance, whenever one reflectively introspects one's ongoing experiences or during explicit self-recognition of one's face in the mirror. By contrast, prereflective self-awareness does not involve any form of high-order self-monitoring. One can get a bearing on this phenomenological take on self-awareness by contrasting it with the view of perceptual awareness defended by Brentano. In Brentano's view ([1874] 1973), when I perceive a cat, I am aware that *I* am perceiving a cat. Importantly, he acknowledges that I do not have here two distinct mental states, but rather one single mental phenomenon: my awareness of the cat is one and the same as my awareness of perceiving it. But by means of this unified mental state, I have an awareness of two *objects*: the cat and *my* perceptual experience. Opposing this view, several contemporary philosophers (Legrand 2006; Zahavi 2014) insist on the phenomenological insight according to which my awareness of *my experience* is not an awareness of it as an object, in the sense that I cannot endorse the perspective of an external observer or spectator on it. In prereflective self-awareness, experience is given not as an object but as a fundamentally *first-personal* subjective experience. Clearly much more needs to be said about perceptual awareness in general and the related debates in contemporary philosophy of mind, but for the purposes of this chapter, we will restrict our focus to the phenomenological insight according to which our experiences always involve a kind of implicit, prereflective self-awareness that is a more *basic* form of self-awareness.

This idea has been the focus of much recent empirical research, including investigations that use experimental "tricks" to systematically manipulate sensorimotor signals, promote their integration, or generate conflicts and illusions, and hence study their role in body awareness (for a review, see Blanke, Slater, and Serino 2015). These studies, as well as investigations in neuropsychiatry (for a review, see Jenkinson and Fotopoulou 2014), suggest that primary sensorimotor signals are integrated and organized at different levels of the neurocognitive hierarchy to form several neurocognitively distinct dimensions of minimal, as well as "extended," selfhood. "Body ownership" (the prereflective sense or metacognitive judgment that I am the subject of a voluntary or involuntary movement, or that I am experiencing a certain sensation like touch) and "body agency" (the prereflective sense or metacognitive judgment that I am the cause of a movement and its consequences) (Gallagher 2003; Legrand 2006) are related notions here.

Recently, one of the authors of this chapter (AF) has used an influential theory from computational neuroscience, the Free Energy Framework (Friston 2010), to describe the processes that constitute the minimal self as the "mentalization" of sensorimotor signals (Fotopoulou 2015). Although the term "mentalization" is traditionally used in psychology to refer to our cognitive ability to infer the mental states of ourselves and others, its alternative use in this context is deliberate: it aims to ground this traditional concept in its embodied origin and highlight that self-awareness is not some "add-on" inferential process of "mind reading" but rather a more fundamental process of organization and schematization of bodily signals that directly and necessarily extends to the mentalization of *any*-body (see hereafter). While a description of the Free Energy Framework itself goes beyond the scope of the current chapter, we heuristically define "body mentalization" here as the process by which primary sensorimotor signals are progressively integrated and schematized to form multiple models of our embodied states in given environments. These models are understood not as static body representations in the brain (e.g., "body schema" versus "body image") but as "hypothetical" (probabilistic, inferential), dynamic, and generative processes (they are constantly updated against received error signals). While we cannot do justice to this topic here, we focus in the next section on how mentalization takes place in relation to different bodily signals derived from the individual body, as well as from other bodies in physical proximity. In doing so, we also position this process in a conceptual space that we think is currently occupied by an arbitrary gulf created between experiential minimalism and social constructivism.

1 The Mentalization of the Body and Others: The Terrain between Experiential Minimalism and Social Constructivism

Adopting the view that minimal selfhood emerges from the progressive mentalization of the experience of an active and situated living body within a wider physical and social environment presents us with the crucial question of the role of others in shaping minimal self-awareness. In other words, can we characterize "mineness" and the minimal selfhood without any contrasting *others*? We address this question by first considering the existing neurophilosophical literature, as recently summarized by Zahavi (2014, 2015a, 2015b). On the one hand, according to the experiential minimalism approach (EM), our experiential life is from the beginning characterized by prereflective self-awareness and by its first-personal character or "mineness," which is an innate, ongoing, and more primary form of self-awareness (see also Gallagher 2005; Legrand 2006; Thompson 2007). On the other hand, according to social constructivism (SC), the self is a socioculturally determined acquisition, emerging in the process of social exchanges and mutual interactions. To put it provocatively, one cannot be a self on one's own but only together with others.

Both EM and SC face criticisms. Critics of EM have pointed out that a minimal under-standing of selfhood overlooks the crucial role of the open-ended construction of individual-ity via narratives and language (Gallagher and Hutto 2008). Indeed, given that we are never cut off from the world, who we are crucially depends on the story we and others tell about ourselves. Critics of SC have argued that selfhood cannot be reduced to that which is narrated, and the very mineness of experience is not constitutively dependent on social interactions and intersubjectivity:

I am not disputing that we *de facto* live together with others in a public world from the very start, but I would deny that the very mineness or for-me-ness of experience is constitutively dependent upon social interaction. In short, I am not disputing the de facto co-existence and co-emergence of experi-ential selfhood and intersubjectivity, but am rather denying their constitutive interdependence. (Zahavi 2015a, 148)

There is much controversy in contemporary social cognition literature, including in the recently formed field of "second-person neuroscience" (Schilbach et al. 2013), over the appropriate understanding of the so-called we-experiences in terms of "we-, shared-, or col-lective intentionality," "second-person cognition," or "plural self-awareness," and whether these we-experiences precede or presuppose the self–other distinction (Searle 1990; Reddy 2008; Gallotti and Frith 2013; Tuomela 2013; Schmid 2014). For our restricted purposes here, we will focus on the issue of the constitutive interdependency between minimal experiential selfhood and intersubjectivity from a developmental perspective. For example, Zahavi and Rochat (2015) have drawn on phenomenological insights and developmental studies to sup-port the idea that we-experiences are not prior to, or equiprimordial with, self-experiences. What is primordial is the first-personal presence, the mineness that amounts to a primitive and minimal form of selfhood. We agree with Zahavi and Rochat's claim that this basic expe-riential ownership functions as a precondition for all normative, narrative, and culturally embedded self-interpretations that might occur later in the development. However, we believe that in dismissing the constitutive interdependence of experiential selfhood and intersubjectivity, one runs the risk of throwing the proverbial baby with the bathwater, if intersubjectivity is understood only as the sense of dynamic interactions between two (or more) socially constructed selves. Indeed, a closer look at the development of experiential life at the most primitive levels might reveal the presence of even more primitive forms of embodied relatedness and proximal intersubjectivity, which do not need to posit socially mediated and culturally constructed selves.

In the next sections, we suggest that it is preferable to refocus the very notion of "inter-subjectivity" to take into account more basic and proximal forms of embodiment present in early infancy. We begin with some conceptual clarifications regarding the term "perceptual experience" itself, as well as the "observability" condition that shapes these debates, to argue that many accounts tacitly endorse a visuospatial model of perceptual experience, and that this might be misleading in understanding more basic forms of intersubjectivity. By contrast, we argue that an understanding of other minds is based on a more direct experience of their bodies as proximal sources of sensory signals, encountered in different conditions of congruency and incongruency with other sensorimotor signals originating from within, on, and outside the physical boundaries of the body. It is the regularities, as well as the unavoidable irregularities, of such processes that determine the progressive mentalization of one's own body and the bodies of others, and therefore the progressive sophistication of the self–other distinction. This view, which avoids an unnecessary prioritization of a "detached-visual" model of perception, is best exemplified by considering the role of interoception and particularly affective touch in infancy and beyond.

2 Proximal Intersubjectivity: Against a Detached Perspective

Contemporary debates in philosophy of mind and cognition conceal deep disagreements in the definition of perceptual awareness. To give a full and detailed account of these debates, as fascinating as they are, would require a substantial digression. For our purposes here, we can simply build on the observation that perceptual experience is taken to be the fundamental point of contact with the world, and as such it provides the primary basis on which beliefs, concepts, and knowledge may be formed to relate to the environment. Typically, the relation *subject-object* constitutes a paradigmatic structured relation of a perceptual experience, where the experiencing subject relates to an object *there*. For example, Crane and French (2015) note that "perceptual experience, in its character, involves the presentation (as) of ordinary mind-independent objects to a subject, and such objects are experienced as *present* or *there* such that the character of experience is immediately responsive to the character of its objects." An important correlate of the problem of perceptual experience concerns the *subject-subject* relation, namely, the social perception of other people. Crucially, this model tacitly presupposes a visuospatial perspective of perceptual experience where a subject faces an object from a safe distance *there* (as opposed to *here*) and in a detached manner (no-contact).[2]

For example, it is common to claim that the view others have of the infant functions as a social "mirror" through which the child becomes aware of herself. However, as we shall shortly see, this idea of one's "visibility" through others' perceptual awareness and the related typical expression—to *see* oneself through the *eyes* of others—is highly misleading when examining more primitive forms of prereflective self-awareness (Ciaunica 2015). For example, Sartre famously argued that my *primary* experience of the other is an experience that involves my own self-consciousness, that is, a self-consciousness in which I am prereflectively aware that I am a *visible object* for another. Sartre characterizes my being-for-others as an external dimension of being, and he speaks of the existential alienation provoked by my

---

2. This discussion has obvious implications for spatial cognition and notions of peripersonal and extrapersonal space that sadly we cannot address within the space limitations of this chapter.

encounter with the other (Sartre [1943] 1956, 287). The primary experience of the other is not that I perceive her as some kind of object in which I must encounter a person. Rather, I perceive the other as a subject who perceives *me* as a visible *object*. However, recent developmental studies suggest that this type of self-apprehension through others' eyes (based on a visuospatial model of perspective taking) is *not* the most primitive form of self–other relatedness (Moll and Kadipasaoglu 2013). In the remainder of the chapter, we focus on interoception and touch as perceptually proximal ways to relate to others, and we provide empirical evidence suggesting that at the basic level, self–other relatedness is not primarily experienced in this detached visuospatial fashion.

3   The Touched Self: The Mentalization of Proximal Bodies

Questions regarding minimal selfhood and otherness have also received increasing empirical attention in developmental psychology. The results of infant imitation experiments, for instance, have yielded a plethora of interpretations and debates. Other empirical researchers have focused on multisensory integration and other "contingency detection" paradigms (for a review, see Gergely and Watson 1999). For instance, some studies have now illustrated that infants as young as three to five months show sensitivity to body-related, proprioceptive-visual synchrony and, as motor control develops, also spatial congruency (Rochat and Morgan 1995). In such paradigms, infants tend to respond differentially to experimentally controlled and visually presented feedback of their body parts (e.g., their legs) moving synchronously and in spatial congruency to their own movements, rather than manipulated visual feedback that does not have these properties (e.g., asynchronous or incongruent movements). A recent study has further found that newborns detect visual-tactile synchrony in stimuli directed to their own faces and are able to discriminate synchrony from visual-tactile asynchrony (Filippetti et al. 2013). Indeed, the detection of "amodal" properties like synchrony is considered key to the integration of the senses and the organization of perceptual input into distinct, unitary multimodal schemata, a process we have termed "sensorimotor mentalization" in this chapter. This is in fact the basis of most multisensory integration paradigms in adults: sensitivity to synchrony (the so-called glue of the senses) across sensory input allows perceiving subjects to experience unitary multimodal events and to separate stimulation originating from the self and stimulation arising from others. Accordingly, developmental studies on sensitivity to synchrony have been considered as evidence for the early ability for a rudimentary distinction between self and other. This conclusion seems to favor experiential minimalism (EM), in the sense that such a distinction seems to precede the need for intersubjectivity in the constitution of the self.

However, we believe this approach misses an important dimension of such integration and mentalization processes. Indeed, human infants seem to respond to fundamental rules of information organization; however, in their everyday experiences (and in some experimental settings) such amodal rules apply also to information received from other bodies in their proximity. Put crudely, the bodies of human caregivers provide an almost continuous embodied engagement in infancy, during which rich patterns of synchrony and asynchrony, and other forms of on–off sensory and spatial contingency and congruency, are implemented through a rich repertoire of culturally defined practices of interaction (e.g., hugging, kissing, singing, clapping, stroking, rocking, holding), as well as necessary and frequent routines of embodied engagement required to satisfy the infant's basic biological and psychological needs (e.g., breast-feeding, washing, rubbing-cleaning, skin-to-skin sleeping, body-to-body temperature regulation, and skin hydration; see also the following sections). Several volumes have recently focused on visual signals from other bodies, like the mirroring assumptions and theories of different kinds (for a review, see Gallese 2013). However, we propose here that the early, crucial role of such practices in the formation of the minimal self is most obvious when one considers the special case of interpersonal "touch," even in experimental settings. In the aforementioned study of Filippetti and colleagues (2013), for instance, part of the sensory input (the tactile stimulation) was caused by another individual in both the synchronous and the asynchronous conditions. Therefore what determines the early mentalization of one's own body, as opposed to that of another individual, may somewhat paradoxically be caused by social interactions (cosubjectivities). For instance, feeding, sleeping, calming-down, or entertaining routines typically include endless repetitions of multisensory bundles from at least two bodies (e.g., active and passive touch, proprioceptive and vestibular information, smell, temperature, visual and auditory feedback). During such experiences, the infant is therefore responding to regularities and irregularities between the various sensorimotor "bundles" and thus mentalizing its own body in interaction with that of the caregiver. In this sense, the very first-person experience of my body as mine is constituted by the presence of, and interaction with, other bodies in proximity.

Before we go on to further explain why physical contact and interpersonal touch have a unique, primary role in the minimal self, it is necessary to stress that contrary to SC, the critical variable emphasized here is "other interacting bodies," rather than "other minds." While the behavior of caregivers is determined by their feelings and intentions toward infants, such general mental states are not regarded as the critical element of the role of proximal intersubjectivity in the formation of minimal selfhood. We illustrate this point in reference to a recent empirical study on infant holding. Most parents would recognize that it is far easier to calm and put to sleep a crying baby while standing up and walking around the room with the baby in one's arms than by holding the baby in one's arms while seated. Indeed, a recent study of infants younger than six months found that being held and carried by a walking mother led the infants to immediately stop voluntary movement and crying and exhibit a rapid heart rate decrease, compared with holding by a sitting mother (Esposito et al. 2013). Furthermore, similar motor, vocal, and heart rate "calming" responses were observed in mouse pups, supporting the idea of a conserved embodied component of mammalian mother-infant interaction. It should be obvious from these findings that such mechanisms relate to the importance of embodied caregiver-infant interactions per se (e.g., mobile versus static holding), without the need to refer to the sharing of any high-order, mental, or even spatial concepts such as intentionality, empathy, or perspective.

Similar effects of embodied and primarily tactile interactions between parents and their offspring has long been established in other mammals (Harlow 1958; Panksepp 1998). For example, early postnatal maternal tactile stimulation (linking/grooming) has been shown to modify the known adverse effects of prenatal stress on physiological and emotional reactivity later in life (e.g., Vallee et al. 1999). Unfortunately, relevant systematic research in human infants is sparse (Sharp et al. 2012), although touch has long been prioritized by proponents of "kangaroo care" and similar "skin-to-skin" and "touch-based" approaches to parenting and health care (for a review, see Field 2001). However, for the most part, scholars of human infancy tend to claim that such effects in humans are mediated by parents' mental states and related higher-order psychological concepts (e.g., theory of mind, attachment style). Even in theories that have stressed embodied aspects of the infant-caregiver relationship, for example, "affect attunements" (Stern 1985) or "contingent marked mirroring" (Gergely and Watson 1999), these are quickly embedded in more complex mentalistic conceptualizations of the caregiver's mind and therefore the view that infants' minds are first "read out" by mothers and then responded to accordingly. While we do not deny the role of such forms of relatedness and intersubjectivity, we agree with Zahavi and others in understanding such factors as secondary to more primitive forms of subjectivity and selfhood. Contrary to extreme versions of experiential minimalism, however, we propose that proximal embodied interactions of caregiving entail a more fundamental form of embodied intersubjectivity that contributes to the constitution of the self from the outset. In the example of the study by Esposito et al. (2013), we assume that the mental states and "mind-reading" capacities of mothers toward their babies do not typically vary depending on whether they are walking or sitting, and yet the particular embodied interactions between mother and infant seem to have direct behavioral and physiological effects on infants. We further specify some of the reasons why such embodied interactions are especially relevant and necessary for the formation of minimal selfhood.

4 The Felt Self: The Mentalization of Interoceptive Signals

What kind of bodily signals become "mentalized" to form the basis of minimal selfhood? In empirical research, although scientists have long proposed that bodily self-consciousness relies on an integrated representation of multiple streams of sensory and motor information, there has been a strong bias in the kind of bodily signals studied in this respect. Specifically, most investigations have focused on "multisensory integration" paradigms that study the integration of exteroceptive (e.g., vision, audition, touch) signals, or on sensorimotor integration paradigms that may also include motor, efferent signals, and proprioceptive or vestibular feedback. Remarkably, however, until recently little work on bodily self-awareness concerned interoception. Interoception refers to the perception of the physiological condition of the body, involving representations from multiple modalities such as temperature, itch, pain, cardiac signals, respiration, hunger, thirst, pleasure from sensual touch, and other bodily feelings. It is distinct from the exteroceptive system, which refers to the classical sensory modalities for perceiving the external environment, as well as proprioceptive, vestibular, and kinesthetic input about the movement and location of the body in space (Blanke and Metzinger 2009; Craig 2002; Critchley et al. 2004). Crucially, contrary to classic views of interoception as "the perception of the body from within," the current notion of interoception is tightly linked to homeostasis and emotion. Interoceptive signals are considered crucial in informing the organism about the homeostatic state of the body in relationship to experiences originating both from within the organism (e.g., cardiac and respiratory functions, digestion, hunger, thirst) and from outside it (e.g., taste, smell, affective touch, pain). Accordingly, interoception is thought to rely on separate specialized neuroanatomical systems that are associated with the autonomic nervous system, special spinal cord pathways, and subcortical and cortical brain areas mapping motivational and homeostatic states (Craig 2009; Critchley et al. 2004; Damasio 1994; Panksepp 1998).

A number of researchers have recently argued that interoception is uniquely related to the generation of subjective feelings, informing the organism about its levels of arousal, bodily needs, and the value or valence of stimuli. As such, interoception has been ascribed a central role in the core of self-awareness (Craig 2009; Critchley et al. 2004; Damasio 1994; Seth, Barrett, and Barnett 2011). Important for our purposes here, preliminary evidence also suggests that interoception can uniquely shape the minimal self, as studied in multisensory integration paradigms and neuropsychiatric disorders (for a review, see Seth 2013). For example, participants with lower abilities to detect their own heartbeat seem more susceptible to bodily illusions of synchronous visuotactile stimulation (Tajadura-Jiménez, Longo, Coleman, and Tsakiris, 2012).

How can one reconcile this view with more classical considerations of the constitution of the minimal self and related debates on intersubjectivity as outlined earlier? At first sight, the potential role of interoception in the minimal self may be interpreted as evidence in favor of experiential minimalism. Specifically, one could say that the mentalization of the body and the constitution of the minimal self are mediated by an innate, specialized system that informs the organism about the homeostatic state of the body and particularly of sensations arising from within the organism. The resulting inner feelings of "arousal," "wakefulness," and "wellness," or lack thereof, combined with exteroceptive and motor signals regarding the body, could thus form the basis of subjectivity and the self, and a fundamental source of information regarding the self–other distinction. This is indeed the view that several scientists and scholars have recently put forward (Craig 2009; Damasio 1994; Critchley et al. 2004; Seth 2013). On closer inspection, however, interoception and its properties point to a view that regards intersubjectivity as fundamental and necessary in shaping the mentalization of interoception and not the other way around. The latter claim is supported by two main observations that we unpack in the following two sections, respectively: (a) interoception itself is derived from the outside and other bodies as much as from the inside of the body; and (b) in early infancy, when the motor system is not yet developed, the functioning of several interoceptive modalities depends wholly on embodied interactions with other bodies.

5 The Affectively Touched Self: The Pleasure of Proximal Interactions

As previously discussed, contrary to classical views, contemporary accounts have defined interoception as the set of modalities that inform the organism about the homeostatic state of the body in relationship to experiences originating from within the organism (e.g., cardiac awareness, hunger) or outside it (e.g., taste, smell, affective touch, pain). We use the example of affective touch to illustrate the importance of this reclassification (for similar considerations and findings on the domain of pain, see Krahé et al. 2013, 2015; Decety and Fotopoulou 2014). Indeed, recent neurophysiological, neuroimaging, and behavioral studies suggest that certain single tactile experiences on the skin, such as the reception of gentle, caress-like strokes, are processed by two partly independent neurocognitive systems. As has been known for decades, tactile stimuli are processed in terms of their exteroceptive, discriminatory processes in classical peripheral pathways and somatosensory cortical areas. Recent research has demonstrated, however, that a specialized peripheral and central system seems to code for the affective properties of the same tactile stimulus. Contrary to purely sensory touch, composed of skin mechanoreceptors projecting to the thalamus and primary somatosensory cortex, the neurophysiological system for affective touch (Vallbo, Olausson, and Wessberg 1999) seems to rely on a distinct subgroup of mechanoreceptors, tactile C-fibers, responding only to slow (between 1 and 10 cm/s), caress-like touch and leading to subjective pleasantness (Löken et al. 2009). Crucially, C-tactile afferents take a distinct ascending pathway from the periphery to a different part of the thalamus and then to the posterior insular cortex (Morrison, Bjornsdotter, and Olausson 2011). According to some researchers, the latter pathway is considered as mediating an early convergence of sensory and affective signals about the body, which are then re-represented in the mid and anterior insula, the proposed sites of interoceptive awareness (Craig 2009; Critchley et al. 2004).

Thus affective touch seems to simultaneously capture information about the "inner" body (e.g., "this experience feels good or not") and the external world (e.g., "this is a material with little friction, moving slowly"). Crucially, a recent study found that nine-month-old infants are sensitive to the particular physical properties of affective touch, in the sense that CT-optimal but not nonoptimal velocities of tactile stimulation led to heart rate deceleration in the infants, possibly reflecting relaxation and increases in their behavioral engagement (gaze shifts and duration of looks) with the stroking stimulus. We speculate that this unique parallel (by definition, synchronous and spatially congruent) activation of pathways relating to the internal representation of the body, as well as the external, mostly social world of the infant (see the previous section regarding the presence of social touch in infancy), presumably acts as an early developmental source not only of self–other relatedness and social connection (for the so-called social touch hypothesis, see Morrison, Loken, and Olausson 2010) but also of bodily information regarding the self–other distinction. Paradoxically, given its dual sensory-discriminatory and affective-motivational nature, social touch, an essential part of early mother-infant interactions, may have a unique developmental role in progressively establishing the physical boundaries of the psychological self. Unfortunately, to our knowledge, no systematic developmental studies have focused specifically on the role of affective touch in the formation of the minimal self. However, the aforementioned application of multisensory integration paradigms to the study of infant body perception (e.g., Filippetti et al. 2013) suggests that the specific developmental role of affective social touch can soon be studied in early infancy and childhood.

In addition, indirect confirmation of our suggestion comes from studies on adults. Recent research has shown that the perception of affective touch can provide information about the emotions and thoughts of other individuals, that is, the touch providers (Hertenstein et al. 2006) and the touch receivers (Gentsch, Panagiotopoulou, and Fotopoulou 2015). More specifically as regards the minimal self, a series of recent studies focused on the role of affective touch in the sense of body ownership. The rubber hand illusion is a paradigm involving the illusion of ownership of a foreign hand following synchronous visuotactile stimulation between one's own unseen hand and another hand seen in proximity and in a congruent spatial position to one's real hand. Using this paradigm, three independent studies, one of them from AF's lab, have now found that slow, caress-like touch of CT-optimal velocities and properties enhanced various subjective and behavioral measures of the rubber hand illusion more than fast, emotionally neutral touch (Crucianelli et al. 2013; Lloyd et al. 2013; van Stralen et al. 2014). That is, affective, pleasant touch delivered by another individual seems to play a unique role in the process of multisensory integration that determines how a body part is subjectively experienced as mine. Support for this idea can also be found in several studies on patients with clinical disorders of the minimal self, such as patients with body ownership disturbances, who seem able to at least momentarily accept their disowned arm as theirs following affective touch (for a review, see Gentsch et al., forthcoming).

6   When the Motor System Is Not Yet Developed: The Social Mentalization of the Body

Young infants show considerable movement of the head, including the face and eyes, the limbs and the trunk, especially when in certain neurophysiological states. Nevertheless they lack strength and control in their large antigravity muscles and are helpless in supporting their own weight, and they are unable to initiate and execute complex sequences of purposeful movements. Therefore a young infant cannot position and balance itself, feed itself, thermoregulate, or protect itself from tissue damage (e.g., skin burns, bone fractures, etc.). Freud wrongly assumed that this human need for early nurturance and care by conspecifics is the ultimate motivation for our early social relating. Seminal studies have since established that humans have developed an innate social attachment drive, unrelated to hunger or thermoregulation, and a corresponding lifelong need for social connection (Bowlby 1969; Harlow 1958; Panksepp 1998). Contrary, however, to recent emphasis on mentalistic concepts such as "attachment styles," it is useful to remember that proponents of this view have indeed emphasized the embodied rather than the mentalistic dimensions of this drive, such as the "need for physical proximity" (Bowlby 1969), "contact comfort" (Harlow 1958), and tickling or bodily play (Panksepp 1998). Indeed, we believe that the primacy of our social attachment drive should not obscure the important embodied role of caregivers in regulating the infant's interoceptive states and in turn the foundations of the minimal self. Thus, contrary to current "mentalistic" views of self-formation, we suggest that the origins of the mentalization process itself are not only embodied (as outlined in the previous sections) but also by necessity involve other people's bodies.

Specifically, in the previous sections, we have described as "the mentalization of the body" the process of detecting "amodal" properties like synchrony between the senses and organizing sensory input of both personal and interpersonal origins into distinct, unitary multimodal schemata. We have also stressed that such senses refer not only to exteroception but also to interoception, the senses that inform the organism about the homeostatic state of the body. Moreover, the mentalization of the body involves not only perceptual integration and subsequent inferences but also sensorimotor integration (active inference, in the terminology of the Free Energy Framework). Given the infant's immature motor system, however, what kind of models can she form? The rudimentary motor system of the infant affords several opportunities for her to learn and build generative models in her first unaided sensorimotor interactions with the environment. For instance, an infant can learn that closing her eyes or looking away causes changes in her visual input, which she can then learn to implement when sudden large changes in environmental light occur. In the case of several interoceptive modalities, however, no movement on the part of the infant alone can change her interoceptive state. In the terms of the Free Energy Framework, there can be no prediction errors, or learning in the longer-term, on the basis of active inference.

In this section, we put forward the radical claim that it is exactly the fact that human infants are born without a fully mature motor system that can change its own physiological states that determines the constitutional role of proximal intersubjectivity in minimal selfhood. As infants experience physiological changes as both internal and external conditions, they can engage in reflexive autonomic and motor behaviors (e.g., crying, kicking, exploring, sucking, etc.). These active behaviors, however, return only rudimentary sensory feedback that rarely changes the interoceptive state of the infant (e.g., the infant can quickly learn to perceive its own crying or can start to associate kicking with the tactile input encountered if the legs touch something in the environment, but these experiences and newly acquired schemata will not change his levels of arousal or satiation). Instead, in good-enough caregiving environments, such behaviors are met—not only by facial expressions and other aforementioned mentally "attuned" responses, but also crucially with a variety of proximal, embodied responses, such as soothing touch, holding, feeding, and so on, as described earlier—and can produce changes in the infant's behavior, physiology, and particularly her interoceptive state (e.g., heart rate reductions, satiation, etc.). It follows that feelings of bodily satisfaction, pain, and pleasure, and the lack thereof, are primarily constituted as "mine" only via behaviors that engage the interacting other (see also Gergely and Watson 1999), rather than being fundamental processes of the singular individual and its body, as certain theories assume (e.g., Damasio 2010; Craig 2009).

Two further points are needed to clarify this position. First, the infant's behaviors during proximal interactions (e.g., crying) are not the manifestation of a preexisting set of psychologically differentiated "feelings" or "subjective states," as the experiential minimalism view implies. Rather, we propose that the infant experiences changes in mentally "undifferentiated" states of physiological arousal, or alertness, as well as "SEEKING" in the active domain (SEEKING is conceived as a primitive motivational system in neuroscience, linked with the neurobiology of the neurotransmitter dopamine and described as an "objectless" urge for action toward the environment, an active version of the notion of arousal; see Panksepp 1998). The infant progressively learns to associate such physiological states with particular behaviors and responses (and later with language and cultural responses; indeed, culture is present in the minimal self only as embodied practices between people, as explained earlier). For instance, crying no longer is only associated with the initial state of physiological change but also predicts a set of external behaviors, some of which in turn are anticipated to change the initial state, and the overall process "binds" physiology to subjectively experienced states such as "pain" or "unpleasure." We believe this is the embodied basis of subjective "feeling states," which therefore do not preexist embodied encounters. The more rudimentary dimensions of arousal and SEEKING are, in our view, the only aspects of subjectivity that can be put forward in support of experiential minimalism. However, as they do not require any preexisting perspectival notion, or self–other distinction (e.g., there is no reason to assume that changes in physiological arousal are experienced by an infant in a way that is anchored to any notion of a self, i.e., as arousal experienced by me), it is unclear whether they should be seen as the basis of minimal selfhood.

Finally, our position does not assume that the infant mentalizes its own body and thus develops a minimal self, because the adult is able to correctly "read out" the infant's psychological states and "metabolize" them into appropriate emotions and feelings (Fonagy et al. 2004). Successful mind reading (Fonagy et al. 2004; Gallese 2013) or "affective attunement" (Stern 1985) by the adult ensures this process can continue to develop successfully as the child grows, and the interactions indeed need to be "attuned" to a richer and more fixed mental, generative models. However, as we have already stressed (see sec. 3), the critical variables in the initial processes of social mentalization emphasized here are assumed to be bodily and exploratory rather than psychological and fixed. We believe caregivers interact with infants mostly in a dynamic, trial-and-error fashion, trying to discover and learn what each infant needs in each instance and from each caregiver. In the same instance, they are contributing to the infant's self-formation and learning. In simple terms, the self is not only constituted intersubjectively but also constitutes particular emotional intersubjectivities at the same time. If this were so, early care would be far easier than it seems to be, and a good textbook could really tell us what all infants need.

Conclusion

Is the minimal self already relational in its very bodily foundations? At first sight, it is tempting to agree with Zahavi's claim that "we should not accept being forced to choose between viewing selfhood as either a socially constructed achievement or an innate and culturally invariant as a given. Who we are is both made and found" (2015a, 147). Indeed, the minimalist notion of an experiential self is fully compatible with a more complex notion of a socially and normatively embedded self, and researchers have proposed several ways of framing this relation. However, we have tried to argue here in favor of a more radically embodied view of intersubjectivity itself that cuts across the distinction between innate and socially constructed notions of selfhood. Proximal intersubjectivity itself is not an index of socialization, or the product of cultural practices, but rather a fundamental condition of human survival, supported by strong human attachment instincts. Specifically, we have outlined that particular types of affective touch and more generally physical contact and proximal interactions are crucial for the mentalization of the body and the formation of minimal selfhood. We have also made the further, radical claim that in early infancy, when the motor system is immature, proximal intersubjectivity is necessary for the mentalization of interoceptive states and therefore the corresponding core aspects of the minimal self. There is thus no gap between the minimal and the interactive self—as constrained and embedded in cultural frameworks—for there is a deep continuity between the principles that govern the mentalization of sensorimotor signals in the singular individual and the mentalization of *any*-body via physical proximity and interaction. Our approach thus echoes Merleau-Ponty's view according to which one must consider "the relation with others not only as one of the contents of our experience *but as an actual structure in its own right*" ([1960] 1964, 140; our italics).

Acknowledgments

Aikaterini Fotopoulou was supported by a European Research Council Starting Investigator Award (ERC-2012-STG GA313755) and a Volkswagen Foundation European Platform for Life Sciences, Mind Sciences, and the Humanities grant.

Anna Ciaunica was supported by a Foundation for Science and Technology Fellowship Grant (FCT) (SFRH/BPD/94566/2013).

References

Blanke, O. 2012. Multisensory brain mechanisms of bodily self-consciousness. *Nature Reviews: Neuroscience* 13:556–571.

Blanke, O., and T. Metzinger. 2009. Full-body illusions and minimal phenomenal selfhood. *Trends in Cognitive Sciences* 13 (1): 7–13.

Blanke, O., M. Slater, and A. Serino. 2015. Behavioral, neural, and computational principles of bodily self-consciousness. *Neuron* 88 (1): 145–166.

Bowlby, J. 1969. *Attachment and Loss*, vol. 1: *Attachment*. Basic Books.

Brentano, F. [1874] 1973. *Psychology from an Empirical Standpoint*. Trans. A. C. Rancurello, D. B. Terrell, and L. L. McAlister. Routledge.

Ciaunica, A. 2015. Basic forms of pre-reflective self-consciousness: A developmental perspective. In *Pre-reflective Self-Consciousness: Sartre and Contemporary Philosophy of Mind*, ed. S. Miguens, G. Preyer, and C. Morando, 422–438. Routledge.

Clark, A. 2013. Whatever next? Predictive brains, situated agents, and the future of cognitive science. *Behavioral and Brain Sciences* 36 (3): 181–204.

Craig, A. D. 2002. How do you feel? Interoception: The sense of the physiological condition of the body. *Nature Reviews: Neuroscience* 3:655–666.

Craig, A. D. 2009. How do you feel-now? The anterior insula and human awareness. *Nature Reviews: Neuroscience* 10:59–70.

Crane, T., and C. French. 2015. The problem of perception. In *Stanford Encyclopedia of Philosophy*, spring 2016 edition, ed. E. N. Zalta. http://plato.stanford.edu/archives/spr2016/entries/perception-problem.

Critchley, H. D., S. Wiens, P. Rotshtein, A. Öhman, and R. D. Dolan. 2004. Neural systems supporting interoceptive awareness. *Nature Neuroscience* 7:189–195.

Crucianelli, L., N. K. Metcalf, A. K. Fotopoulou, and P. M. Jenkinson. 2013. Bodily pleasure matters: Velocity of touch modulates body ownership during the rubber hand illusion. *Frontiers in Psychology* 4:703.

Damasio, A. R. 1994. *Descartes' Error: Emotion, Reason, and the Human Brain*. G. P. Putnam's Sons.

Damasio, A. R. 2010. *Self Comes to Mind: Constructing the Conscious Brain*. Heinemann.

Decety, J., and A. Fotopoulou. 2014. Why empathy has a beneficial impact on others in medicine: Unifying theories. *Frontiers in Behavioral Neuroscience* 8:457.

De Jaegher, H., and E. A. Di Paolo. 2007. Participatory sense-making: An enactive approach to social cognition. *Phenomenology and the Cognitive Sciences* 6:485–507.

Esposito, G., S. Yoshida, R. Ohnishi, Y. Tsuneoka, M. del C. Rostagno, S. Yokota, S. Okabe, et al. 2013. Infant calming responses during maternal carrying in humans and mice. *Current Biology* 23 (9): 739–745. doi:10.1016/j.cub.2013.03.041.

Field, T. 2001. *Touch*. MIT Press.

Filippetti, M. L., M. H. Johnson, S. Lloyd-Fox, D. Dragovic, and T. Farroni. 2013. Body perception in newborns. *Current Biology* 23 (23): 2413–2416.

Fonagy, P., G. Gergely, E. L. Jurist, and M. Target. 2004. *Affect Regulation, Mentalization, and the Development of Self*. Karnac.

Fotopoulou, A. 2015. The virtual bodily self: Mentalisation of the body as revealed in anosognosia for hemiplegia. *Consciousness and Cognition* 33:500–510.

Friston, K. 2010. The free-energy principle: A unified brain theory. *Nature Reviews: Neuroscience* 11:127–138.

Gallagher, S. 2000. Philosophical conceptions of the self: Implications for cognitive science. *Trends in Cognitive Sciences* 4 (1): 14–21.

Gallagher, S. 2003. Bodily self-awareness and object-perception. *Theoria et Historia Scientiarum: International Journal for Interdisciplinary Studies* 7 (1): 53–68.

Gallagher, S. 2005. *How the Body Shapes the Mind*. Oxford University Press.

Gallagher, S., and D. Hutto. 2008. Understanding others through primary interaction and narrative practice. In *The Shared Mind: Perspectives on Intersubjectivity*, ed. J. Zlatev, T. P. Racine, C. Sinha, and E. Itkonen, 17–38. John Benjamins.

Gallese, V. 2013. Mirror neurons, embodied simulation and a second-person approach to mind reading. *Cortex* 49:2954–2956. doi:10.1016/j.cortex.2013.09.008.

Gallotti, M., and C. Frith. 2013. Social cognition in the we-mode. *Trends in Cognitive Sciences* 17 (4): 160–165.

Gentsch, A., E. Panagiotopoulou, and A. Fotopoulou. 2015. Active interpersonal touch gives rise to the social softness illusion. *Current Biology* 25 (18): 2392–2397.

Gentsch, A., L. Crucianelli, P. Jenkinson, and A. Fotopoulou. Forthcoming. The touched self: Affective touch and body awareness in health and disease. In *Affective Touch and the Neurophysiology of CT Afferents*, ed. H. Olausson, J. Wessberg, I. Morrison, and F. McGlone. Springer.

Gergely, G., and J. S. Watson. 1999. Early social-emotional development: Contingency perception and the social-biofeedback model. In *Early Socialization*, ed. P. Rochat, 101–136. Erlbaum.

Harlow, H. F. 1958. The nature of love. *American Psychologist* 13:673–685.

Hertenstein, M. J., D. Keltner, B. App, B. A. Bulleit, and A. R. Jaskolka. 2006. Touch communicates distinct emotions. *Emotion* 6 (3): 528–533. doi:10.1037/1528-3542.6.3.528.

Hohwy, J. 2007. The sense of self in the phenomenology of agency and perception. *Psyche* 13 (1): 1–20.

Jenkinson, P. M., and A. Fotopoulou. 2014. Understanding Babiski's anosognosia: 100 years later. *Cortex* 61:1–4.

Krahé, C., A. Springer, J. A. Weinman, and A. Fotopoulou. 2013. The social modulation of pain: Others as predictive signals of salience—a systematic review. *Frontiers in Human Neuroscience* 7:386.

Krahé, C., Y. Paloyelis, H. Condon, P. M. Jenkinson, S. C. R. Williams, and A. Fotopoulou. 2013. Attachment style moderates partner presence effects on pain: A laser-evoked potentials study. *Social Cognitive and Affective Neuroscience* 10:1030–1037.

Legrand, D. 2006. The bodily self: The sensorimotor roots of pre-reflective self-consciousness. *Phenomenology and the Cognitive Sciences* 5:89–118.

Lloyd, D. M., V. Gillis, E. Lewis, M. J. Farrell, and I. Morrison. 2013. Pleasant touch moderates the subjective but not objective aspects of body perception. *Frontiers in Behavioral Neuroscience* 7:207.

Löken, L. S., J. Wessberg, I. Morrison, F. McGlone, and H. Olausson. 2009. Coding of pleasant touch by unmyelinated afferents in humans. *Nature Neuroscience* 12 (5): 547–548.

Menary, R. 2007. *Cognitive Integration: Mind and Cognition Unbounded*. Palgrave.

Merleau-Ponty, M. [1960] 1964. The child's relations with others. Trans. W. Cobb. In *The Primacy of Perception and Other Essays on Phenomenological Psychology, the Philosophy of Art, History, and Politics*, ed. J. M. Edie, 96–155. Northwestern University Press.

Metzinger, T. 2003. *Being No One: The Self-Model Theory of Subjectivity*. MIT Press.

Moll, H., and D. Kadipasaoglu. 2013. The primacy of social over visual perspective-taking. *Frontiers in Human Neuroscience* 7:558.

Morrison, I., M. Bjornsdotter, and H. Olausson. 2011. Vicarious responses to social touch in posterior insular cortex are tuned to pleasant caressing speeds. *Journal of Neuroscience* 31 (26): 9554–9562. doi:10.1523/JNEUROSCI.0397-11.2011.

Morrison, I., L. S. Loken, and H. Olausson. 2010. The skin as a social organ. *Experimental Brain Research* 204 (3): 305–314.

Panksepp, J. 1998. *Affective Neuroscience: The Foundations of Human and Animal Emotions*. Oxford University Press.

Prinz, W. 2012. *Open Minds: The Social Making of Agency and Intentionality*. MIT Press.

Reddy, V. 2008. *How Infants Know Minds*. Harvard University Press.

Rochat, P., and R. Morgan. 1995. Spatial determinants in the perception of self-produced leg movements by 3- to 5-month-old infants. *Developmental Psychology* 31 (4): 626–636.

Sartre, J.-P. [1943] 1956. *Being and Nothingness*. Trans. H. E. Barnes. Philosophical Library.

Schilbach, L., B. Timmermans, V. Reddy, A. Costall, G. Bente, and T. Schlicht. 2013. Toward a second-person neuroscience. *Behavioral and Brain Sciences* 36:393–414.

Schmid, H. B. 2005. *Wir-Intentionalität: Kritik des ontologischen Individualismus und Rekonstruktion der Gemeinschaft*. Karl Alber.

Schmid, H. B. 2014. Plural self-awareness. *Phenomenology and the Cognitive Sciences* 13 (1): 7–24.

Searle, J. R. 1990. Collective intentions and actions. In *Intentions in Communication*, ed. P. R. Cohen, J. Morgan, and M. Pollack, 401–415. MIT Press.

Seth, A. K. 2013. Interoceptive inference, emotion, and the embodied self. *Trends in Cognitive Sciences* 17 (11): 565–573. doi:10.1016/j.tics.2013.09.007.

Seth, A. K., A. B. Barrett, and L. Barnett. 2011. Causal density and integrated information as measures of conscious level. *Philosophical Transactions of the Royal Society A* 369:3748–3767.

Sharp, H., A. Pickles, M. Meaney, K. Marshall, F. Tibu, and J. Hill. 2012. Frequency of infant stroking reported by mothers moderates the effect of prenatal depression on infant behavioural and physiological outcomes. *PLoS One* 7 (10): e45446. doi:10.1371/journal.pone.0045446.

Stern, D. N. 1985. Affect attunement. In *Frontiers of Infant Psychiatry*, vol. 2, ed. D. Call, E. Galenson, and R. L. Tyson, 3–14. Basic Books.

Tajadura-Jiménez, A., M. Longo, R. Coleman, and M. Tsakiris. 2012. The person in the mirror: Using the enfacement illusion to investigate the experiential structure of self-identification. *Consciousness and Cognition* 21 (4): 1725–1738.

Thompson, E. 2007. *Mind in Life: Biology, Phenomenology, and the Sciences of Mind*. Harvard University Press.

Tuomela, R. 2013. *Social Ontology: Collective Intentionality and Group Agents*. Oxford University Press.

Vallbo, A. B., H. Olausson, and J. Wessberg. 1999. Unmyelinated afferents constitute a second system coding tactile stimuli of the human hairy skin. *Journal of Neurophysiology* 81 (6): 2753–2763.

Vallee, M., S. MacCari, F. Dellu, H. Simon, M. Le Moal, and W. Mayo. 1999. Long-term effects of prenatal stress and postnatal handling on age-related glucocorticoid secretion and cognitive performance: A longitudinal study in the rat. *European Journal of Neuroscience* 11 (8): 2906–2916.

Varela, F., E. Thompson, and E. Rosch. 1991. *The Embodied Mind: Cognitive Science and Human Experience*. MIT Press.

van Stralen, H. E., M. J. van Zandvoort, S. S. Hoppenbrouwers, L. M. Vissers, L. J. Kappelle, and H. C. Dijkerman. 2014. Affective touch modulates the rubber hand illusion. *Cognition* 131 (1): 147–158.

Zahavi, D. 2005. *Subjectivity and Selfhood: Investigating the First-Person Perspective*. MIT Press.

Zahavi, D. 2014. *Self and Other: Exploring Subjectivity, Empathy, and Shame.* Oxford University Press.

Zahavi, D. 2015a. Self and other: From pure ego to co-constituted we. *Continental Philosophy Review* 48 (2): 143–160.

Zahavi, D. 2015b. You, me and we: The sharing of emotional experiences. *Journal of Consciousness Studies* 22 (1–2): 84–101.

Zahavi, D., and P. Rochat. 2015. Empathy ≠ sharing: Perspectives from phenomenology and developmental psychology. *Consciousness and Cognition* 36:543–553.

9 Thin, Thinner, Thinnest: Defining the Minimal Self

Dan Zahavi

My initial work on the relationship between experience, self, and self-consciousness dates back to the late nineties (Zahavi 1999, 2000a), where I started defending the view that all three notions are interconnected and that a theory of consciousness that wishes to take the subjective dimension of our experiential life seriously also needs to operate with a minimal notion of self. Further elaboration of this early work led to the book *Subjectivity and Selfhood* (2005). During the decade that followed, I continued to refine the position and also started to respond to various criticisms that the view encountered (Zahavi 2007, 2009, 2011a, 2011b, 2012), eventually bringing these different efforts together in the book *Self and Other*, which was published in 2014. The criticisms and suggested revisions offered by Matthew Ratcliffe as well as Anna Ciaunica and Aikaterini Fotopoulou in their contributions to the present volume can to some extent be seen as representing a new phase in the discussion, not only because they engage with the recent arguments of *Self and Other*, but also because their criticism differs from the criticism offered in the past by, for instance, advocates of a no-self view, narrativists, or phenomenal externalists. Rather than denying the existence of the minimal self, their concern is with its proper characterization and interpersonal constitution. I appreciate their revisionary ideas, and I am grateful to the editors for having urged me to offer a response, thereby giving me the opportunity to clarify a few aspects of my own view.

1 Ratcliffe

In his chapter "Selfhood, Schizophrenia, and the Interpersonal Regulation of Experience," Ratcliffe does not set out to deny or dismiss the existence of the "minimal self," nor does he want to denigrate its significance. In fact, he readily agrees that the minimal self is integral to experience and inextricable from the very structure of experience, and that it is more fundamental than richer conceptions of self, including narrative accounts. No, his main aim, as stated in his introduction, is to argue that the minimal self has "to be reconceptualized in *interpersonal* terms," and that the "most basic sense of self is ... developmentally dependent on … other people," for which reason the minimal self cannot really be distinguished from the interpersonal self.[1]

What is Ratcliffe's central argument? He asks whether minimal selfhood involves a sense of being the locus of a specific type of experience, or whether an awareness of simply being the locus of some (unspecified) experience might suffice, and he defends the former option. In his view, one has to be prereflectively aware of experiencing *x* in a certain specific way, say, perceptually or imaginatively or in recollection, and so on, to qualify as a minimal self. Why is that? As Ratcliffe argues, without a proper demarcation between perceiving and remembering or perceiving and imagining, our sense of our own temporal and spatial location would break down. But without a sense of one's spatiotemporal location, it is not obvious that one could continue to experience oneself as a singular, coherent locus of experience. The final move in the argument is then to insist that the ability to make the required discriminations is interpersonally constituted. That is, "the sense of being in one kind of intentional state rather than another depends on a certain way of experiencing and relating to others, both developmentally and constitutively" (sec. 1).

Ratcliffe is certainly right when he claims that "interpersonal processes can enrich, diminish, or transform the nature of what one perceives, remembers, or imagines" (sec. 2). I also think it is correct that a fuller appreciation of the distinction between different intentional acts has ramifications for our self-understanding, and that this appreciation is facilitated (and perhaps even enabled) by interpersonal interaction. Perhaps he is even right—though I am somewhat less persuaded by this—that "modalities of intentionality depend on certain kinds of interpersonal relation for their development and sustenance" (sec. 5). However, I would insist that all of this is irrelevant to the matter at hand.

A crucial element in my defense of minimal selfhood has been reflections on the first-personal character of phenomenal consciousness. Roughly speaking, the idea is that subjectivity is a built-in feature of experiential life. Experiential episodes are neither unconscious nor anonymous; rather, they necessarily come with first-personal givenness or perspectival ownership. The what-it-is-likeness of experience is essentially a what-it-is-like-for-me-ness (Zahavi and Kriegel 2016). More specifically, this for-me-ness is taken to reside in the basic prereflective or reflexive (not reflective!), that is, self-presentational or self-manifesting, character of experience. The experiential self is consequently, and very importantly, not some experiential object. It is not as if there is a self-object in addition to all the other objects in one's experiential field. Rather, the claim is that all of these objects, when experienced, are given in a distinctly first-personal way. In short, if we want to "locate" the

1. Ultimately Ratcliffe wants to argue that this reconceptualization has ramifications for psychopathology, since the widespread assumption that schizophrenia is fundamentally related to a disturbance of the minimal self must now be given an interpersonal, relational twist. In the following comments, however, I do not engage with this aspect of his paper (but see Parnas, Bovet, and Zahavi 2002, where we argue that schizophrenic autism involves a simultaneous disturbance of self, intentionality, and intersubjectivity).

experiential self, we should look not at what is being experienced but at how it is being experienced. It is consequently no coincidence that the idea of a minimal self grew out of considerations concerning the relation between phenomenal consciousness and self-consciousness.

And here is the point. When talking about minimal selfhood, I am talking of something that is part and parcel of any experiential episode qua its experiential givenness, regardless of whether the episode in question is (ontologically) constituted or (epistemically) recognized as a particular intentional act type, or not. Indeed, to claim that the experiential episode would only be self-manifesting after such determination seems quite odd. Equally odd would be the claim that this fundamental reflexive character of phenomenal consciousness is interpersonally constituted such that infants who had not yet engaged in sufficient interpersonal relations, as well as all nonsocial organisms, would lack phenomenal consciousness and minimal selfhood. A crying newborn is not a zombie bereft of experiences but a creature whose crying is expressing an experience of distress. The crying newborn is a subject of experience whatever else it might be. This is also not something Ratcliffe is eager to deny. But the move he makes is somewhat surprising, since it ultimately changes the nature of his challenge. Ratcliffe argues that even if selfhood might be possible without intersubjectivity, we should avoid making the mistake of thinking that the "same kind of self-experience is preserved in unadulterated form throughout social development, as an underlying kernel" (sec. 1). We consequently have to construe self-development as a transformative process. This is a both interesting and important question. Do adult language users really have nothing in common experientially with infants and nonhuman animals? What we experience will undoubtedly change through development, but will development also affect the most fundamental structures of phenomenal consciousness? Will it also change and transform the most basic structures of prereflective self-consciousness and inner time consciousness? I have my doubts, but regardless of what the answer might be, it should be obvious that the challenge is now different. What Ratcliffe is disputing is now no longer the existence of a nonsocial minimal self but any claim to the effect that the minimal self is unchanged by development. Contrary to the (more) minimal self of an infant, the (less) minimal self of an adult is interpersonally constituted.

Indeed, I think the best way of making sense of Ratcliffe's argument is as follows: Given his emphasis on spatiotemporal location, the minimal self that he thinks is interpersonally constituted is a less minimal self than the one I am concerned with. If so, our respective views might be quite compatible, since we are simply targeting different notions of self. Not surprisingly, however, I would then claim that my thinner and more minimalist self is a condition of possibility for Ratcliffe's interpersonally constituted minimal self.

2 Ciaunica and Fotopoulou

In *Self-Awareness and Alterity* (1999), as well as in various writings from around that time, I defended the interdependency of self and alterity. Indeed, as I argued in the article "Alterity in Self," given the temporal and bodily character of experience, even the minimal sense of selfhood entails and depends on alterity (2000b, 126). Have I since changed my view? No, not at all, since what I made clear back then, and have repeated since, is that we need to distinguish different types of otherness. There is an alterity internal to myself, there is the alterity of the world, and there is the alterity of other subjects. Denying that the alterity of other subjects is constitutively involved in minimal selfhood is not to deny that there might be other forms of alterity, which are indeed constitutively involved. Why this sudden reference to these older texts? Because it should make clear why I find it somewhat puzzling to be criticized by Ciaunica and Fotopoulou in their chapter "The Touched Self: Psychological and Philosophical Perspectives on Proximal Intersubjectivity and the Self" for being committed to a detached visuospatial model of selfhood and social understanding, and for having failed to realize the role of worldly engagements for even minimal forms of self-awareness. Denying that the minimal self is interpersonally constituted, denying that one only becomes a subject of experience, that our experiential life is only imbued with its prereflective self-presentational character, and that phenomenal episodes only acquire for-me-ness, in virtue of one's relations to others, in no way entails that the minimal self is a self-enclosed self. As subject of intentional experience, it is inherently open to the world and others. Furthermore, nothing in the endorsement of the minimal self rules out that there are prereflective forms of sociality or that there are other dimensions of selfhood that are intrinsically interpersonally co-constituted. The second and third parts of *Self and Other* were precisely devoted to an extensive argument for this idea.

Ciaunica and Fotopoulou also insist that my position leads to positing an arbitrary gulf between experiential minimalism and social constructivism, and that my minimal understanding of selfhood overlooks the crucial role of the open-ended construction of individuality via narratives and language. I am also puzzled by these criticisms. The third part of *Self and Other* is explicitly devoted to elaborating an interpersonal dimension of self, that is, the self in its relation to and interaction with others, and I argue there that this dimension can serve as a crucial bridge between the minimal self and the normatively enriched narrative self (Zahavi 2014, 208, 238). Furthermore, as I have already made clear, the minimal notion of self doesn't overlook the open-ended construction of individuality. On the contrary, one of the reasons for introducing it was precisely to make comprehensible how such an open-ended construction could take place. Thus the minimal notion of self was never intended or presented as an exhaustive account of selfhood. Indeed, the label minimal (or thin) was partially employed to highlight how limited the notion is and how much more must be said to account for the fully and distinctly human self (Zahavi 2014, 50). In their conclusion, Ciaunica and Fotopoulou write that their approach "echoes Merleau-Ponty's view according to which one must consider 'the relation with others not only as one of the contents of our experience *but as an actual structure in its own right.*'" It is hard to see, however, why such a view, which I wholly endorse, should spell trouble for experiential minimalism. Nothing in the latter view commits one to the claim that the only role for sociality and otherness is qua content, and that others do not have an impact on the very structure of our subjectivity, as well.

Ciaunica and Fotopoulou write that interoception (the inner feelings of arousal, wakefulness, wellness, etc.) is crucial for self-experience and subjectivity, and fundamental for the very self–other distinction, and they then mount an argument to the effect that several interoceptive modalities are dependent on and changed by the embodied interaction with others. More specifically, they argue that states of physiological change such as crying, for example, become associated with and tied to particular behaviors and responses, as well as subjectively experienced states, through the social environment and embodied interpersonal interactions. Subjective "feeling states" are then taken to be the outcome of this so-called process of mentalization, for which reason such states cannot be said to preexist embodied encounters. But this whole line of argument is puzzling. Since Ciaunica and Fotopoulou concede that experiential states are constituent parts of—rather than products of—the process of mentalization, they cannot be claiming that the phenomenality of these very states is interpersonally constituted. But are they then at all targeting the position I am defending? That is what they take themselves to be doing, since they explicitly argue that the preexisting experiential states are too rudimentary to support experiential minimalism. The states allegedly don't come with any self–other distinction, don't require any preexisting perspectival notion, and are not anchored to any notion of self. But are the features whose presence Ciaunica and Fotopoulou want to deny really the features that I want to ascribe to phenomenal states, or are we simply targeting different levels? What they seem to have in mind with their reference to mentalization (and here their use differs somewhat from how the term is used not only in the theory of mind literature but also by attachment theorists like Fonagy or Gergely) is a fundamental process of organization and schematization. It is not about having phenomenal states but about coming to experience them as inner, as private, as mine rather than yours, and so on. As I make clear in chapter 2 of *Self and Other*, however, it is not a requirement for having a first-personal experiential life that one is able to appreciate, let alone conceptualize, the distinctly subjective or "inner" or "private" givenness of one's own experiences (Zahavi 2014, 27–30). This is undoubtedly a late achievement, which in all likelihood requires one to compare and contrast one's own perspective with that of others ("mine" meaning "not yours"). What I have in mind when referring to the subjective or first-personal character of experience is a feature of experience that it possesses in virtue of being the phenomenally conscious state it is. To claim that this feature is interpersonally constituted is to say that phenomenality as such is interpersonally constituted, and that is indeed a quite radical claim. When Ciaunica and Fotopoulou write that "feelings of bodily satisfaction, pain, pleasure, and lack thereof are primarily constituted as 'mine' only via behaviors that engage the interacting other" (sec. 5), the question is, in short, precisely how radical they want to go. Are they making a fairly uncontroversial claim, or are they aligning themselves with the social constructivism of Wolfgang Prinz, who famously declared that selves are sociocultural constructs rather than natural givens, and that human beings who are deprived of the required social interaction and denied socially mediated attributions of self would also lack me-ness, be self-less and without consciousness, and therefore remain "unconscious zombies" (Prinz 2003, 526)? I think Ciaunica and Fotopoulou's text makes it clear that they shy away from endorsing the latter position. But if so, I suspect we are talking at cross-purposes. If not, Ciaunica and Fotopoulou would have to argue that the features I am concerned with are either insufficient (though perhaps necessary) for selfhood or features that experiential states could lack while still being phenomenally conscious states. Whereas the former option quickly risks reducing the discussion to a terminological squabble (Zahavi 2014, 47, 62, 89), the second option is more theoretically interesting and challenging (Zahavi 2014, 25–41). But it is not one that Ciaunica and Fotopoulou explore or defend in any detail.

To repeat, when I claim that phenomenal consciousness is first-personal, what I mean is that we are acquainted and presented with our own experiential life in a way that differs from the way in which we are acquainted and presented with the experiential life of others. This first-personal experiential givenness is manifest in the very having of the experience. It is a givenness that obtains even when we are not explicitly aware of it, and even when we lack the conceptual skills to articulate or appreciate it. Indeed, all of this follows directly from the core claim, namely, that phenomenally conscious episodes by necessity are experientially manifest. A conscious mental state is not merely conscious of something, its object; it is simultaneously self-disclosing or self-revealing. This is what makes it different from any purported nonconscious representational state. This view has had many prominent advocates in the history of Western philosophy. It is also a view that has been defended by proponents of reflexivist or self-illumination theories in Indian philosophy. Needless to say, it is not a view that is universally accepted, but if one wants to criticize it, one has to engage with the relevant debate in philosophy of mind. To fail to do so, and to base one's objections on findings in developmental psychology concerning the child's capacity to discriminate perspectives, is to miss both the target and the point of the argument.

References

Parnas, J., P. Bovet, and D. Zahavi. 2002. Schizophrenic autism: Clinical phenomenology and pathogenetic implications. *World Psychiatry: Official Journal of the World Psychiatric Association (WPA)* 1 (3): 131–136.

Prinz, W. 2003. Emerging selves: Representational foundations of subjectivity. *Consciousness and Cognition* 12 (4): 515–528.

Zahavi, D. 1999. *Self-Awareness and Alterity: A Phenomenological Investigation.* Northwestern University Press.

Zahavi, D. 2000a. Self and consciousness. In *Exploring the Self,* ed. D. Zahavi, 55–74. John Benjamins.

Zahavi, D. 2000b. Alterity in self. *Arob@se* 4 (1–2): 125–142.

Zahavi, D. 2005. *Subjectivity and Selfhood: Investigating the First-Person Perspective.* MIT Press.

Zahavi, D. 2007. Self and other: The limits of narrative understanding. In *Narrative and Understanding Persons*, ed. D. D. Hutto, 179–201. Cambridge University Press.

Zahavi, D. 2009. Is the self a social construct? *Inquiry* 52 (6): 551–573.

Zahavi, D. 2011a. Unity of consciousness and the problem of self. In *The Oxford Handbook of the Self*, ed. S. Gallagher, 316–335. Oxford University Press.

Zahavi, D. 2011b. The experiential self: Objections and clarifications. In *Self, No Self? Perspectives from Analytical, Phenomenological, and Indian Traditions*, ed. M. Siderits, E. Thompson, and D. Zahavi, 56–78. Oxford University Press.

Zahavi, D. 2012. The time of the self. *Grazer Philosophische Studien* 84:143–159.

Zahavi, D. 2014. *Self and Other: Exploring Subjectivity, Empathy, and Shame.* Oxford University Press.

Zahavi, D., and U. Kriegel. 2016. For-me-ness: What it is and what it is not. In *Philosophy of Mind and Phenomenology: Conceptual and Empirical Approaches*, ed. D. O. Dahlstrom, A. Elpidorou, and W. Hopp, 36–53. Routledge.

10 The Emergence of Persons

Mark H. Bickhard

Within classical metaphysical frameworks, there doesn't seem to be much that persons could be other than some sort of substance or entity. Entity-based metaphysics, however, encounter fatal problems, certainly for minds, and arguably for persons as well. Furthermore, although entity-based metaphysics, in the form of particle-based frameworks, still dominate in philosophy, they are arguably not coherent (Bickhard 2009; Seibt 2009, 2010), and they are demonstrably inconsistent with contemporary physics (which is based on quantum field processes, not particles; Bickhard 2009). The ontology of persons is thus doubly problematic: not only is there a question of what that ontology might be, but there is a background question of what *kind* of an ontology is even plausible.

Does this entail that persons don't exist, or are epiphenomenal? Not necessarily, and certainly no such entailment exists if metaphysics that offer alternatives to entity-based frameworks are considered. I will be arguing that persons are emergent kinds of phenomena, developing this point within a process metaphysics, not an entity metaphysics.

1 Process

The model that I will be developing requires genuine ontological emergence, and that, so I argue, requires a process metaphysics. In particular, a particle framework makes emergence not possible, while a process framework makes emergence almost quotidian (Bickhard 2009, 2015a).

First, how does a particle metaphysics preclude emergence? Emergence is a property of organization: new organization is supposed to yield new (causally efficacious) properties. But within a particle framework, organization is neither a substance nor an entity and thus is not even a candidate for having any causal efficacy (Bickhard 2000, 2009; Campbell 2015).

This background assumption manifests itself in multiple arguments against emergence. One of Kim's arguments, for example, is that new organization might produce new causal *regularities*, but those are just the result of the basic causal interactions among those particles in that configuration (Kim 1991; Campbell 2015); all the genuine causality is at the level of the particles. The assumption that "configuration" is not even a candidate is clear.[1] Furthermore, so long as the fundamental metaphysics of the world is assumed to be constituted solely by substances and particles, then configuration should *not* be a candidate for having causal power.

But what is wrong with a particle metaphysics? Even if it precludes emergence, perhaps the world is *in fact* constituted out of particles. A particle model, however, suffers from fatal problems. First, in a world of nothing but point particles, nothing would happen: point particles have a zero probability of hitting one another.[2] Second, our best physics entails that particles do not exist: the world is composed of quantum fields—processes—and all that is left of a particle framework is that field interactions are quantized. But that quantization is the same kind of quantization as is found in a guitar string: a whole (quantized) number of wavelengths in the guitar string oscillations (and there are no guitar sound particles). A hybrid framework of point particles interacting *via* fields (a frequent contemporary assumption) is not consistent with contemporary physics, but it already involves fields—processes— and that is the crucial step (Bickhard 2009).

Nevertheless, it is appropriate to ask what support there is for a process metaphysics. If a process framework is incoherent or false in itself, then it does not matter if it might enable a metaphysics of emergence. First is the argument by elimination: entity or particle metaphysics are incoherent. Second is the physics of the world as quantum fields. Fields are processes, and quantum fields are in process even in "empty" space. These processes cannot be modeled in terms of particles (though, again, they will be quantized).

One of many empirical manifestations of such quantum field activity, even in a vacuum, is the Casimir effect. The vacuum is filled with excitations with oscillatory properties that are ephemeral but nevertheless have consequences; the vacuum is *not* "empty." If two very flat metal plates are brought close together, the oscillatory processes between the plates are constrained in the wavelengths that can occur, just as the guitar string being pinned at two points constrains the wavelengths of the guitar string oscillations. But the activity outside the plates is not so constrained, so there is more vacuum activity outside the plates than there is between them, and there is a net force pushing the plates toward each other (Mostepanenko, Trunov, and Znajek 1997; Sciama 1991). This and many other phenomena confirm the quantum field process framework and make no sense within a particle view.

2   Emergence

How, in turn, does a process metaphysics enable a model of metaphysical emergence? The simple point is that organization is the locus of emergence, and organization cannot be

1. For a discussion of Kim's more well-known preclusion argument, see Bickhard 2009.

2. "Particles" with finite extension encounter multiple problems. For example, their encounters would involve instantaneous transmission of force across their extent; there is extreme difficulty (perhaps impossibility) of explaining any kind of attraction among particles (hooks and eyes?), and so on.

delegitimated as a potential locus of causal efficacy in a process framework without eliminating *all* causal efficacy from the world. Fields have whatever consequences they have for the rest of the world necessarily in terms of their organizations. Organization is a locus of "causality" for processes,[3] and new organization is a legitimate candidate for new causality: emergence.[4]

Rescuing emergence from the impossibility arguments based on particle metaphysics, however, does not provide any model of any particular kind of emergence. For my purposes here, what is needed is a model of *normative* emergence. In particular, I will outline a model of the emergence of normative function and, based on that, a model of the emergence of representation.

2.1 Function and Representation

Unlike substance-based metaphysics, for which the default condition is stasis, the default for process is change. Instead of needing to explain change, thus, the explanatory challenge is to explain (the possibility of) stability.

Two sorts of stability are important for my current purposes. First, there is stability of process organization that remains stable because some above-threshold input of energy would be required to disrupt the organization; the organization is in some sort of "energy well." An atom is a good example: it is a furious process of QED and QCD processes that, if not disrupted, can remain in a nucleus/electron organization for cosmic time periods.

A second form of stability is more subtle and more important for issues of normative emergence. Some sorts of process organizations, such as an atom, can be isolated, go to thermodynamic equilibrium, and remain in that condition indefinitely. Others, however, are ontologically far from thermodynamic equilibrium, and, if they go to equilibrium, they cease to exist. A canonical example is a candle flame: it is far from equilibrium, and it cannot be isolated without that far-from-equilibrium condition ceasing to exist, and thus the flame ceasing to exist. The flame is a necessarily *open* process: it *is* the flow of air into a hot region, which combusts with wax vapor, thus maintaining the temperature of that region, and inducing convection, which brings in still more air and dissipates waste products.

The candle flame is an example of self-organizing process—it inherently organizes into the flow of air with combustion at its core. Not all self-organizing processes contribute to their own existence as a flame does: the flame contributes by maintaining above-combustion-threshold temperature, inducing convection, and so on. The candle flame is, in that sense, *self-maintaining* (Bickhard 2009); it is such self-maintaining process organizations that I will be primarily concerned with.

A further kind of complexity constitutes what I have called *recursive self-maintenance* (Bickhard 2009). A candle flame is self-maintenant, but only with certain ranges of

3. Whatever "causality" is (Bickhard 2011, 2015a).

4. Such a notion of emergence does *not* support the British emergentist view of emergent properties as being in-principle not derivable (McLaughlin 1992).

conditions: too much wind, too little oxygen, running out of wax—all constitute conditions in which the process organization will cease, and the flame has no alternative kinds of process it could engage in to maintain its condition of being self-maintenant should it encounter such conditions.

A bacterium, however, is also self-maintenant, but it can in addition adjust its activities—under some changes in condition—so as to *maintain the property* of self-maintenance: it is *recursively* self-maintenant. For example, if a bacterium is swimming up a sugar gradient, it will tend to keep swimming. If, however, it is swimming down a sugar gradient, it will tend to tumble for a moment and then resume swimming. Swimming contributes to self-maintenance if it is oriented up a sugar gradient but impairs self-maintenance if oriented down a sugar gradient, and the bacterium can adjust its activities so as to self-maintain self-maintenance.

2.1.1 Function Processes that contribute to the (self-)maintenance of a far-from-equilibrium process organization are, in that sense, *functional* for and relative to the continued existence of that system. This is the primitive form of the emergence of normative function. This kind of functionality is normatively functional in the sense that it contributes to the system's ability to maintain itself against the entropic tendencies to which it is subject. It is normative in the sense that it can do so in better or worse ways and can be, in fact, dysfunctional—for example, a bacterium that continues swimming even though oriented down a sugar gradient.

This differs in significant ways from etiological models of the nature and emergence of normative function (Bickhard 2009). I will not address those alternative models here, except to point out that this model just limned focuses on "serving a function" as the primary locus of functional emergence, rather than "having a function," which is the standard focus (Millikan 1984, 1993). This, of course, issues a promissory note to explicate *having* a function in terms of *serving* a function, rather than the other way around.

This explication introduces a consideration that is important for later discussion, so I will outline it here. The basic idea is that processes in an organism will tend to serve certain functions—contribute to the overall far-from-equilibrium "health" of the body—if various other conditions exist and are themselves maintained, and often, though not always, maintained in certain locations in the body. In that sense, a given functional process *presupposes* that those other enabling conditions obtain and, in some cases, are maintained in particular locations, perhaps by particular organs—for example, a kidney filtering blood. It is the network of resulting functional presuppositions that constitutes some organs and some processes as having functions: they *have* the functions of *serving* the functions that they are presupposed by other processes of the organism to serve. This *functional presupposition* relation is of central importance, not only here but also for understanding other phenomena, such as representation.

2.1.2 Representation In particular, having a function is modeled in terms of the more or less constant presuppositional relationships involved, with having a function constituted in being presupposed to serve some particular function(s). But more limited functional processes can also involve functional presuppositions.

Consider the evolution of complex agents. Such agents must be able to select what to do next—what interactions to engage in—from among some functional indications of what interactions are appropriate or possible in current circumstances: it is not functional to try to get something to eat from the refrigerator if you are in the middle of a forest. Such indications thus serve the function of providing a basis for the organism selecting and guiding its interactions. And those functions involve their own functional presuppositions.

Specifically, an indication that some particular interaction is possible will tend to be correct if appropriate conditions hold in the environment—appropriate in the sense that they would support or enable that interaction—and not correct if those conditions do not hold. That is, the indications will be *true* if their presuppositions hold, and not true—*false*—if those presuppositions do not hold. Such indications have truth values and thus constitute an emergent basis for representation.

This is a minimal form of representing, but it offers resources for more complex representing, resources that evolution has made strong use of. In particular, such indications can (1) branch and (2) iterate. Consider a frog that has opportunities to flick its tongue in one direction for a fly, another direction for a different fly, and yet another direction for a worm. The frog will presumably have indications of all three possibilities and will select one—perhaps the worm because it is larger—on the basis of other criteria. Such multiplicity of indications constitutes a kind of branching of anticipations of possibilities.

The frog may also indicate that, if it were to move left a small amount, then a different worm and a different fly would come into range. Such connections constitute a kind of iteration of conditional possibilities: the second worm (or fly) could be accessible, conditional on making the move left. That is, some possible interactions would create the conditions that would support other possibilities.

Branching and conditional iterations of indications of interactive possibilities can elaborate into complex webs and do so in still more complex organisms, such as primates and, especially, humans.

To illustrate how such webs can constitute the resource for more complex forms of representing, consider a child's toy wooden block. The block offers multiple possible visual scans, manipulations, throwing, dropping, and so on. A subweb of indications of manipulations and scans, for example, will have two important properties: (1) it is internally completely reachable, in the sense that any interaction in the subweb is reachable from any other location in the web; and (2) that internally reachable subweb remains invariant in its organization under a wide range of other interactions that the child could engage in, such as dropping the block, leaving it on the floor while going into the other room, putting it away in the toy box, and so on. The subweb, however, is not invariant under all other interactions, such as crushing or burning.

But it is the important invariance under various kinds of location and change of location that constitutes such a subweb as representing a small manipulable object.[5] In this general manner, webs of indications of interaction possibilities can yield more complex representing.

2.1.3 Situation Knowledge and Apperception The overall web of indications of potential interactions can be highly complex and serves all action and interaction of complex agents. It is called the organism's *situation knowledge*—the interactive knowledge of its situation.[6]

Situation knowledge is subject to constant change and updating. Every interaction on the part of the organism will change situation knowledge, as will simply the passage of time, actions of other agents, and so on. The processes of maintaining and updating situation knowledge are processes of *apperception*. Apperceiving a situation is thus constituted as the construction or updating of situation knowledge concerning what is possible, what could be anticipated, in that situation.

2.1.4 Representation? It has become almost orthodox in recent years to take an antirepresentationalist stance. It might be useful, therefore, to say a bit more about this model being a model of representation—of proposing that representation exists at all. Such a proposal does not comport with that orthodoxy.

Antirepresentationalist positions are commonly supported by arguments against symbol manipulation notions and information semantic notions of representation. The general logic is that these frameworks do not and cannot work, and, therefore, that there is no representation (at least not organism-level representation; e.g., Hutto and Myin 2012). It should be noted, however, that this is an invalid form of argument: it is an argument by elimination, and not all alternatives have been eliminated. In particular, symbolic and information semantics are not the only frameworks for modeling representation.[7]

5. This is basically Piaget's (1954) model of the representation of small objects, stated in the terms of this model. Such borrowing is possible because both models are action based. For a discussion of how this model can address representations of abstractions, such as of the number three, see Bickhard 2009. There are also partial convergences of this general framework with Gibson's notion of affordance (Bickhard and Richie 1983), but, among other divergences, Gibsonian affordances cannot branch or iterate—they cannot form webs.

6. Note that this is a model of knowledge as pragmatic—knowing how—rather than of a classical "justified true belief" notion of knowledge. I argue against the existence, and conceptual coherence, of classical proposition-based models of knowledge (e.g., Bickhard 2009).

7. Many antirepresentationalist stances, including within enactivism, would hold that, for example, affordances suffice. But indications of an affordance for an organism—for example, the affordance of the frog flicking its tongue and eating—can be false (perhaps it's a pebble, not a fly) and therefore have

In fact, the interactivist model has been arguing for decades that symbol manipulation and information semantic models, as well as many other models of representation, all make a common underlying error: that all representation is constituted as some form of encoding correspondence (e.g., Bickhard 1980, 1993, 2009). Encoding models have been the only real framework on offer for millennia, so it is understandable that eliminating these might be taken as eliminating representation.[8]

But these are not the only possible approaches to representation. There are general pragmatic approaches, and, in particular, the interactivist model just outlined. The very existence of this model demonstrates the invalidity of the arguments by elimination. Furthermore, this model, arguably, transcends and avoids the myriad problems of encodingist models (Bickhard 2009).

2.1.5 Transcending Encodingist Aporia   For example, one problem inherent to encodingist correspondence models is the problem of the possibility of error. If the special kind of correspondence—causal, lawful, informational, structural, and so on—that is supposed to constitute representation exists, then the representation exists *and is correct*; it is true. If the correspondence does not exist, then the representation does not exist. There is no third model possibility to model the representation existing but being false. There have been multiple attempts to solve this problem in recent decades—for example, Dretske (1988) and Fodor (1987, 1990)—but they do not succeed (Bickhard 2009).

The interactive model, in contrast, has no difficulty in modeling representations that are in error: they are those indications of interactive possibilities that are in fact not possible.

More deeply, a problem that is not addressed in most of the literature is that of *organism-detectable error*, not just error per se. The attempts to address error per se are from the perspective of an external observer with a perspective on both the organism and the environment; the external observer is supposed to be able to determine what the organism is (*factually*) in a representational correspondence with and to determine what the representation is *supposed* to represent. That external observer, then, can compare the two to determine if the representation is correct. As mentioned earlier, none of these attempts succeed.

The deeper problem is to model how the organism itself could detect that it is in error. According to classical correspondence models, the animal would have to step outside itself

---

the normative truth-valued character of representation. Also, as mentioned, Gibson's affordances cannot form webs, but once it is recognized that indications of interactive potentialities *can* form webs, the presumed bar to representation disappears. For further discussion regarding enactivism, see Bickhard 2016.

8. Note that the argument against encoding*ism* is not an argument against encodings per se. It is an argument that encodings must be a derivative form of representation, and therefore the assumption that *all* representation is constituted as encodings must be false (Bickhard 2009).

to compare what is actually in the environment with what the animal is representing as being in that environment to determine if the representation is correct. But no animal can step outside itself, so it is not possible (within these modeling constraints) to model organism-detectable error.

But if it is not possible in principle for an animal to detect (however fallibly) that it is in error, then error-guided behavior and learning are not possible. We know that error-guided behavior and learning occur, therefore organism-detectable error is possible, and any model that precludes it is thereby refuted.

This problem is the classical problem of radical skepticism. It has not been resolved in centuries of attempts, which is perhaps why no one in contemporary literature addresses it. But if the conclusion of the radical skeptical argument were correct, then error-guided behavior and learning would not be possible. So something must be wrong with the argument. I contend that the radical skeptical problem is itself artifactual: it is created by the underlying encodingist correspondence model of representation.

In particular, the interactive model has no difficulty at all in modeling the possibility of organism-detectable error, thus of the in-principle possibility of error-guided behavior and learning. If the organism selects an indicated interaction to engage in, and that interaction does not proceed as indicated, then the indication is false, and it is detected as being false by the (functional) detection of the failure of the indication.

The key shift here is from a past-oriented correspondence model—a "spectator" model, in Dewey's term (Dewey [1929] 1960; Tiles 1990)—to a future-oriented pragmatic model. We cannot peer backward into the past down the input stream, but we can determine if the future unfolds in the manner indicated (Bickhard 2009, in preparation).

3   Social Situations

Apperceptive updating with respect to most situations is based on prior interactions and their outcomes, which yield differentiations of what kind of environment is involved, how it is changing, and what can be anticipated within it. Visually scanning a water glass, for example, sets up multiple possibilities of drinking, using it as a paper weight, and so on.

A special problem arises, however, when two or more complex agents are in each other's presence. Each person's interactive characterization of the situation will include a characterization of the other(s). But an important aspect of the interactive potentialities of other people will depend on their characterization(s) of the situation, including their characterization(s) of the first person(s). But this problem iterates: I must characterize you, including your characterization of me, as characterizing you, and so on. Any solution to the problem must constitute a kind of "fixed point," in the sense that each person's situation knowledge is consistent with that of every other person, so no further "meta-characterization" iterations are needed.

Such a situation constitutes a *coordination problem*, in the sense of Schelling (1963): each person has an interest in arriving at a solution to the problem, and many possible such solutions would resolve the issue, so the "problem" is to arrive at a coordination—a commonality—regarding how the overall situation can be interactively characterized. A solution to a coordination problem, in turn, constitutes a *convention* in the sense of Lewis (1969). Since this is a "convention" concerning how the social situation is to be (conventionally) characterized, I call it a *situation convention* (Bickhard 1980, 2009). Situation conventions that are repeatable, such as driving on the right-hand side of the road, constitute *institutionalized conventions*. But not all situation conventions are institutionalized, such as the commonalities of understanding in a conversation at a given moment that permit participants to commonly understand further utterances, resolve anaphora, and so on. These specific situation conventions—commonalities of understanding—may never repeat.[9]

The domain of situation conventions, including institutionalized conventions, their interrelationships, processes of change, and so on, constitutes social ontology. These ontologies range from, for example, simple turn-taking games with infants, to language coordinations (see below), to the vast realm of institutions and institutionalized forms of interaction that constitute society at large. These are all forms of coordinating and coordinated interactions (Bickhard 1980, 2008, 2012, 2013, in preparation).

3.1 Language
The model of situation conventions supports a related model of the nature of language. Utterances are commonly thought to be composed of (re-)encodings of mental contents into strings of words. It is now recognized that this does not work for deictic or indexical utterances, and the general position encounters serious problems. I argue elsewhere that these problems are unsolvable (Bickhard 1980, 2009)—such models presuppose an incoherent encodingism—and I propose a different way of modeling language.

Utterances, according to this model, are interactions, just like all other engagements with the environment, but interactions with a special locus of interactive "object." In particular, utterances are interactions with situation conventions (Bickhard 1980, 2007, 2009, 2015b). They induce changes in situational common understandings that (when successful) modify them into new situational common understandings. The apperception of utterances modifies the situation conventions in which the utterances occur.

Among other consequences, this model entails that an important part of the possibilities that constitute a social situation are the possibilities of further language processes. That is, an important part of social ontology is constituted in language potentialities. Social realities are—in significant part, though not entirely—linguistic.

9. This involves some changes in Lewis's detailed model: Lewis only considered what I am calling institutionalized conventions, and his model cannot handle situation conventions that do not repeat (Bickhard 1980, 2009).

4 Persons

Persons are animals, and animals are agents: persons are agents. But persons constitute a special emergent form of agent-hood. Persons have an intrinsic social ontology.

Compare social insects: there is an emergence of sociality at the level of a nest or hive, but each individual insect is not itself socially constituted. The insects *collectively* constitute social processes but are not themselves *individually* so constituted. A human infant, in contrast, *develops* as an agent—in major part becoming an agent that can interact with, participate in, and thereby co-constitute the social ontologies of family, society, and culture. The ontology of the person as agent is itself socially constituted. The person would be a different, perhaps a very different, agent if he or she had developed in a different culture.

There appears to be some degree of this social ontology in primates, but humans are unique in the extent to which they are socially constituted. Society and culture are emergents in the collective individuals involved, and the persons who constitute societies and cultures share the ontology that they co-constitute (Bickhard 2012, 2013).

In this sense, cultures and societies create their own emergence bases via development of the constituting agents—persons—from infancy through adulthood.

Persons thus are not entities, at least no more so (and no less so) than a hurricane is an entity; they are not substances. Persons are complex organizations, involving multiple levels of emergence, including normative emergence, of processes. They are inherently sociocultural. They inherently have a social and language-based ontology.

Persons thus do not just live in social realities; they are *constituted* in interactive processes and potentialities with those realities. Persons *are* sociocultural agents.

5 Some Notes on Ethics

Such a process model of persons has a number of consequences for related domains. I will briefly indicate some of those consequences for ethics. This is not the place to develop this approach, but outlining the general approach to ethics that is enabled by such a model can illustrate some similarities to, and differences with, major approaches in the domain.

One form of argument for ethics is that violations of ethics are violations of the ontology of what it is to be human. For Aristotle, this involved violations of the natural purpose of human beings (Irwin 2007). At least one possible interpretation of Kant would have violations of morality be violations of one's respect for others as reasoning beings, with respect for one's own reason being necessarily a respect for all reasoning beings (Hill 2012; Kant 2002).

Few today would accept Aristotle's notion of human purpose, and Kant's framework encounters undermining problems of multiple sorts. For Kant, one such problem is simply that there is no necessary entailment from respect for my own reason to respect for yours.[10] A problem in common to Aristotle and Kant is that the ontologies proposed, within the framework of which ethics and morality are supposed to be derived, are not correct: humans do not have the "purposes" that Aristotle proposed, and humans are not constituted in the difficult hybrid between an ontology of reason and an ontology of animal nature that Kant proposed.[11]

The model outlined earlier, however, offers an alternative ontology within which ethics might be developed. In particular, to live one's life in a way that distorts or "violates" the basic social ontology of being a person is to distort or violate one's own ontology. This is not the same kind of issue as respecting other reasoning beings because I am a reasoning being; that is based on a conceptual commonality, a commonality of conceptual category.[12] The ontology of persons as sociocultural beings, however, is not just a conceptual commonality: I am *constituted in* my interactions and potential interactions with others, and that fact is beyond any conceptual framework applied to that ontology. Violating my own ontology in this sense is intrinsically in error, and its being in error does not depend on any rules or purposes. This consideration of the ontology of being a person does interestingly converge with character-based models of ethics, and therefore in part with virtue ethics, but it offers a different framework within which such person characteristics are understood.

Much more needs to be developed for this to yield an ethics. For example, what constitutes "distort or violate," and why and in what way would that be normatively negative? These issues are pursued elsewhere, but perhaps this outline suffices to indicate that the model offers a different framework within which to explore the realm of ethics.

Conclusion

Persons have a social ontology and cannot be modeled except in terms of that ontology. But social ontologies require their own supporting metaphysical frameworks, in particular:

1. A model of cognition that makes sense of the coordination problems, and therefore the fixed-point conventional solutions, posed by situations involving other people;
2. A model of normative function that enables such a model of interactive cognition;
3. A model of metaphysical emergence;
4. A process metaphysics that makes sense of genuine metaphysical emergence.

Persons are social-cultural-historical-linguistic agents and are so as a matter of developmental emergence. This is a complex ontology that, arguably, cannot be understood without the

10. Whatever "respect" is.

11. And inherited from Aristotle.

12. Thus, to not respect you as a reasoning being might be to disrespect my own *kind* of ontology, but it is not necessarily to distort or disrespect my *own* ontology as a reasoning being—it is not to disrespect my own reason.

full supporting framework involving process and emergence. Recognizing such an ontology, in turn, enables different approaches to modeling further emergents, such as that of ethics.

References

Bickhard, M. H. 1980. *Cognition, Convention, and Communication*. Praeger.

Bickhard, M. H. 1993. Representational content in humans and machines. *Journal of Experimental and Theoretical Artificial Intelligence* 5:285–333.

Bickhard, M. H. 2000. Emergence. In *Downward Causation*, ed. P. B. Andersen, C. Emmeche, N. O. Finnemann, and P. V. Christiansen, 322–348. University of Aarhus Press.

Bickhard, M. H. 2007. Language as an interaction system. *New Ideas in Psychology* 25 (2): 171–187.

Bickhard, M. H. 2008. Social ontology as convention. *Topoi* 27 (1–2): 139–149.

Bickhard, M. H. 2009. The interactivist model. *Synthese* 166 (3): 547–591.

Bickhard, M. H. 2011. Some consequences (and enablings) of process metaphysics. *Axiomathes* 21:3–32.

Bickhard, M. H. 2012. A process ontology for persons and their development. *New Ideas in Psychology* 30 (1): 107–119.

Bickhard, M. H. 2013. The emergent ontology of persons. In *The Psychology of Personhood: Philosophical, Historical, Social-Developmental, and Narrative Perspectives*, ed. J. Martin and M. H. Bickhard, 165–180. Cambridge University Press.

Bickhard, M. H. 2015a. The metaphysics of emergence. *Kairos* 12:7–25.

Bickhard, M. H. 2015b. The social-interactive ontology of language. *Ecological Psychology* 27 (3): 265–277.

Bickhard, M. H. 2016. Inter- and en-activism: Some thoughts and comparisons. *New Ideas in Psychology* 41:23–32.

Bickhard, M. H. In preparation. *The Whole Person: Toward a Naturalism of Persons—Contributions to an Ontological Psychology.*

Bickhard, M. H., and D. M. Richie. 1983. *On the Nature of Representation: A Case Study of James Gibson's Theory of Perception*. Praeger.

Campbell, R. J. 2015. *The Metaphysics of Emergence*. Palgrave Macmillan.

Dewey, J. [1929] 1960. *The Quest for Certainty*. Capricorn Books.

Dretske, F. I. 1988. *Explaining Behavior*. MIT Press.

Fodor, J. A. 1987. *Psychosemantics*. MIT Press.

Fodor, J. A. 1990. *A Theory of Content and Other Essays*. MIT Press.

Hill, T. E. 2012. *Virtue, Rules, and Justice: Kantian Aspirations*. Oxford University Press.

Hutto, D. D., and E. Myin. 2012. *Radicalizing Enactivism: Basic Minds without Content*. MIT Press.

Irwin, T. 2007. *The Development of Ethics*. Vol. 1, *From Socrates to the Reformation*. Oxford University Press.

Kant, I. 2002. *Groundwork for the Metaphysics of Morals*. Ed. T. Hill Jr. Trans. A. Zweig. Oxford University Press.

Kim, J. 1991. Epiphenomenal and supervenient causation. In *The Nature of Mind*, ed. D. M. Rosenthal, 257–265. Oxford University Press.

Lewis, D. K. 1969. *Convention*. Harvard University Press.

McLaughlin, B. P. 1992. The rise and fall of British emergentism. In *Emergence or Reduction? Essays on the Prospects of Nonreductive Physicalism*, ed. A. Beckermann, H. Flohr, and J. Kim, 49–93. Walter de Gruyter.

Millikan, R. G. 1984. *Language, Thought, and Other Biological Categories*. MIT Press.

Millikan, R. G. 1993. *White Queen Psychology and Other Essays for Alice*. MIT Press.

Mostepanenko, V. M., N. N. Trunov, and R. L. Znajek. 1997. *The Casimir Effect and Its Applications*. Clarendon Press.

Piaget, J. 1954. *The Construction of Reality in the Child*. Basic Books.

Schelling, T. C. 1963. *The Strategy of Conflict*. Oxford University Press.

Sciama, D. W. 1991. The physical significance of the vacuum state of a quantum field. In *The Philosophy of Vacuum*, ed. S. Saunders and H. R. Brown, 137–158. Clarendon Press.

Seibt, J. 2009. Forms of emergent interaction in general process theory. *Synthese* 166 (3): 479–512.

Seibt, J. 2010. Particulars. In *Theory and Applications of Ontology: Philosophical Perspectives*, ed. R. Poli and J. Seibt, 23–57. Springer.

Tiles, J. E. 1990. *Dewey*. Routledge.

III   Cultural Affordances and Social Understanding

11   The Significance and Meaning of Others

Shaun Gallagher

In support of a pluralist approach to social cognition, I want to introduce a distinction found in debates about the nature of interpretation, namely, the distinction between *meaning* and *significance*. First I review the hermeneutical debates and several ways of thinking about the relation between significance and meaning. I then explore some implications of this distinction for contemporary debates in social cognition and defend the idea that even if different situations of social cognition may involve a variety of interpretive approaches, interpretive significance remains primary.

1   Meaning and Significance in Hermeneutics

In the wake of Hans-Georg Gadamer's magnum opus on philosophical hermeneutics, *Truth and Method* ([1960] 2004), several critics entered into debates with Gadamer concerning questions about objectivity. Emilio Betti, a legal historian, published his critical comments in *Die Hermeneutik als allgemeine Methodik der Geisteswissenschaften* (1962), and Gadamer responded to Betti in the foreword to the second edition of *Truth and Method*. The literary theorist E. D. Hirsch then joined the debate on the side of Betti and developed his own hermeneutical theory in two books, *Validity in Interpretation* (1967) and *The Aims of Interpretation* (1995). I focus on these authors, although I will briefly mention one other debate that has relevance: the well-known debate between Gadamer and Habermas.

Gadamer ([1960] 2004) develops a dialectical or dialogical theory of hermeneutics. Our understanding of a text, for example, involves contributions from both the interpreter and the text as they enter into a dialogical relation. The interpreter, embedded in her own culture and historical period, as well as her own interests and prejudices, contributes to the resulting interpretation as much as the text does. "The real meaning of a text as it addresses itself to an interpreter … is always co-determined by the historical situation of the interpreter" ([1960] 2004, 280). Rather than thinking that the employment of a method designed to neutralize the interpreter's personal horizon could lead directly to the truth of the text, interpretation involves a *fusion of horizons (Horizontverschmelzung)* and opens up new possibilities of meaning: "In the process of understanding there always occurs a true fusion of perspectives in which the projection of the historical perspective really brings about a sublation of the same" (290). This view reflects a dialogical conception of truth where truth and meaning, rather than residing inertly in the text, are generated in the process of interpretation.

Betti (1962), drawing on the romantic hermeneutics of Schleiermacher and Dilthey, criticized Gadamer's hermeneutics as being overly subjectivist. For Betti, careful hermeneutical method aims to eliminate biases in interpretation so as to arrive at an objective understanding and the truth that resides in the text, which is equated with the original intention of the author. What the author intended, or how the intended and contemporary audience understood the text, should be the ruling criterion. To get at the meaning of the text, one needs to do careful philological and historical research into the life and times of the author and the way that the text's vocabulary was understood at the time of writing. The use of such hermeneutical rules, for example, is reflected clearly in originalist readings of the U.S. Constitution (e.g., Brest 1980). In this view, the meaning of the text is fixed and can be considered either as under the control of the author at the time of writing or as what a reasonable audience would understand by the text at that time.

Gadamer's response to this critique was to distinguish his own project from the kind of prescriptive hermeneutics that Betti was discussing. Gadamer argues that he is describing how interpretation happens, not how it ought to happen. His concern is philosophical, "not what we do or what we ought to do, but what happens to us over and above our wanting and doing" ([1960] 2004, xxvi). This position suggests that even if in hermeneutical practice we seek to attain objectivity or seek to find the truth of the *mens auctoris*, larger forces are at work that may undermine what we can accomplish. At least in part, the larger forces are the effects of history, which differentiate the horizon of the interpreter and the horizon of the original author or text. Gadamer addresses the question: How is understanding possible? The answer will have implications for what we consider valid hermeneutical method, and what can be accomplished by using such a method. Gadamer suggests, however, that we cannot decide those issues until we have a good conception of what is possible. In this regard, it may not be possible for an interpreter, for example, a legal historian like Betti, to understand the text better than the author himself (which is Schleiermacher's claim), or for an interpreter to work out an interpretation that corresponds exactly with the author's intention; rather, as Gadamer puts it, interpreters will always understand differently.

In his review of *Truth and Method*, Hirsch (1965) reinforces Betti's critique and makes a useful distinction between meaning and significance. Hirsch considers the title of Gadamer's work to be ironic, since he considers it "a polemic against [the] 19th-century preoccupation with objective truth and correct method" (Hirsch 1965, 488) as found in romantic hermeneutics. Hirsch takes Gadamer to be defending the idea that "every putative re-cognition of a text is really a new and different cognition in which the interpreter's own historicity is the *specifica differentia*" (488). Hirsch decries the Gadamerian claim that the meaning of a text is generated not in the mental processes of the author but in the interpretation and is the product of the interpreter's grasping as much as it is expressed by the text. To be clear, Hirsch's claim to the contrary notwithstanding, Gadamer does not contend that "textual meaning can somehow exist independently of consciousness altogether" (Hirsch 1965, 491); for Gadamer, it clearly depends to some degree on the consciousness of the interpreter. But even in this modified reading, Hirsch still has a complaint to the effect that the author's mental processes are discounted in a wholesale manner, an idea expressed in a well-known statement by Gadamer: "The meaning of a text goes beyond its author not just sometimes but always. Understanding is not a reproductive but always a productive activity. One understands differently when one understands at all" ([1960] 2004, 280). Hirsch takes this to mean that "'the' meaning of the text is a never-exhausted array of possible meanings lying in wait for a never-ending array of interpreters. But if this is so it follows that no single interpretation could ever correspond to the meaning of the text" (1965, 492). Meaning in this view is entirely indeterminate.[1]

To save the idea that meaning is in fact determinate and interpretation can be valid, Hirsch distinguishes between "the meaning of a text," which remains unchanged, and "the meaning of a text to us today," that is, the significance or relevance it has for us, which may change with each reader or generation of readers.

Now the meaning of a text is that which the author meant by his use of particular linguistic symbols. Being linguistic, this meaning is communal, that is, self-identical and reproducible in more than one consciousness. Being reproducible, it is the same whenever and wherever it is understood by another. But each time this meaning is construed, its meaning to the construer (its relevance) is different. Since his situation is different, so is the character of his relationship to the construed meaning. But it is precisely because the meaning of the text is always the same that its relationship to a different situation is a different relationship. (Hirsch 1995, 498–499)

In *Validity in Interpretation* (1967), Hirsch rephrases this distinction as the distinction between meaning and significance (see also Hirsch 1984). Significance is not unimportant for Hirsch. Indeed, it can be considered a "more valuable object of interpretation" precisely because of its relevance to present circumstances.

As I have suggested elsewhere (Gallagher 1992), the Gadamerian response to Hirsch would be that if there actually is an unchanging meaning that belongs to a text, there is no access to it that doesn't go by way of significance. Indeed, since it is not clear what justifies the claim that a text has an unchanging meaning independent of interpretation, it may be a matter of our interpretational bias to think that it does—a bias that may be productive in some way, as Gadamer might say, but a bias nonetheless. Which is to say that anytime an

---

1. Here one might think of Quine's notion of the "indeterminacy of meaning (or translation)" and Davidson's concept of "radical interpretation." It would take us too far afield and complicate the analysis to a high degree to consider the relation between Gadamer's hermeneutics and these theories. For some of these complications, see the debate between Dreyfus (1980), Rorty (1980), and Taylor (1980), and essays by McDowell (2002) and Stueber (1994).

interpreter makes a claim on the meaning, it will turn out to be a claim about its significance for the interpreter, even if the interpreter takes validity and objectivity to be the most significant aspects of interpretation. This doesn't mean, for Gadamer, that some interpretations aren't better than others, but the measure of that involves the productivity of interpretation rather than the reproduction of a textual meaning.

My aim here is not to resolve this debate but to take this distinction between meaning and significance and use it to sort out some issues in recent debates about social cognition. In its simplest terms, my argument is that theory-theory, like Hirsch's hermeneutics, defends the idea that we can gain insight into the other's meaning, understood in folk psychological terms of intentions, beliefs, desires, and so on; in effect, mind reading is like reading a text to discover the author's intention. In contrast, interaction theory follows the Gadamerian line by placing importance on the significance that emerges in the interaction between agents; intersubjective understanding, by way of interaction itself, is more like having a conversation with the text. Before turning to such issues, however, I will mention, briefly, one other debate focused on Gadamer's hermeneutics. In this debate, Habermas (1982), concerned with social and political issues, raises some important questions about how interpretation can be shaped by various social, economic, and cultural forces seemingly ignored by Gadamer. This debate was wide ranging, but the point most pertinent to my discussion here is a proposal by Habermas to address such issues by developing a "depth hermeneutics." He suggests that to fully comprehend the role of social, economic, and cultural forces (which may involve race, class, and gender issues, for example), we need not only a hermeneutical mode of interpretation of the sort that Gadamer proposes but a scientific mode of explanation that reveals the substructural forces at work in any understanding. Building on Dilthey's (1989) distinction between understanding and explanation, a critical depth hermeneutics would employ both understanding to explicate the surface meaning, and explanation to work out the usually hidden and deeply entrenched forces that shape and misshape intersubjective interpretations. Employing the meaning-versus-significance distinction, one can extrapolate and suggest that understanding, following Gadamer's hermeneutical approach, is best thought as grappling with significance; explanation, following Dilthey's view, delivers a more scientific, causal account, but, as Habermas contends, one that is required in order to grasp, with the assistance of critical reflection, the full and deeper meaning, which may involve distortions that shape intersubjective and communicative practices. In this view, the deeper meaning is equivalent not to the author's intention, or to the original audience's understanding, but to a realization of how certain socioeconomic forces shaped such intentions and understandings and their subsequent interpretations.

2 Meaning and Significance in Social Cognition

Accepting the distinction between meaning and significance at face value, that is, as Hirsch defines it, can help us understand some issues concerning social cognition and the ways that we understand others. That hermeneutical practice is not simply about textual interpretation was made clear by Dilthey (1989), who argues that hermeneutics can be used as a method in the human and social sciences. Heidegger (1962) takes it even further, conceiving of hermeneutics as a way to characterize human understanding more generally. A hermeneutical approach to social cognition, however, does not decide any issues in advance; the issues would be decided by taking a particular hermeneutical approach—for example, a Gadamerian approach rather than a more conservative (Betti, Hirsch) or critical (Habermas) approach. With respect to how we understand others, one can match up different hermeneutical approaches to several of the established theoretical approaches to social cognition. Thus Chenari (2009), for example, argues, that Schleiermacher was a proto-theory-theorist, and Dilthey, who emphasized the role of empathy, foreshadowed simulation theory. I take a different strategy and simply ask how we might put the meaning-significance distinction to work in contexts of social cognition.

I will assume a pluralist view of social cognition. That is, I will assume that we have a number of ways of understanding others (Fiebich, Gallagher, and Hutto, in press; Gallagher 2015), ranging from embodied interactive processes (that contribute not only to understanding others' intentions but to the formation of the other's intentions), to direct social perception of intentions and emotions, to communicative and narrative practices, to the mind-reading processes associated with simulation and theoretical inference. If we think that both meaning and significance, as defined by Hirsch, are important factors for our understanding of others, this may have implications about how we conceive of processes involved in social cognition.

It seems clear that if we think that the meaning of someone's behavior or action involves something that is not easy to get to, and that it takes a careful (scientific-like) attempt to get to and explain that meaning, then theory-theory (TT) clearly has a role to play in social cognition. Meaning, in this case, would be portrayed as residing in an agent's original intention, and that intention, as a mental state, and as conceived by many theory-theorists, would not be directly accessible (this is sometimes referred to as the unobservability principle; see, e.g., Krueger 2012). Thus TT holds that we engage in mind reading—that is, inferring the other person's mental states as a way to explain or predict his or her behavior—on the basis of folk psychology, understood as a (science-like) theory. According to these versions of TT, the minds of others are not readily accessible; the intersubjective interpreter has to consider behavioral evidence and formulate a hypothesis consistent with folk psychological principles to infer their mental states. Even young infants are portrayed as little scientists, forming hypotheses and experimenting as a way to make their folk psychological theories more precise (e.g., Gopnik, Meltzoff, and Bryant 1997).

Such an approach resembles Hirsch's notion that for an interpretation to reach meaning, the interpreter needs to take a hypothetico-deductive approach to interpretation. Hirsch develops this in hermeneutical terms of textual interpretation, but the idea is easily applied to the setting of intersubjective understanding. If we take folk psychology to summarize cultural norms and conventions, such norms and conventions can ground our pre-apprehensions of (our hypotheses about) the other person's meaning. "This is what gives the interpreter's pre-understanding a sporting chance of being correct. For the [other's] meaning has a shape and scope that is governed by conventions which the interpreter can share as soon as he is familiar with those conventions" (Hirsch 1965, 506). A folk psychological hypothesis is like any other: "The best hypothesis is the one that best explains all the relevant data. This ... suggests that the much-advertised cleavage between thinking in the sciences and the humanities does not exist. The hypothetico-deductive process is fundamental in both of them, as it is in all thinking that aspires to knowledge" (507).

Seeburger makes clear what is at stake for Hirsch in his concern for validity. It involves the ability to test claims about meaning:

> That, in turn, requires that it be possible to appeal to standards of evidence, verification, and validation which provide a distinction between "correct" and "incorrect" claims, and provide ways for deciding, in given cases, which claims are which. Finally, deciding between correct and incorrect claims is possible only in relation to a stable, determinate object, independent of those claims, to which the claims refer. (Seeburger 1979, 250–251)

The determinate object in this case is the meaning of the other person's behavior, rather than the significance it might have for the interpreter.

As Seeburger goes on to show, Hirsch appeals explicitly to only one criterion as a way to distinguish meaning from significance, namely, voluntariness. In the context of textual hermeneutics, what the author (or speaker) intentionally states is equivalent to the meaning; any involuntary aspects that may be part of context (e.g., verbal mistakes or slips) may have significance but should not determine meaning content. This view is problematic in at least two ways, however. First, as Seeburger suggests in reference to this specific example, since involuntary verbal slips may have deeper meaning (in a Freudian interpretation, for example), voluntariness cannot be the determining criterion for meaning. Here one can point to Habermas's notion of depth hermeneutics to make a similar point. In this approach, which Habermas models on both psychoanalysis and neo-Marxist economic analysis as modes of depth analysis, intentions that may be under the control of the other person as determinants of her behavior do not tell the whole story. Meaning is further shaped by deeper, nonconscious social and cultural forces that may explain why people act the way they do. Not all explanation will take form in the folk psychological space of reasons; there is, so to speak, a landscape of power relations that require theoretical explanation, as well. Although this approach is still consistent with TT, it seems somewhat distant from our everyday attempts to understand others, in the sense that we do not usually engage in anything like in-depth theoretical investigations to understand the deep meaning of their actions. It is possible, of course, that some people's interpretations are informed by a folk psychology that includes references to unconscious motives or the role of economic circumstances that qualify the voluntariness of another's behaviors. We may recognize that a person acts not fully out of her own intention but because cultural or institutional practices push her in a certain direction, or because a heteronomy of established values lure her to desire this rather than that. But again, this would undermine voluntariness as a criterion that could distinguish meaning from significance, since such unconscious or economic factors likely do not come under the agent's voluntary control.

A second problem is that, in some respects, mind reading based on folk psychology, rather than using voluntariness (i.e., the fact that the agent intends to do what she does) as a criterion to test for the other's meaning, takes the voluntariness of the other's intentions to be included in the mental states that need to be explained. Using voluntariness as an explanatory criterion to test the validity of one's interpretation seems to presuppose that one already knows what the other's intention is and that it is voluntary; but the other person's intention is part of what one is trying to understand.

To address this last issue, one might suggest that we need a hybrid theory that combines theoretical inference with simulation. Simulation, especially the low-level automatic kind associated with mirror neuron activation, may provide a more immediate, empathic comprehension of the fact that the other's action is intentional (e.g., Brincker 2015). Alternatively, a hybrid of direct social perception and theoretical inference may fulfill the same requirement (for discussion, see, e.g., Lavelle 2012; Carruthers 2015; Gallagher 2015). Views that champion the role of direct social perception argue that we can directly see the intentions and emotions of others in their bodily comportment. If we have immediate access to the fact that the other person is acting intentionally, and perhaps even access to her specific intention, we are, in Hirsch's view, already accessing the true meaning of the other person's actions.

For Hirsch, however, in line with TT approaches, a full comprehension of the other's meaning involves more than direct perception; although it conceives of social cognition in observational terms, it necessarily involves inferential work to explain a meaning intrinsic to the text or, in this case, hidden in the mind of the other. Seeburger (1979) argues that any interpreter's application of voluntariness as a criterion for meaning is already context dependent, already framed by the interpreter's interests; interpretation is thereby limited to significance. In regard to social cognition, this amounts to saying that whenever anyone understands another's actions, it is always from a perspective informed by pragmatic and cultural factors that shape the interpreter's perception. This is consistent with the Gadamerian view and, in the context of social cognition, with interaction theory (IT).

IT holds that we are not primarily observers of the other's behavior, but in most everyday encounters, we interact with others in a second-person (I-you) or first-person-plural (we) relation. Understanding happens on occasions when an interpreter enters into a dynamical (dialogical or interactional) process with the other that generates meaning as a relational phenomenon irreducible to the actions or the sum of actions that belong to the individual participants. This applies to communicative actions as well as what are not strictly communicative actions, such as joint actions oriented toward a pragmatic goal. What Gadamer says of communicative actions also applies to noncommunicative actions:

"The real event of understanding ... goes continually beyond what can be brought to the understanding of the other person's words [or actions] by methodological effort and critical self-control. It is true of every conversation that through it something different has come to be" (1990, 158). In this context, the notion of direct social perception is not a purely observational phenomenon that feeds a further inferential process (e.g., Carruthers 2015); rather, just as perception is normally enactive, that is, action oriented, so social perception is *inter*action oriented. I see your action in terms of what it affords as a response on my part. That is, understanding the action of the other is understanding its significance for me (or for us), and for my potential actions or future interactions with you. The significance of your action, for me, may be best expressed in terms of what it affords in terms of my response or our future interactions. Importantly, however, it is not necessarily a matter of my narrow interests, which may lead to attempts to manipulate the other's behavior; rather, in the context of interaction, it is often a matter of joint interests and joint intentions as we engage in joint actions. One basic principle of this type of social interaction is that the outcome is a new structure or new dynamical system that always goes beyond the intentions and actions of the individuals involved (De Jaegher, Di Paolo, and Gallagher 2010).

In some regard, this is precisely the way that Hirsch implicitly understands significance. Seeburger suggests that just as Hirsch associates meaning with voluntariness, so he implicitly associates significance with purposiveness. In a general sense, we can think of significance as being about pragmatic or normative purpose (defined in terms of affordances and immediate context), which may support interaction or joint actions. According to IT, for most of our everyday interactions, significance suffices; moreover, significance is constituted in the interactions themselves. That is, most of our understandings of others are generated in contexts of interacting with them, where the meaning of their actions and responses depends to some degree on the meaning of our own actions and responses, and such understanding just is the pragmatic grasp of the other's action as it relates to one's own action. Social understanding operates primarily within what Jerome Bruner (1986) calls the landscape of actions rather than the mental landscapes of beliefs and desires that are in some way occluded by behavior.

Even in cases where this kind of enactive perception of the significance of the other's action is not clear, or in cases where the other's action is puzzling or unexpected and we need to go beyond what may have become a disaffordance for any possible interaction, to make sense of the situation, it is not always mentalistic interpretation, the kind of mind reading based on theoretical inference or simulation, that provides the best route. Rather, communicative and narrative practices may be the most productive way to gain understanding (Gallagher 2006; Gallagher and Hutto 2008; Hutto 2007). Most simply, we can ask other people what they're doing or what they're thinking. And in cases where asking may be impossible or undependable, as Bruner makes clear, the landscape of action, as well as the space of reasons, may be better accessed by narrative understanding, since it is often the complex situation and the individual's history that allows me to make sense of her intentions and feelings and reasons for acting one way rather than another. Narratives are more local and tied to the particulars of the situation and the person. What I know of someone's peculiar personality and history, or even what I know of the particular social role that someone is playing in the specific circumstances of our encounter, will allow me to intuit or work out his or her meaning. These types of details typically fit better into the frame of a narrative than into a theoretical explanation that starts and sometimes remains on a general level. Moreover, the situation may involve a degree of diversity that undermines any attempt at one-to-one simulation (Gallagher 2007). Rather than a simulative imagination (or what Goldman [2006] calls "E-imagination"), a narrative or "N-imagination" may be more productive.

Whether a narrative understanding of another person is a matter of understanding the significance or the meaning of her actions, intentions, emotions, thoughts, ways of life, and so on, may depend on how we conceive of narrative. It is not the case, however, that a person's self-narrative is a fixed and determinate meaning; self-narrative tends to adjust itself to circumstance and, importantly, to depend on others. Developmental studies show that one's self-narrative originates in our encounters with others and incorporates some parts of the other's story (Nelson 2003; see Gallagher and Hutto 2008). That type of self–other integration in the formation of narratives also reflects our everyday interactions with others. In this respect, it is often the significance of one's actions for others that rules how those narratives are shaped. How one's actions are perceived by others, rather than what one originally intends by them, tends to dominate how narratives are shaped. Indeed, a person's narrative focus or insistence on what he intended is usually motivated by another's misunderstanding.

Conclusion

A pluralist approach to social cognition suggests that there are a number of ways to understand others (Fiebich, Gallagher, and Hutto, in press). Which approach may be the most productive will depend on the who, what, when, and why of circumstance—how much we know about the other person, what our interaction with them is, how close or far from a goal we are, what normative factors constrain our actions, and so forth. I have argued that in most everyday encounters significance rather than meaning (in Hirsch's definition) takes priority in intersubjective affairs, and this is reflected primarily in the fact that in most everyday encounters that call for understanding we interact with others rather than just observe them. That there are some circumstances in which we are observers looking to understand or even to develop an explanation of another's behavior is not to be denied. In most circumstances, however, we are interacting with others in ways that generate meaning that is relevant to us or to the situation (i.e., significance). Indeed, what others intend or feel or start to believe or desire is often motivated and sometimes generated in our interactions with them.

References

Betti, E. 1962. *Die Hermeneutik als allgemeine Methodik der Geisteswissenschaften*. Mohr.

Brest, P. 1980. The misconceived quest for the original understanding. *Boston University Law Review* 60:204.

Brincker, M. 2015. Beyond sensorimotor segregation: On mirror neurons and social affordance space tracking. *Cognitive Systems Research* 34:18–34.

Bruner, J. 1986. *Actual Minds, Possible Worlds*. Harvard University Press.

Carruthers, P. 2015. Perceiving mental states. *Consciousness and Cognition* 36:498–507.

Chenari, M. 2009. Hermeneutics and theory of mind. *Phenomenology and the Cognitive Sciences* 8 (1): 17–31.

De Jaegher, H., E. Di Paolo, and S. Gallagher. 2010. Can social interaction constitute social cognition? *Trends in Cognitive Sciences* 14 (10): 441–447.

Dilthey, W. 1989. *Selected Works*, vol. 1: *Introduction to the Human Sciences*. Ed. R. A. Makkreel and F. Rodi. Princeton University Press.

Dreyfus, H. 1980. Holism and hermeneutics. *Review of Metaphysics* 34 (1): 3–23.

Fiebich, A., S. Gallagher, and D. Hutto. In press. Pluralism, interaction, and the ontogeny of social cognition. In *Routledge Handbook of Philosophy of the Social Mind*, ed. J. Kiverstein. Routledge.

Gadamer, H.-G. [1960] 2004. *Truth and Method*. Trans. J. Weinsheimer and D. G. Marshall. Bloomsbury.

Gadamer, H.-G. 1990. The universality of the hermeneutical problem. In *The Hermeneutic Tradition: From Ast to Ricoeur*, ed. G. Ormiston and A. Schrift, 147–158. SUNY Press.

Gallagher, S. 1992. *Hermeneutics and Education*. SUNY Press.

Gallagher, S. 2006. The narrative alternative to theory of mind. In *Radical Enactivism: Intentionality, Phenomenology, and Narrative*, ed. R. Menary, 223–229. John Benjamins.

Gallagher, S. 2007. Simulation trouble. *Social Neuroscience* 2 (3–4): 353–365.

Gallagher, S. 2015. The new hybrid theories of social cognition. *Consciousness and Cognition* 36 (November): 452–465. doi:10.1016/j.concog.2015.04.002.

Gallagher, S., and D. Hutto. 2008. Understanding others through primary interaction and narrative practice. In *The Shared Mind: Perspectives on Intersubjectivity*, ed. J. Zlatev, T. Racine, C. Sinha, and E. Itkonen, 17–38. John Benjamins.

Goldman, A. I. 2006. *Simulating Minds: The Philosophy, Psychology, and Neuroscience of Mindreading*. Oxford University Press.

Gopnik, A., A. N. Meltzoff, and P. Bryant. 1997. *Words, Thoughts, and Theories*. MIT Press.

Habermas, J. 1982. *Zur Logik der Sozialwissenschaften*. Suhrkamp.

Heidegger, M. 1962. *Being and Time*. Trans. J. Macquarrie and E. Robinson. Harper and Row.

Hirsch, E. D. 1965. Truth and method in interpretation. *Review of Metaphysics* 18 (3): 488–507.

Hirsch, E. D. 1967. *Validity in Interpretation*. Yale University Press.

Hirsch, E. D. 1984. Meaning and significance reinterpreted. *Critical Inquiry* 11 (2): 202–225.

Hirsch, E. D. 1995. *The Aims of Interpretation*. University of Chicago Press.

Hutto, D. 2007. The narrative practice hypothesis: Origins and applications of folk psychology. *Royal Institute of Philosophy* 60 (Supplement): 43–68.

Krueger, J. 2012. Seeing mind in action. *Phenomenology and the Cognitive Sciences* 11 (2): 149–173.

Lavelle, J. S. 2012. Theory-theory and the direct perception of mental states. *Review of Philosophy and Psychology* 3 (2): 213–230.

McDowell, J. 2002. Gadamer and Davidson on understanding and relativism. In *Gadamer's Century: Essays in Honor of Hans-Georg Gadamer*, ed. J. Malpas, U. von Arnswald, and J. Kertscher, 173–193. MIT Press.

Nelson, K. 2003. Narrative and the emergence of a consciousness of self. In *Narrative and Consciousness*, ed. G. D. Fireman, T. E. J. McVay, and O. Flanagan, 17–36. Oxford University Press.

Rorty, R. 1980. A reply to Dreyfus and Taylor. *Review of Metaphysics* 34 (1): 39–46.

Seeburger, F. F. 1979. The distinction between "meaning" and "significance": A critique of the hermeneutics of E. D. Hirsch. *Southern Journal of Philosophy* 17 (2): 249–262.

Stueber, K. 1994. Understanding truth and objectivity: A dialogue between Donald Davidson and Hans-Georg Gadamer. In *Hermeneutics and Truth*, ed. B. R. Wachterhauser, 172–189. Northwestern University Press.

Taylor, C. 1980. Understanding in human science. *Review of Metaphysics* 34 (1): 25–38.

12   Feeling Ashamed of Myself Because of You

Alba Montes Sánchez and Alessandro Salice

According to most accounts, shame is an emotion that typically focuses on the self, in the sense that its intentional object is the individual who feels it. But if this is on the right track, how is it possible for anyone to feel ashamed of what someone else does or says? Imagine that you are meeting some friends for drinks after work, and you bring along a colleague who happens to be Sicilian. At one point, one of your friends makes a slightly xenophobic comment against Sicilians, implying they are all "Mafiosi," and *you* feel ashamed (this example is a modification of the one offered by Scheler; cf. Scheler 1957, 81). Now, if the intentional object of shame is the very emoting subject, then in which sense can one speak about shame here? Obviously you didn't make the comment, and therefore you seem to have no reason for being ashamed, and yet you *are* feeling that emotion. This issue has not been addressed in great detail in the philosophical literature on shame, but it deserves further analysis, first because it puts some pressure on the received view, and second because it sheds light on an underexplored variety of shame, and thus on certain aspects of shame in general.

We tackle the problem in three steps. In section 1, we begin by thematizing what—despite several differences in the details—could be labeled the "standard" account of shame and, specifically, the rather uncontroversial claim that shame is about the self who feels shame. We then introduce the notion of *Fremdscham* by discussing the foregoing example.[1] This form of shame could be taken to represent a challenge to the standard account. A strategy to dispel this challenge could be to reduce Fremdscham to other kinds of emotions, especially to embarrassment (sec. 1.1), an indignation-like reaction (sec. 1.2), "standard" shame (according to Scheler's idiosyncratic interpretation) (sec. 1.3), and "fictional" shame (sec. 1.4). We do not deny that these feelings are possible, but we claim that there is a further phenomenon that cannot be reduced to them. Although the main gist of this chapter is that, given the necessary conceptual adjustments (and, in particular, if considered as a salient by-product of group identification), Fremdscham is not a challenge to the standard account, we do believe

---

1. We use the German word *Fremdscham* for brevity, but we do not mean to do an ordinary-language analysis of what German mother-tongue speakers mean by this term. More on this later.

that this *is* an original form of shame, and we therefore end the first section by vindicating its phenomenological credentials.

In section 2, we offer our positive account of Fremdscham by claiming that it can be qualified as a genuine and yet peculiar form of shame. It is a *genuine* form of shame because, in tune with the standard account, this affective response can be argued to be about the self. But it is also a *peculiar* form of shame because it appears to be triggered by the subject having group identified (i.e., by the subject conceiving of him- or herself as a member of the group to which the shameful individual also belongs). Although the cause or the event prompting shame is not an action or a situation brought about by the subject who is feeling shame, the event or cause can be regarded as salient for the subject, insofar as it is elicited by someone who is seen as an in-group member. Hence it seems plausible to maintain that Fremdscham is about the self, but it is about the self as member of a group. In other words, we suggest that it is possible to defend the claim that Fremdscham is about the ashamed self, if the theory of Fremdscham is supplemented by a theory of group identification.

In section 3, we address the possibility that one might feel ashamed of someone with whom the ashamed self apparently does not enter forms of group membership in any substantial sense. We introduce this possibility through an example that, if sound, seems to show that one can feel Fremdscham for persons who are not in-group members. How can our account accommodate such cases? Our tripartite strategy develops along the following lines: (1) this example depicts a related emotion, namely, indignation, but not Fremdscham in our sense; (2) there is an extremely thin and minimal sense of the "we," according to which all human beings (simply in virtue of their being human) form a "we"; (3) to group identify is to form a representation with a peculiar direction of fit; more specifically, to adopt a "we"-perspective or (what we here use as a synonym) to have a social self is based on the mere *presupposition* that there is a group to begin with (but this presupposition can be mistaken). Although each of these lines of argument per se might be seen as too weak to fully account for such cases, we argue that, considered together, they provide us with sufficient evidence to dismiss possible objections to our account of Fremdscham based on them.

1   Shame and Hetero-Induced Shame (Fremdscham)

Shame is generally experienced as a distressing, often very painful, emotion that makes the subject feel faulty and unworthy, exposed, vulnerable, and judged. A widespread way of cashing this out is to say that shame arises in connection to a negative self-evaluation, which might allow one to speak of shame as an "emotion of self-assessment" (Taylor 1985). This negative self-assessment can be due to both active and passive aspects of selfhood: to actions and omissions of all kinds (telling a lie to a friend, failing to defend one's values so as to maintain status in a particular social group), to things that befall us (victims of abuse typically feel it), to character traits, physical features, social background, and so on.

What appears to be a unifying feature of all these cases is that the intentional object of the emotion is not mainly the situation or action that gives rise to the shame episode but the self of the person ashamed. The shameful situation is such that it changes one's self-experience from one where selfhood is a tacit dimension of one's experience of the world to one where one's self becomes conspicuous, the object of focus (see Sartre 1972, 298–302; Zahavi 2014, 212–214). In fear, for example, one is chiefly directed at the world, at the threatening object, while in shame, the situation causes one to focus on oneself and feel exposed and diminished. Shame can be said to be "reflexive," in the sense of being directed back at myself, and that is why it has been labeled as a "self-conscious emotion" (see Tangney 2005; Lewis 1995). In shame, I focus on myself and see myself as small, faulty, or inadequate.

In the literature, there is a controversy around the issue of whether shame is also a "social emotion," that is, whether or not it is a response to being exposed to the censoring gaze of a real, an imagined, or an internalized audience (cf. Deonna, Rodogno, and Teroni 2011). For our current purposes, we do not need to enter into this debate; our proposal concerning Fremdscham is compatible with both positions. Elsewhere, with the aim of doing justice to the plausible intuitions of both camps, Zahavi (2014, 2012) and Montes Sánchez (2014) have characterized shame as an emotion of *social self-consciousness*: an emotion that relies on a form of self-consciousness that is only available to a social being, an emotion where we are aware of a dimension of ourselves (the intersubjective or interpersonal dimension) that is constituted and revealed to us in our encounters with others. This dimension escapes our control, as Sartre (1972) claimed, because it is impossible to fully determine how other people perceive us. This idea suggests that even if, to feel shame, an audience is not always necessary, the dimension of selfhood at stake in shame would not be possible without intersubjectivity. This is a claim about the self of shame: shame is not possible for a monadic, isolated self (Zahavi 2014, 2012; Montes Sánchez 2014). Put another way, the self of shame is intrinsically social. Our proposal regarding Fremdscham addresses an additional aspect of the sociality of this emotion.

Let us now turn to the issue of the intentional object of shame. As we have stressed, most accounts endorse the view that this is the self of the person feeling shame. Shame is mostly characterized as an emotion that focuses on oneself (as does pride), and not on others or the world (as does love or fear). But here we must draw an important distinction: what elicits our shame is an occasion for it, or its cause, but *not* its object. I feel ashamed *of* myself *because* I lied to my friend, or *because of* my appearance, or many other possible reasons, but I always feel shame *of myself* (Deonna et al. 2011, 83). Now, the cause and the intentional object of this emotion are strongly connected. The cause gives rise to the distinctive anxiety of shame precisely because I perceive it as reflecting back on me. As Hume (1978, 286) remarked, if the affective phenomenon is not intentionally directed at myself, I may perhaps feel other unpleasant emotions, but not shame.

However, is it right to assume that the object of this emotion is always the ashamed self? Can I not feel ashamed in a situation where someone else, and not me, is in a shameful position? It seems relatively clear that we *can*, and examples of it are common. Consider again our example of the xenophobic comment made by your friend in front of your Sicilian colleague. The remark seems to go unnoticed, as no one gives any sign of feeling awkward (not even your Sicilian colleague, who is used to this sort of comment), and yet you feel ashamed.

What this example seems to illustrate is that shame can arise in connection with someone who indeed is *not* the person feeling shame, that is, when the negative evaluation does not appear to fall on the ashamed subject. One could try to argue that the phenomenon tracked by these scenarios has a vicarious nature, meaning that shame would be caused by a process of emotional contagion (Goldie 2000, 181; Scheler 1973). However, as especially Scheler is at pains to highlight, this situation is interesting precisely because this is not the case: the object of shame seems to be another person, and yet you have *not* picked up her emotion by contagion (as can happen when we come to a party and soon find ourselves in the same festive mood as the others). In the example, neither your xenophobic friend nor your Sicilian colleague shows *any shame*, and yet you feel it. Again, hetero-induced shame is puzzling and interesting because it does not require anyone else feeling shame, and it typically includes shame-inducing subjects who are themselves shameless.

This example illustrates a clear instance of what in Spanish is called *vergüenza ajena*, an expression that means "shame that doesn't belong to me but to someone else" and refers to a feeling of shame about another person. So does the German *Fremdscham*. Italian has a compound expression to refer to this phenomenon, *vergognarsi per qualcuno*, where the preposition *per* signals that the felt shame is directed toward another person. In what follows, we employ the German expression *Fremdscham* to refer to this form of shame for purely syntactic reasons: being a noncompound singular term, we can avoid complexities in the construction of sentences, but we do not mean to engage in an ordinary-language-philosophy analysis of what German mother-tongue speakers mean when they use the term *Fremdscham*. We simply use it as a shorthand term to refer to hetero-induced shame or, more accurately, to the specific form of it that we delineate later in the chapter.

At this juncture, it is important to stress that there is more than one way in which others can induce our shame.[2] As Zahavi (2012, 313n8) notes, one can feel shame *with* or *for* somebody. He clarifies this point through an example that is similar to ours: you are taking a walk with a friend, who happens to be black, and by chance you encounter your father, who makes a racist comment against black people. In such a situation, you may feel shame *with* your friend or *for* your father. Feeling shame *with* somebody might be a matter of emotional contagion or, more interestingly, might be a collective emotion. The latter option deserves a much more careful investigation, and we believe it could reinforce our underlying claim that

2. Note that when others humiliate or shame us, one of the possible responses (aside from anger or indignation) is shame, but this shame is unambiguously about oneself. What we mean to address here are different ways in which one's own shame might be thought to be about others.

the form of selfhood that is at stake in shame (and other self-conscious emotions) is essentially social, but we cannot address it in this chapter. Here we focus on Fremdscham insofar as it is shame one feels *for* somebody, that is, an individual, not a collective emotion, but induced by someone else. In what follows, we use the term *Fremdscham* in this more restricted sense.

Although the various expressions in different languages that we have presented seem to track a genuine phenomenon—and even if one is careful to rule out emotional contagion, as Scheler does, and to stress that this phenomenon can be nonvicarious—it is important to address four other possible explanations that might endanger the irreducibility of the phenomenon we are after. Before offering a positive account of hetero-induced shame, we want to resist the ideas that (1) Fremdscham can be cashed out in terms of embarrassment, (2) Fremdscham can be reduced to a sort of indignation about the shamelessness of others, (3) there is no principled difference between Fremdscham and shame, and (4) Fremdscham can be reduced to fictional shame.

Although we resist the attempts to reduce Fremdscham to these other emotive reactions, we do agree that such reactions are all possible. Admittedly, sometimes, or perhaps even often, German mother-tongue speakers refer to these feelings with the word *Fremdscham* (and Spanish mother-tongue speakers with the expression *vergüenza ajena*). Still, our point is that explaining Fremdscham in terms of these other emotions simply does not do justice to the peculiarity of hetero-induced shame.

1.1 Fremdscham Is Not Embarrassment

A first, quick objection can be raised against the example of your father's racist comment: it is not about shame but about embarrassment. Both emotions are closely related, and it is not rare for embarrassment to slide into shame, but they are different in important ways that would make it easier for one to feel embarrassment when one is no more than a witness to a ridiculous or inappropriate situation. Embarrassment is shallower (it does not affect the self so deeply) and much more fleeting; it is always linked to public appearances, to social awkwardness; and it always requires an audience. Rom Harré (1990, 197) proposes that the difference between shame and embarrassment lies in the type of code one is breaching. In his view, shame is connected to breaches of moral norms or of an honor code that would deeply impact the evaluation of one's character, while embarrassment would purely be about breaches of convention, of a code of manners, which simply make one look a bit foolish. Thus one might argue that merely witnessing something can be socially awkward, that the whole situation would go against the conventions of smooth and polite social exchanges, and therefore call for embarrassment, regardless of who caused it. If one accepts this, Scheler's extension of the object of shame to other selves could be interpreted as a consequence of his failure to distinguish between shame and embarrassment, and accounts along Scheler's line of thought could be subject to the accusation of calling "shame" what should appropriately be called "embarrassment." Although this may be a correct description of *some* cases, accepting it as a general explanation of Fremdscham would imply that the boundary between shame and embarrassment is perfectly distinct, clear-cut, and impermeable, which it is not, and would also imply that all examples of hetero-induced shame should be described as cases of embarrassment, and not shame, which is also dubious. Let us explain why.

First of all, as Susan Miller (1985, 28) claims, emotion terms are not so easily applied to experiences as concepts such as "chair" or "table" are applied to objects, and thus it is problematic to assume that emotion terms designate clear-cut areas of experience. Boundaries are blurry, and there are often areas of confusion and overlap between closely related emotions. This is the case with shame and embarrassment, which belong to the same emotional family; they resemble each other, and embarrassment can sometimes slide into shame. As Zahavi (2012, 305) argues, Harré's clear-cut distinction between the two emotions fails to explain how the exact same situation can cause one or the other, depending on the person. Harré's distinction tracks a significant difference, but he seems to ascribe it to the wrong instance: what matters is not the socially established code (of morals or manners) that is being breached but the way in which the situation affects the person's sense of self, which is not determined *univocally* by the social code at play. The difference should be cashed out in phenomenological terms, that is, in terms of how the subject experiences the situation, what it means for her, and how she evaluates it, and not exclusively in terms of social norms and conventions. At that level of description, it is an open question when "shallow" issues of social presentation become "deep" issues that affect one's sense of self: "appearances" are often less superficial and more far-reaching that one may think. If, in the given example, you were to construe your friend's remarks as a cruel insult, they might very well cause you shame, not mere embarrassment. But if this seems implausible, consider another possibility: you invite a group of colleagues to a classical music concert where your son is the main soloist. When he appears onstage, he is visibly drunk, behaves rudely, and gives a dreadful performance.[3] Would it be plausible to argue here that you are feeling mere embarrassment and not shame? We think not.

1.2 Fremdscham Is Not Indignation about the Shamelessness of Others

One could argue that the phenomenon of "being ashamed of someone else" is equivocal (see Scheler 1987, 19): it does not refer to a painful awareness of a self put on the spot. Rather, it refers to an aversive reaction that we experience because someone else has brought discredit or disgrace on us or is offending us with her shamelessness. According to this account, Fremdscham would be an indignation-like response to somebody shamelessly doing something disgraceful. Imagine, for example, that while browsing TV channels, you happen on a particularly outrageous reality show and, finding it unbearable to watch, rapidly change the channel. It might be possible to argue that if you feel Fremdscham in such a case, this does not differ so greatly from being shocked or horrified by somebody's impudence. In this

3. We thank Dan Zahavi for suggesting this example.

interpretation, Fremdscham would appear to be more akin to indignation, in that it would imply a condemning verdict of shamelessness. Indeed, one could describe scenarios like this one as cases where the subject isolates herself from the shameful feature or situation, condemns it, and reassures herself of her superiority: she remains keenly aware that she is *not* the shameful subject. The main difference between indignation and Fremdscham here would be that indignation refers to injustice, while Fremdscham refers to shamelessness (and thus Fremdscham would be connected to scenarios of shame, not to scenarios of guilt), but the intentional structure of Fremdscham would resemble indignation much more than regular shame.

Perhaps this is what the word *Fremdscham* often idiomatically refers to. It is even plausible to think—although we do not make a strong claim about this here—that a systematic connection exists between some varieties of hetero-induced shame and indignation about shamelessness. Indeed, Fremdscham implies a desire to distance oneself from something perceived as worthy of shame, and therefore something like a condemnation, but this verdict on the situation should not be confused with the intentional structure of the emotion, which is still that of shame. One can feel indignation before a shameless deed, and this emotion can get mixed or be felt at the same time as Fremdscham, but this is not what Fremdscham amounts to. Even if there were a systematic connection between these emotional responses, the systematic co-occurrence of two emotions does not mean that one can be reduced to the other.

Furthermore, that one experiences a desire or an impulse to distance oneself from the cause of shame is not characteristic of Fremdscham only; it is characteristic of all forms of shame. Such a desire is present as well in the shame that one feels unambiguously about oneself, and it is consistent with its phenomenological descriptions. Indeed, it lies at the core of them: in shame, I feel tainted by an *unwanted* identity (Ferguson, Eyre, and Ashbaker 2000), burdened by an aspect of myself that escapes my control (Sartre 1972; Zahavi 2014; Montes Sánchez 2014), and I desperately want to *escape* from it, but I cannot, because "I am riveted to myself" (Lévinas [1935] 2003, 64). And in this sense, as we argue in section 2, I am riveted to my group, too. Or at least in such cases I experience myself as riveted to it while wishing I was not. This tension is what makes shame so unpleasant. Fremdscham is no different in this from regular shame: it is simply that my wanting to reject my identification with the group also implies rejecting other members of the group, but the structure appears to be analogous to that of regular shame.

1.3 Fremdscham Has Specificity vis-à-vis Shame

In line with our view about Fremdscham, Scheler argues that shame (and not just embarrassment) can be felt for others in a genuine, nonvicarious fashion. Section 2 draws substantially on this insight, but Scheler's explanation of Fremdscham displays some problematic aspects due to his view about shame. More precisely, he holds that this emotion is not self-conscious, that is, it is not specifically about the one who is feeling shame. According to Scheler, self *and* other can function both as the intentional object of shame and as the shame-inducing audience. I can be ashamed of myself *before* myself or *before* the other. And I can be ashamed *of* myself or *of* the other. So for Scheler it follows that the intentional object of shame is "the individual self in general [*das individuelle Selbst überhaupt*]" (Scheler 1987, 18, trans. mod.; for the German version, see Scheler 1957, 81). For Scheler, there are states of affairs that call for me to feel shame, independently of who is in them. In his view, "Shame is a protective feeling of the individual and his or her value against the whole sphere of what is public and general" (Scheler 1987, 17). Shame for him is a mechanism against the contamination of higher, positive self-values by lower ones; it protects the sphere of humanity and individuality. This wider sphere of higher spiritual values that, in his view, distinguishes humans from animals can be put at risk in myself or in another.

Admittedly, there must be something right in the view that shame is linked to our humanity and that we have a sense of commonality that allows us to feel shame for others. However, the way Scheler talks about the object of shame being "the individual self in general" seems to blur important distinctions between self and others that are present in Fremdscham, and it does not sit easily with the sense of self-individuation that accompanies shame (for other problematic aspects, see the end of sec. 3). How would Scheler explain that sometimes we do not react with Fremdscham to violations of those spiritual human values he speaks about, but rather respond with outrage, contempt, or even ridicule? Is it simply because we are indecent and have no protective sense of shame? This might sometimes be the case; however—as we have already shown—self-involvement seems to be necessary for shame, including Fremdscham. In Fremdscham, shame's object is still the self of the person ashamed, who feels, so to speak, exposed by proxy. Without a sense of my own investment and exposure in the situation, I might feel perhaps indignant or scandalized, or even amused at how the other is making a fool of himself, but not ashamed.

1.4 Fremdscham Is Not "Fictional" Shame

One could further argue that when I feel shame for another, I put myself imaginatively in her shoes and feel her shame. As such, I could take on the role of the shameless agent and feel ashamed because I deem her actions shameful and would feel ashamed of myself, or I could imagine myself in the role of the shamed individual and react with the shame I would feel if someone humiliated me in this way. But I might also take on entirely or partially the other's perspective, as is often the case in enjoying works of fiction, for example, if I read about Anna Karenina's shame and come to feel it too, although my values do not coincide with those of a Russian aristocrat in the nineteenth century. These cases, however, are in tune with the standard account of shame and do not seem to capture what is typical about Fremdscham; fictional shame would still be intentionally directed at myself, only here myself coincides with another individual—either the shameless or the humiliated one. Consider an analogous situation: you imagine being Napoleon contemplating the desolation of Austerlitz; in this case, it is *Napoleon's* perspective that you are adopting toward the desolation of Austerlitz, not *your* perspective (Williams 1976, 43). The problem would thus be displaced: it would no longer concern the theory of shame per se but rather would concern fictional emotions or imagined emotions.

We do not deny that this happens often: we put ourselves imaginatively in somebody else's shoes and feel what we imagine are his or her emotions. This is arguably one of the main reasons why fiction can be so gripping. But can all examples of hetero-induced shame be explained thus? We do not think so. Recall the example of my son disgracing himself in the concert. Is it plausible to think that I feel shame because I am putting myself in his shoes? This seems fairly unlikely: if the soloist was a complete stranger, I would be more likely to feel contempt for the musician and angry toward the organizers who charged me for seeing the show. In our view, we must seek the explanation somewhere else.

2 Fremdscham and Group Identification

Despite these reductivist attempts toward hetero-induced shame, it should not go unmentioned that some authors have made gestures toward recognizing the genuineness of this affective phenomenon. The interesting element of these gestures is the link that is established between the ashamed self and the individuals who, with their behavior, trigger the emotion of shame.

Walsh (1970), for instance, claims that shame can arise by association, through members of one's family, nation, profession, and so on. Our initial Scheler-inspired example of the xenophobic remark could thus be interpreted as a result of my association with my friend. The common examples of people feeling ashamed of their family members can also be explained in this way. As Cavell stresses, "Shame is felt not only towards one's own actions and one's own being, but towards the actions and the being of those with whom one is identified—fathers, daughters, wives … the beings whose self-revelations reveal oneself" (Cavell 2002, 286). The crucial word here appears to be "identified": *I* feel exposed by proxy. It seems intuitively plausible to say that most of us have intimate others with whom we can identify closely, in ways that make us feel ashamed *of ourselves* through their exposure. But the problem with both Walsh and Cavell is that they are vague about the mechanisms of association and identification: they give the reader no satisfying account of such mechanisms.

But how is that at all possible? How can others impact my self-conscious emotive life in such a fundamental way?[4] To illustrate our idea, we first refer to an established paradigm in social psychology, generally labeled "social identity theory," according to which individuals not only have an individual self but also have one or more social selves. One's social self does

---

4. This is the case not only for negatively colored emotions like shame but also for positive ones like pride. For example, consider the pride a parent feels when her daughter receives some important results. The parent is not the agent in this scenario, but he or she feels a perspicuously self-directed emotion. For more on this, see Salice and Montes Sánchez 2016.

not have to be conceived as something different from one's personal identity; rather, one's social identity could be said to be a part of one's personal identity. The social self can be intended as the representation one has of oneself as a member of a group and can be seen as the result of a process of self-categorization (Brewer and Gardner 1996). This process seems to be elicited by the individuals starting to perceive themselves as being similar to others in a certain respect or as sharing certain properties with others. The properties can be of a manifold nature: social properties such as "being an analytic philosopher" or "being a Buddhist" can be used to characterize oneself, for example, within an academic or a religious context. But empirical evidence from experiments in social psychology (Tajfel 1970; Tajfel et al. 1971) seems to show that much more minimal properties (in the sense of properties negligible in usual contexts) like "preferring Beethoven to Mozart" or "preferring Paul Klee to Wassily Kandinsky" in certain contexts could acquire salience and form the basis of self-categorization, thus giving rise to so-called "minimal groups."[5]

What seems to be important for our current purposes, however, is that the mere understanding of belonging to a class or group, that is, to the group identified by a given property (e.g., the class of analytic philosophers, the class of the Beethoven enthusiasts, etc.), is not yet per se to identify with that group. The subject who is aware of exemplifying certain properties does not have to articulate her experiences by using the first-person-plural pronoun or, said another way, to frame the situation she is in according to a we-perspective. In a sense, the subject does not live through the group from within, and the mere idea of belonging to a specific class of individuals does not necessarily affect the way in which one behaves, feels, or thinks. But this is exactly what it means to see oneself as a member of a "we" or to adopt a we-perspective: a wealth of psychological literature identifies marked predispositions to altruism toward in-group members, to emotional sharing, to sympathy, to collective actions, to we-talk, and so on (Turner 1987, 50), as quintessential by-products of group identification.

At the same time, the route that leads individuals to group identify does not have to be self-categorization: in certain cases, dyadic forms of intentionality (e.g., as instantiated in face-to-face communication) can lead the individuals involved to conceive of themselves as *us* (Zahavi 2015). And perhaps other processes as well could be conducive to group identification (Bacharach 2006, 76). We will come back to the notion of group identification in section 3, but for present purposes, suffice it to say that once group identification has occurred, the subject's self-consciousness undergoes a specific transformation: the *self* is now sublated under a "we," a "we" to which others are also sublated and to which the self feels attached.

How does all of this bear on the issue of hetero-induced shame? To illustrate the role of group identification or we-perspective taking, let us slightly modify (or perhaps clarify) the example we began with. Assume the example extends through time in the following way: at

---

5. But see Oakes 1987 on the difficulties of characterizing "salience."

time *t* the friends gather, have a chat, and make xenophobic comments about Sicilians (no one feels ashamed). At time *t'* the same group of friends convenes again, but this time a Sicilian is also present. At some point, friend *a* makes the same comment made at *t*, and friend *b* feels ashamed. It now seems plausible for the scenario to be cashed out in the following terms: the presence of the Sicilian person assigns salience to the fact that person *b* shares certain properties with other persons (e.g., person *b* shares a certain xenophobic attitude toward Sicilians), and this triggers group identification to the effect that person *b* starts to frame the relevant situation in we-terms. Given that person *a* is seen by person *b* as a member of the same group to which *b* also belongs, and *b* considers xenophobic behavior shameful, *b* feels ashamed. Although the cause or situation that moved person *b* to feel shame is *not* brought about by *b*, the emotion of shame is nevertheless about *b*—however, it is about *b* as a member of group *G*.

If this is on the right track, then one could argue that, on the one hand, Fremdscham does align to shame (insofar as both emotions share an analogous intentional structure), and hence recognizing its authenticity does not call into question the standard account of shame. On the other hand, one would have to appreciate the peculiarity of this form of shame and anchor it in the fact that the event causing the emotion is not brought about by the ashamed self and that it presupposes group identification.

3   Fremdscham with and without Group Membership

The proposed description of Fremdscham makes explicit a premise that remained tacit throughout the entire discussion and can now be targeted: the individuals that figure in the example so far discussed are *friends*, meaning that the example involves individuals who are members of a group (and, indeed, of a particularly interconnected one). In other words, the example we began with assumes that there is an actually obtaining link between the ashamed self and the subject who, with his joke, causes the emotion. But is this assumption justified? After all, one could conceive of real-life scenarios that seem to testify in favor of the possibility of an individual feeling ashamed of someone with whom he or she does not entertain any form of group membership. Here is an example that seems to clearly invalidate our account of Fremdscham:

You are walking on the street and see a beggar sitting on the sidewalk. Suddenly, the man who was walking just in front of you spits on the beggar. On witnessing this, you feel ashamed. This man is a complete stranger, someone you had never seen before and who has no association with you, and yet you feel Fremdscham.[6]

If this example is sound, how can our account accommodate cases like this? After all, one might be tempted to argue that Fremdscham does *not* build on group identification, given

6. This is a variation of an example suggested to us by Henning Nörenberg.

that the relevant subjects mentioned in the example are not members of the same group. This possible objection could grant that in certain cases, group identification accompanies Fremdscham, but per se this element is not a necessary condition for Fremdscham to be elicited. So how to square the necessity claim made in section 2 with this example? The suggestion is that three possible arguments could help mount a sufficiently solid defense of our account by making it hospitable to those scenarios. We do not take each of these arguments to be compelling enough to block the objection. However, we do believe that, if considered together, they acquire sufficient force to dismiss it and produce adequate evidence to accommodate this kind of case. As we will see, some of the reasons related to this caution are associated to metaphysical considerations that we are not in a position to settle in this chapter.

The first strategy is to question the cogency of the example, for one could argue that the scenario depicted is not about Fremdscham (though a native German speaker might designate it with that word). It would fall within the possibilities covered earlier: it would be a case of indignation about shamelessness, not of hetero-induced *shame*. However, although this idea could be employed in an ad hoc way to reject some possible counterexamples, the same argument we provided earlier to distinguish shame from embarrassment could be brought to bear on this issue, again putting on us the burden of the proof. Luckily, this first strategy can be complemented by a more detailed presentation of the account developed in section 2.

What provides force to the objection is the assumption that whenever one group identifies, one forms or enters a group. Since the example does not involve the existence of group, there cannot be any group identification. But—to put it bluntly—this assumption seems to us to be off track: to group identify does not yet equate to form or enter a group. If the outcome of group identification is to form a certain representation (a representation of oneself as group member, or a social self), and if representations have directions of fit (hereafter DoF; for the standard treatment of this notion, see Searle 1983), then mere group identification does not already count as forming a group or entering a group. More precisely, the social self is not a representation with a double DoF: representations or acts with a double DoF are those that bring about the very facts they are about. For example, if one has the authority to adjourn the meeting, and if one adjourns the meeting, then one brings about the fact that the meeting is adjourned. The social self does not have such DoF, because one could group identify without any group existing in the first place. Indeed, the above-mentioned experiments on the minimal-group paradigm show that individuals can accomplish the identification process with other individuals without even knowing them (in addition, there are arguments against the very idea of representations with double DoF; see Laitinen 2013). This could well be the case in the example we just offered. Now, if the process of group identification does *not* create groups, then in which relation does this process stand with groups? In other words, what kind of DoF could the social self have?

Let us first focus on the world-to-mind DoF that characterizes conative states like intentions, wishes, desires, and so on. To be sure, there are many ways in which the social self can be related to mental states of a conative nature. For example, in certain cases, the fact that I group identify might involve my positive attitude toward the group I count myself in—and perhaps even my wish or desire that the group does well and flourishes. Or group identification might even be *preceded* by my wish or desire to be a member of the group. But, obviously, it is one thing to wish to be a member of a group, and another thing to act on behalf of the group and to articulate one's experiences in we-terms. This seems to suggest that if the social self has a DoF, then it is not of the world-to-mind kind.

Accordingly, one is perhaps better advised to choose a different line and argue that the social self is characterized by a mind-to-world DoF typical of cognitive states like, for instance, perceptions. Just as a perception has to be qualified as a hallucination if (among other conditions) the allegedly perceived fact or object does not exist, so similarly group identifying misfires if there is no group there to identify with. If the social self has this kind of DoF, then the objection to our account of Fremdscham could now be reformulated in the following terms: since there is no group in the examples that we have offered, group identification misses its target (the subject's experience of group identifying would thus be a kind of hallucination, so to speak), and yet the person is feeling Fremdscham. Therefore Fremdscham does not presuppose group identification.

One preliminary consideration that might speak against assigning a mind-to-world DoF to the social self is that, in forming this representation, the subject seems to *perform* something (although perhaps not always consciously or spontaneously), whereas perceptions or other cognitive states are not "performed" in the same sense—they just occur. But even if one is not persuaded by this argument, then another element in the revised version of the objection might be called into question. One could suggest that group identification indeed is tracking an actually existing group: this is the class of all human beings, a universal "we" that encompasses all other groups. If the social self were to have a mind-to-world DoF, then the claim that Fremdscham relies on group identification could still be secured under the condition that there were such an encompassing and universal "we." In addition, to operate with the embarrassment/Fremdscham distinction, this could be a second possible strategy: to group identify is to adopt a representation with a mind-to-world DoF, and when one feels hetero-induced shame, this representation tracks the fact that the subject belongs to an encompassing "we."

However, some might find this hypothesis suspect: How large does this "we" have to be? Why should it be restricted to human beings? What about other animals?[7] Should we just include particularly intelligent animals, or only our pets, or all animals? Or even: what if panpsychism is correct; that is, what if consciousness is a primitive constituent of reality to the effect that it has to be assigned to almost everything?

7. See Derrida's (2006) remarks on feeling ashamed before his cat.

We leave these questions open, but we would suggest a third and final reply for those who have qualms about accepting the existence of a universal "we." A promising option seems to be assigning the so-called "ø [Null]" or, according to Searle's (2011) most recent view, "Presup" DoF to the social self. Remorse, for instance, could be described as a mental state with such DoF insofar as it can be modeled as deep regret or guilt for something that is already *presupposed* to be a wrong committed (the same holds for apologies). Why do we hold that the social self might have a Presup DoF, and how does this impact our picture of Fremdscham? The relevant consideration here seems to be that group identification, to be elicited, has to presuppose the existence of a group (the group with which the individual then identifies), and this presupposition can be either correct or incorrect. Whereas in the two kinds of examples we have presented, the mental acts that the subject performs in group identifying (or even the experience the subject is living through when she group identifies) are indistinguishable and of exactly the same kind, not all of these mental acts track actually existing groups. They draw on presuppositions that can misfire. Thus only in the first and second examples we propose (i.e., the xenophobic remark and my son's concert) group identification tracks an actually existing group. By contrast, in the third scenario, the presupposition misfires, as there is *no* group to begin with.

If this account of group identification is persuasive enough, then our theory of Fremdscham would remain unaffected by the objection. Or rather, the counterexamples would provide even more evidence for our claim that Fremdscham is a by-product of group identification. Against this background, one could indeed argue that in both kinds of scenarios the element triggering Fremdscham *is* group identification, with the simple addendum that only in the first (but not in the second) scenario, group identification relies on a presupposition, which actually tracks a group.

Conclusion

In many respects, Fremdscham is a fascinating form of shame. What makes it so compelling is that it reveals to us how intrinsically and deeply related our self is to other selves. When an individual understands him- or herself as being member of a "we" (and this happens all the time, generally without activating any particularly demanding mental process), then the individual self undergoes a peculiar transformation: suddenly one feels attached to others and sees them as "belonging" to oneself in such a profound manner as even to trigger self-conscious emotions like shame (or pride, for that matter). By feeling this emotion, the self remains in the focus of the emotion itself, and yet this is not a monadic self, or a self that, in a sense, remains oblivious to the social dimension in which it is embedded. Rather, this is a "transformed" self; this is the self *insofar as it is a member of us*.

Acknowledgments

We thank Mikko Salmela, as well as the editors of this volume, for their comments on this chapter. Preliminary versions were presented at the Vienna-Copenhagen Workshop "'We': Distributive and Collective" (Vienna, March 30, 2015), the University of Hokkaido (Japan, June 30, 2015), and the International Conference of the EPSSE (Edinburgh, July 17, 2015), where we received much-appreciated feedback. Our research was funded by three projects: Marie-Curie Initial Training Network, TESIS (FP7-PEOPLE-2010-ITN, 264828); "The Genomic History of Denmark" (KU2016); and "The Disrupted 'We': Shared Intentionality and Its Psychopathological Distortions" (KU2016).

References

Bacharach, M. 2006. *Beyond Individual Choice, Teams, and Frames in Game Theory*. Princeton University Press.

Brewer, M. B., and W. Gardner. 1996. Who is this "we"? Levels of collective identity and self representations. *Journal of Personality and Social Psychology* 71 (1): 83–93.

Cavell, S. 2002. The avoidance of love: A reading of *King Lear*. In *Must We Mean What We Say? A Book of Essays*, 267–353. Cambridge University Press.

Deonna, J., R. Rodogno, and F. Teroni. 2011. *In Defense of Shame: The Faces of an Emotion*. Oxford University Press.

Derrida, J. 2006. *L'animal que donc je suis*. Galilée.

Ferguson, T. J., H. L. Eyre, and M. Ashbaker. 2000. Unwanted identities: A key variable in shame-anger links and gender differences in shame. *Sex Roles* 42 (3–4): 133–157.

Goldie, P. 2000. *The Emotions: A Philosophical Exploration*. Oxford University Press.

Harré, R. 1990. Embarrassment: A conceptual analysis. In *Shyness and Embarrassment: Perspectives from Social Psychology*, ed. W. Ray Crozier, 181–204. Cambridge University Press.

Hume, D. 1978. *A Treatise of Human Nature*. 2nd ed. Ed. L. A. Selby-Bigge and P. H. Nidditch. Clarendon Press.

Laitinen, A. 2013. Against representations with two directions of fit. *Phenomenology and the Cognitive Sciences* 13 (1): 179–199.

Lévinas, E. [1935] 2003. *On Escape: De l'évasion*. Ed. J. Rolland, trans. B. Bergo. Stanford University Press.

Lewis, M. 1995. *Shame: The Exposed Self*. Simon & Schuster.

Miller, S. B. 1985. *The Shame Experience*. Analytic Press.

Montes Sánchez, A. 2014. Intersubjectivity and interaction as crucial for understanding the moral role of shame: A critique of TOSCA-based shame research. *Frontiers in Cognitive Science* 5:814.

Oakes, P. J. 1987. The salience of social categories. In *Rediscovering the Social Group: A Self-Categorization Theory*, ed. J. C. Turner, M. A. Hogg, P. J. Oakes, S. D. Reicher, and M. S. Wetherell, 117–141. Blackwell.

Salice, A., and A. Montes Sánchez. 2016. Pride, shame, and group identification. *Frontiers in Psychology* 7 (557). doi:10.3389/fpsyg.2016.00557.

Sartre, J.-P. 1972. *L'être et le néant*. Gallimard.

Scheler, M. 1957. Über Scham und Schamgefühl. In *Schriften aus dem Nachlass*, vol. 1, *Zur Ethik und Erkenntnislehre*, ed. M. Scheler, 65–154. A. Francke.

Scheler, M. 1973. Wesen und Formen der Sympathie. In *Gesammelte Werke*, vol. 7, ed. M. Frings. Francke.

Scheler, M. 1987. Shame and feelings of modesty. In *Person and Self-Value: Three Essays*, ed. M. Frings, 1–85. Martinus Nijhoff.

Searle, J. R. 1983. *Intentionality: An Essay in the Philosophy of Mind*. Cambridge University Press.

Searle, J. R. 2011. *Making the Social World: The Structure of Human Civilization*. Oxford University Press.

Tajfel, H. 1970. Experiments in intergroup discrimination. *Scientific American* 223:96–102.

Tajfel, H., M. G. Billig, R. P. Bundy, and C. Flament. 1971. Social categorization and intergroup behaviour. *European Journal of Social Psychology* 1 (2): 149–178.

Tangney, J. P. 2005. The self-conscious emotions: Shame, guilt, embarrassment, and pride. In *Handbook of Cognition and Emotion*, ed. T. Dagleish and M. J. Power, 541–568. John Wiley & Sons.

Taylor, G. 1985. *Pride, Shame, and Guilt: Emotions of Self-Assessment*. Clarendon Press.

Turner, J. C. 1987. A self-categorization theory. In *Rediscovering the Social Group: A Self-Categorization Theory*, ed. J. C. Turner, M. A. Hogg, P. J. Oakes, S. D. Reicher, and M. S. Wetherell, 42–67. Blackwell.

Walsh, W. H. 1970. Pride, shame, and responsibility. *Philosophical Quarterly* 20 (78): 1–13.

Williams, B. 1976. *Problems of the Self: Philosophical Papers, 1956–1972*. Cambridge University Press.

Zahavi, D. 2012. Self, consciousness, and shame. In *The Oxford Handbook of Contemporary Phenomenology*, ed. D. Zahavi, 304–323. Oxford University Press.

Zahavi, D. 2014. *Self and Other: Exploring Subjectivity, Empathy, and Shame*. Oxford University Press.

Zahavi, D. 2015. You, me, and we: The sharing of emotional experiences. *Journal of Consciousness Studies* 22 (1–2): 84–101.

13   The Extent of Our Abilities: The Presence, Salience, and Sociality of Affordances

John Z. Elias

While the concept of affordances is central to broadly embodied approaches to cognition, it is often relied on without sufficient explication. Here I seek to contribute to its clarification in the course of exploring its scope. The concept in its original and basic sense refers to potentialities for action, constituted by the relationship between animals and their physical environment (Gibson 1977; Greeno 1994). Recent research has sought to extend the concept into the social and cultural sphere (Ferri et al. 2011; Rietveld 2012). Such an extension, though promising, poses its problems and complications.

My focus here will be mainly on the ability component, or complement, of affordances, as shaped and informed socially for human beings. We—in virtue of our sociability and plasticity—are particularly open to altering and developing our capacities and abilities, thereby extending the range of available affordances, expanding the scope of our space of potentiality. The distinctively dynamic and expansive nature of abilities for human beings, however, raises questions concerning the ontology of affordances, given their relativity to abilities, their *being* relative to abilities. These questions are particularly pressing, since much of the concept's power comes from the claim that affordances are *real*, that they *exist* in some sense (Turvey 1992; Chemero 2003).

I begin, briefly, with the basic core, to form a solid grasp of the concept before expanding outward into the social realm. I then raise some ontological issues posed by the changeability and variability of our (human, social) abilities. The resolution of these issues, I suggest, involves taking the temporal dimension of abilities and affordances seriously, particularly in terms of interaction across multiple temporal scales. Such a temporal perspective perforce encompasses the persistence of social and cultural patterns over time. I end by addressing abilities as they extend into—and are extended by—social interaction and coordination, and introduce the notion of *joint affordances* specifically, in contrast to the sociality of affordances more generally.

Of course, a full consideration of much of what is touched on here exceeds my present scope and space. Indeed, this paper forms a portion of what is ultimately a much larger project, which involves tracing in detail the substance and application of the concept of affordances, from basic embodied engagements with the immediate physical environment to the possibilities of its application in the social and, perhaps, linguistic realm.

1 An Embodied Modal Realism

Put simply, as a starting point at least, an affordance is a possibility or opportunity for action. Such a possibility depends on an animal's abilities as well as what's available in its environment; an affordance, therefore, is necessarily complementary, a matter of mutuality between animal and environment (Stoffregen 2003). Thus a tree affords climbing for a cat that can climb; it does not afford so for a dog that cannot. It also affords shading for either, among other things. And the tree still affords climbing even when the cat is shading itself: that is, the affordance *climbability* persists whether or not the cat is in fact climbing or otherwise occupied by the act of climbing. Which is to say, affordances exist independently of their being currently perceived or attended to—so this isn't some form of subjective idealism, in which their existence consists in their being perceived or intended (Sanders 1997). Yet their existence does depend on perceptual capacities more generally, since perception is an essential part of an animal's ability to act in and on the world.

However, everyone would agree—given one's abilities and what is available in the environment—that there will be certain things one can and cannot do; this is hardly news. What keeps the idea from being trivial, what gives it substance, is its boldly ontological character, the claim that affordances actually *exist*, as *actual* possibilities for action. Thus the presence of a sharp stick *presents* the possibility for stabbing: it is a *real possibility* within the situation; absent such a stick, such a possibility would not exist. (Here, incidentally, one may wonder about the human ability to consider, or *represent*, possibilities not currently present, to reach beyond immediately available affordances, and to what extent other animals remain enclosed within their sphere of affordances; see, e.g., Millikan 2006.)

The ontological status of affordances, their existence as real possibilities, or potentialities, for action, goes hand in hand with the core ecological claim of *direct perception* (Gibson 1979). We, and animals generally, perceive potentialities for action *directly*. Nothing intercedes between the presentation of the world and the possibilities that it presents. That is, we do not (and do not need to) project possibility onto a barely present world, like bright bulbs onto near darkness. Potentiality inheres in the world, in its very revelation to us. This is not to deny, though, that we are able to adopt a detached, reflective stance, from which to consider the world more neutrally, apart from its immediate relevance to actions and activities (indeed, one may wonder whether this capacity is peculiar to human beings). The claim, rather, is that perception, in its primary, primitive mode, is perception of affordances, of potentialities for action and interaction. In our immediate engagement with the world, the world shows up as already caught up in the circuit of actions and activities (e.g., Noë 2004).

Affordances, then, constitute a space of possibility, the set of possible actions actually *present* in a situation, or, perhaps more aptly, *comprising* the situation itself. This opens the possibility of what might be called an *embodied modal realism*, the position that we have genuine access to real possibilities and opportunities for action and interaction, again as a matter of mutuality between our embodied abilities and aspects of our environment. So just as we can direct our attention toward objects present around us without having to *re-present* them, we may direct ourselves toward the possibilities present to us, around the space of possible actions available to us. In other words, an ecological approach enables a conception of directedness toward—and responsiveness to—potential actions, which we might aim at, target, or track, within an existing space of possible actions, without, again, having to *represent* those potentialities. This view stands in stark contrast to the standard picture of internal intentions, intentions that represent the conditions to be satisfied by the actions to which they give rise (Searle 1980).

2   The Variability of Abilities and the Ontology of Affordances

It is important, however, not to speak in a too fixed and individualistic way about affordances, as if they were static objects to be simply picked up by individual creatures. For one thing, our abilities—and therefore affordances—change over time. A seven-year-old may not yet be able to ride a bicycle. We might say, though, that she is *capable* of riding one, insofar as she is fairly healthy and normal. So *capability* may be thought of as the ability to acquire an ability. And such capability is itself made up of more basic, enabling abilities, such as the ability to coordinate the use of one's limbs, the ability to sit astride a bicycle to begin with, and so on. And acquiring certain abilities makes you *capable* of acquiring more in turn. The notion of *know-how*, furthermore, may provide additional discrimination. While I may be said to be *able*, say, to form a D-minor-7 chord on a guitar, I may not yet *know how* to do so. The relations and gradations between capabilities, abilities, and know-how (not to mention *competences* and *skills* as well), indicate continuities with which our ordinary language—the digital discreteness of our words—struggles (hence controversies concerning the import and boundaries of these categories; e.g., Alter 2001; Noë 2005).

Children, of course, are *taught* to ride bicycles; beginning guitarists *learn* how to play chords. As social creatures, we are especially susceptible to acquiring new abilities and learning further skills, thereby altering the contours of our space of possibilities. From the beginning, we are inculcated and enculturated into an ever-widening range of practices and traditions, which more often than not are social through and through. Keeping in mind the relativity of affordances to abilities, however, this suggests difficult and potentially problematic distinctions between *actual possibilities* (i.e., potentialities present in the current situation) and *possible possibilities* (i.e., potentialities potentially emergent but not yet present). A chair may not yet afford sitting for an infant, but it affords sitting for the *form of life* into which the infant is growing (Rietveld and Kiverstein 2014). Should affordances therefore be relativized to an entire form of life? But then what about individuals and their individual differences, their various and varying abilities and inabilities? Are affordances *for* a whole form of life or *for* individuals and local groups? And if ability is indeed a matter of degree, of being more or less able, should we speak then of proximities to possibilities, of affordances near and far? Or of affordances more or less present or strong?

The problems posed by the relativity of affordances to abilities can usefully be illuminated in connection to psychopathology. As an example of the application of affordances, de Haan and colleagues (2013) distinguish between the *landscape* of affordances, which can be understood as the total set of affordances available to a particular form of life, and the field of *relevant* affordances for an individual, a subset of that larger landscape. So the idea seems to be that—for someone suffering from depression, for instance—their field of relevant or salient affordances will be diminished. But if a depressed person is genuinely *unable* to function in certain respects, and if affordances are determined in part by abilities, then would only their field of *relevant* affordances be diminished, or their *existent* set or landscape? That is, is the problem one of not noticing—of not being sensitive or alive to—affordances that nevertheless still exist, or is the very existence of affordances, of potentialities for action, at stake? Here a potential tension arises between the reality or objectivity of affordances—as exemplified by their existence independent of current or passing interests—and their dependence on abilities.

As mentioned earlier, one response is to relativize affordances to a whole form of life: the existence of affordances would then be determined and secured relative to the abilities and practices of an entire community. Yet the problem of variable abilities could then be expanded to the community itself: just as an individual may be more or less able or unable, so too can a community. One can imagine a particular tradition, as a set of practices, gradually fading out of existence: a certain style of music, say, or a special carpentry technique, no longer transmitted and sustained. Furthermore, such an approach does not address the issue of the role of individuals in constituting affordances. Shouldn't we allow for the possibility of particular individuals having access to affordances, to potentialities for action, unavailable to others in their community, in virtue of their uniquely cultivated skills?

Another tack is to take up the temporality of abilities and affordances and treat them in terms of multiple temporal scales, ranging from short to long. Returning to the case of depression: if a diminished set of salient affordances becomes entrenched such that, over time, the depressed person's abilities more generally begin to narrow and atrophy out of lack of exercise, their total set of affordances would presumably gradually diminish as well. In normal, well-functioning cases, occasional depletions of interest or motivation are more or less momentary matters of passing concern, posing little or no threat to the long-term viability of abilities. But for the sufferer of depression, depleted interest or the absence of interest might persist long enough to begin to erode his or her abilities.

This suggests complex interactions across temporal scales, from the large down to the small, a view potentially encompassing intention and motivation as well. At the outset, I asserted that the existence of affordances is not epistemic, in that they do not depend on merely momentary motivations and currently occurrent intentions; they exist, rather, in relation to abilities. However, admitting that abilities come in degrees and may wax and wane over time, we might then consider modulating influences on abilities, occurring over various temporal scales. And while we would not want to say that we are only able to do what we're willing or motivated to do, this much, at least, is true: ability, minimally, is modulated by motivation (e.g., Nicholls 1984). So while motivation and ability may be distinct, they are mutually and intimately interrelated. One's degree of ability depends, in part, on motivation, both at larger scales, in the motivation to acquire and cultivate particular abilities to begin with, and at the scale of specific situations, where motivations may inform what one is willing and able to do. Though again, one is still able to ride a bicycle even if one isn't especially motivated to at the moment; in many or most cases, any differential impact of motivation on ability may well be negligible. Nevertheless there is a significant difference between motivation having, at times, a negligible effect on ability versus having no effect at all.

Thus treating ability as a matter of degree allows us to see the modulation of motivation on ability. But this would seem to call into question the very stability of abilities, a stability that, again, secures the existence of affordances. Securing the stability of abilities, and hence the existence of affordances, requires reference to the reality of different temporal scales (e.g., Pattee and Rączaszek-Leonardi 2012). Abilities persist over periods of time: while they may not be static or fixed, they do exist at comparatively longer temporal scales. Indeed, a certain stability seems inherent to the very idea: to be said to have an ability is to be able to manifest it across situations, over time, and not just as a one-off, isolated occurrence (e.g., Honoré 1964). It wouldn't make sense—or at least it would strain our usual notion—to be merely momentarily able to do something (this isn't to say, however, that one couldn't have an ability without demonstrating it regularly; that is, there's a distinction between having versus exhibiting an ability). Unless one had a rather magical conception of the acquisition and diminution of abilities, one would not be prepared to attribute an ability, properly speaking, to someone who doesn't genuinely have it over some course of time (notwithstanding the fact that a person could suddenly lose an ability, due, say, to injury). This long view, incidentally, could include the persistence of abilities in and through the community, in the transmission and replenishment of communal practices and traditions from generation to generation.

Intentions and motivations, on the other hand, may be more or less momentary, whether extending over the expanse of long-term projects or narrowing to the pinpoint of a particular situation. Concerning the modulating influence of motivation on ability, then, the question becomes one of making sense of interactions across temporal scales. Such a question, of course, is large and complex (see, e.g., Spivey and Dale 2006 for studies of multiscale dynamics of cognitive activities). For present purposes, I might resort to the ancient image of a flowing river or current to help ground the point. A large and stable current (the Gulf Stream, say) persists over long spans of space and time. Within that larger current, however, various shifts and alterations may occur, stirred by numerous factors (ships, storms, etc.). Momentary reversals or redirections of flow may even occur within these smaller spans of space and time. And these local, small-scale variations, moreover, can impact events within their domain: a boatman, for instance, needing to beat more strongly against the current, or a school of fish carried along by a transient stream. Yet still with these short-term, small-scale fluctuations, the larger current maintains itself in its overall flow. Similar, then, is the momentary influence of varying motivations on long-term abilities. Such that, say, someone under the shock of devastating news may be said to be truly *unable*, at that moment, to get on and ride a bicycle, or to do much else, for that matter, while all the while still having that ability in the long term, to be manifested again in a matter of time.

Thus a sense of temporal scale is crucial, one that telescopes and expands according to explanatory demands, encompassing the expansiveness of our abilities, as well as our ability to bring them to bear on specific situations, in bringing relevant possibilities into more or less sharp relief. Given these considerations, however, the distinction between the *presence* and *salience* of affordances, between their *relevance* and very *existence*, may begin to be blurred, if not broken down. For if abilities are indeed influenced, in part, by desires, intentions, motivations, and so on, is there then room to say that affordances, in certain situations, will not only be more or less *salient* but more or less *present*—in some sense—depending on current concerns and interests? In other words, do affordances themselves *vary* along with ability, as modulated over time? For if we admit abilities come in degrees, should we admit too that affordances, somehow, are a matter of degree? If so, then presence, perhaps, would just *be* salience, given the pressures and purposes of the present moment. From this perspective, salience is built into the very relativity of affordances to ability, as modulated from moment to moment, over time.

Yet to speak of degrees of being would seem to be an ontological rabbit hole. Things, we would want to say, either exist or not (e.g., Quine 1948). Given their existence, though, their various aspects or properties can vary over various dimensions. An object may be more or less heavy or light. Yet the object exists, whatever its weight may be. It must first exist, to be heavy to begin with. Similarly, perhaps, with affordances. An affordance either exists or not, is either there or not (relative, again, to abilities); if it exists, it is more or less salient, more or less relevant—given the situation. So salience is not simply a subjective projection, as if interest or concern were a spotlight of sorts, projected upon preexistent affordances from some independent source (from some mental or subjective realm, perhaps). Salience is not separate from the presence of affordances: an affordance, insofar as it is present, will have a certain salience, depending on the situation. And with salience comes *solicitation*, thus circling back to the role of motivation in informing situations (Rietveld 2008).

In this section, I have addressed the shapability and variability of abilities while exploring some implications for the ontology of affordances. The variability of abilities seemed to pose a threat to the existence of affordances. This threat was defused, and their ontological status secured, by reference to the long-term scale of abilities, their persistence over longer periods of time, in comparison to momentary motivations and situation-specific interests, which, in the short term, modulate abilities. I have touched on these issues, however cursorily or provisionally, to cover and secure ground for considering affordances in the social realm. For our sociality and plasticity go hand in hand (e.g., Davidson and McEwen 2012), making us particularly prone to altering and extending our abilities.

3   Abilities and Affordances in the Social Sphere

Having explored the extent and variability of our abilities, I now address their sociality and social extensiveness more directly. Among other things, I draw a distinction between socially informed yet individual abilities, those that belong to individuals however much they may be socially acquired, and abilities of a more specially social sort, abilities that arise necessarily in and through intersubjective coordination (e.g., Sebanz, Bekkering, and Knoblich 2006).

The thoroughly social nature of human abilities and affordances introduces considerations of *normativity*. A preliminary distinction may be drawn between *regulative* versus *constitutive* constraints on affordances. On the regulative side, norms having to do with standard, conventional, or appropriate uses of objects are often in place. Though chairs in a conference room afford standing on, most conference attendees (the well-behaved ones, at least) use the chairs appropriately to sit. Such regulative constraints are imposed on already existing affordances (hence the heightened salience of specifically *canonical* affordances, as a consequence of regulative norms; Costall and Richards 2013). On the constitutive side, communities and cultures cultivate the abilities they value, raising and training people into habits, practices, and traditions. Affordances, then, emerge in virtue of such enculturation and cultivation. So the distinction, again, is between regulative norms operating on preexisting affordances, and constitutive norms that engender and create the affordances themselves, which guide and determine the acquisition and cultivation of abilities, without which the respective affordances would not exist.

Yet these constraints feed into one another, to the point, perhaps, where the distinction itself becomes one not of kind but of degree. Given an established set of affordances, constituted in virtue of abilities cultivated in a particular community, regulative norms differentially emphasize certain affordances, certain potentialities for action, over others. Take the example of utensils. We become better able to hold and use utensils in specific ways, due to our being brought up and taught to use them in those ways. Though we could still wield forks and knives in ways that would be deemed ill-mannered and inappropriate, our enculturation encourages us toward their more well-mannered and appropriate use. But if these encouragements and enforcements hold over time, our abilities more generally will conform to them in the long term. So we see a similarity with the aforementioned role of motivation in modulating abilities, though here the modulating forces are social and cultural norms. Thus they shape the sphere of affordances: while a wider range of affordances are present within a situation, the specifically canonical ones are typically the most salient, and hence soliciting.

As evident from the discussion thus far, most, if not all, affordances for human beings are socially informed, insofar as most, if not all, of our abilities are acquired and cultivated socially (Zukow-Goldring 2012). However, there are abilities that arise in virtue of social interaction and coordination, which require the active co-presence of others. Next I discuss what I call *joint affordances* specifically, to distinguish them from social affordances or the sociality of affordances more generally.

I will start with a simple example. A piano may not be movable by a single person. But in the presence of another person, that possibility—that is, the affordance *movability*—becomes *present* as an actual potentiality (imagine moving a piano: it really does become movable in the presence of another, with possibilities for action opening up in relation to another). This appearance of the piano as *movable* is not merely a matter of additional physical strength: the piano movers have to have the ability to *coordinate* their actions as well. This involves an openness, a permeability, to the perspectives of others that is not a matter of theoretical elaboration or attribution but rather an aspect of our automatic responsiveness to others (e.g., Surtees and Apperly 2012). We have access to how the situation is informed for them, from their perspective. And that we can reliably anticipate and respond to the actions of others is also an essential aspect of our ability to coordinate. This suggests an especial expansiveness of abilities in the social realm, in situations constituted by coordinations among people, at various scales. Indeed, this provides an additional dimension to the extensiveness of our abilities, by going beyond the bounds of individual abilities. And so, in virtue of the dependence of affordances on abilities, this begins to open up another dimension of potentiality, of real possibilities for action, possibilities that actively depend on the presence of other persons.

Say, however, in the foregoing scenario, that one of the would-be movers refuses to move the piano or otherwise *intends* not to participate. Where does this leave the affordance, then? If both don't *intend* to move the piano together, they would not, it would seem, be *able* to move the piano. This would seem to indicate a dependence on intention, and so a necessary epistemic dimension at odds with the basic ontic claim of the reality of affordances.[1]

Remembering the relativity of affordances to abilities helps yield a response to this problem. Again, if affordances are relative to abilities and exist only in relation to them, then insofar as an ability is not present—is not possessed by an agent or agents—then the respective affordance would not be present. In the case of the piano movers, they must be properly coordinated to be able to move the piano accordingly. If they are not so coordinated, then they are not able to move the piano, and the affordance *movability* would not be present for them in the situation. Anything that interferes with or undermines their coordination would therefore interfere with the existence or emergence of the affordance; absent coordination, they are unable to move the piano. Coordination, in this case, constitutes ability.

---

1. I would like to thank Zuzanna Rucińska for bringing this question to my attention.

To reiterate and summate: affordances are relative to abilities, exist only in relation to them. Absent relevant abilities, the relative affordances do not exist. The piano movers are able to move the piano only when coordinated in a certain way, only when in a certain state of coordination. Insofar as they are not in the relevant state of coordination, they are not able to move the piano. Therefore the affordance *movability* would not be present to them, would not exist for them. Only coordination of a particular kind would enable the existence of the relevant affordance.

However, crucially, it is possible to enter into states of coordination—to be bound up in coordinative dynamics—without explicitly intending to do so. Though this would be unlikely in the case of piano moving, entering into coordinated states, finding oneself in such states, without any intention, explicit or otherwise, to do so, is a quite prevalent and permeating aspect of human interaction. Consider, for instance, the coordination of pedestrian traffic on busy city streets, which we simply enter into by walking into it. We are often engaged in coordinated activity without having formulated or entertained any intention to do so. Indeed, coordination is often a function of the very structure of social spaces, which impose strictures on our activities (Heft 2007). Coordination, in and of itself, therefore does not necessarily depend on intention. This holds the epistemic factor constant and preserves the ontic character of joint affordances, their existence independent of intention and epistemic considerations.

This example of a *joint affordance*, then, serves to illustrate that some affordances are constitutively social in the sense that they actively depend on the presence of others, and exist necessarily in coordination with others. Among other things, the notion may allow for a way of making sense of at least some collective actions, in that it simply becomes a matter of jointly attending to, and engaging with, the same possibility for action—the same affordance—just as we might jointly attend to the same object in a room. So problematic entities such as collective intentions, and concomitant questions concerning how they are instantiated and distributed (Pettit and Schweikard 2006), potentially drop out of the picture. That is, collective intentionality (at least in certain cases) would just be directedness toward the same possibility or potentiality for action, again just as different people can unproblematically direct their attention toward the same shared object.

4   Further Extensions and Future Directions

Thus the variability of our abilities is bound up with the plasticity of our sociality. Not only are our individual abilities shapable and changeable because of our sociality, but specifically social abilities arise in virtue of our capacity to interact and coordinate socially. A further distinction can be drawn between the kind of co-located interpersonal coordination described in the previous section, and abilities that arise within a larger social system, against a social and institutional backdrop, which depend on others who are not necessarily co-present. That is, these abilities are not contingent on the active co-presence of others but are rather a matter of social organization, a function of the very structure of social space (Barker 1968). It is only within such contexts that something like a mailbox can be said to *afford* the mailing of letters (Gibson 1979).

Of course, much more can be said about this, as well as all the rest. As I mentioned in the introduction, this paper is part of a larger overall project, one that attempts to substantiate the ontology of affordances in the course of extending their application to social practices. In line with my emphasis on temporality here, perhaps I can conclude by characterizing affordances in terms of dynamical agent-environment systems, with abilities and aspects of the environment understood as constraints on the potential trajectories of such systems (e.g., Riccio and Stoffregen 1988). These constraints or parameters are themselves modifiable over time, in terms of both the development of abilities and the shaping and informing of the material world. Moreover, they are mutually modifiable: with new abilities come new ways of changing the environment, which in turn give rise to further abilities, and so on (here there may be a connection to explore with the notion of the *virtual field* in enactivism; see, e.g., Di Paolo 2015). So, given the situation, the local conditions of the moment, and what an animal (or animals) can do, certain paths through state space are available, which open (or close) further paths in turn. Affordances, then, can be conceived as possible states of agent-environment systems, or as particular locations within the overall state space of these systems. Again, such an approach necessarily takes temporal extendedness, the reality of the temporal dimension, seriously.

References

Alter, T. 2001. Know-how, ability, and the ability hypothesis. *Theoria* 67 (3): 229–239.

Barker, R. G. 1968. *Ecological Psychology: Concepts and Methods for Studying the Environment of Human Behavior*. Stanford University Press.

Chemero, A. 2003. An outline of a theory of affordances. *Ecological Psychology* 15 (2): 181–195.

Costall, A., and A. Richards. 2013. Canonical affordances: The psychology of everyday things. In *The Oxford Handbook of the Archaeology of the Contemporary World*, ed. P. Graves-Brown, R. Harrison, and A. Piccini, 82–93. Oxford University Press.

Davidson, R. J., and B. S. McEwen. 2012. Social influences on neuroplasticity: Stress and interventions to promote well-being. *Nature Neuroscience* 15 (5): 689–695.

de Haan, S., E. Rietveld, M. Stokhof, and D. Denys. 2013. The phenomenology of deep brain stimulation-induced changes in OCD: An enactive affordance-based model. *Frontiers in Human Neuroscience* 7:653.

Di Paolo, E. 2015. Interactive time-travel: On the intersubjective retro-modulation of intentions. *Journal of Consciousness Studies* 22 (1–2): 49–74.

Ferri, F., C. G. Campione, R. Dalla Volta, C. Gianelli, and M. Gentilucci. 2011. Social requests and social affordances: How they affect the kinematics of motor sequences during interactions between conspecifics. *PLoS One* 6 (1): e15855.

Gibson, J. J. 1977. The theory of affordances. In *Perceiving, Acting, and Knowing: Towards an Ecological Psychology*, ed. R. Shaw and J. D. Bransford, 127–143. John Wiley & Sons.

Gibson, J. J. 1979. *The Ecological Approach to Visual Perception*. Erlbaum.

Greeno, J. G. 1994. Gibson's affordances. *Psychological Review* 101 (2): 336–342.

Heft, H. 2007. The social constitution of perceiver-environment reciprocity. *Ecological Psychology* 19 (2): 85–105.

Honoré, A. M. 1964. Can and can't. *Mind* 73 (292): 463–479.

Millikan, R. 2006. Styles of rationality. In *Rational Animals?* ed. S. Hurley and M. Nudds, 117–126. Oxford University Press.

Nicholls, J. G. 1984. Achievement motivation: Conceptions of ability, subjective experience, task choice, and performance. *Psychological Review* 91 (3): 328.

Noë, A. 2004. *Action in Perception*. MIT Press.

Noë, A. 2005. Against intellectualism. *Analysis* 65 (288): 278–290.

Pattee, H. H., and J. Rączaszek-Leonardi. 2012. *Laws, Language and Life: Howard Pattee's Classic Papers on the Physics of Symbols with Contemporary Commentary*. Springer Science & Business Media.

Pettit, P., and D. Schweikard. 2006. Joint actions and group agents. *Philosophy of the Social Sciences* 36 (1): 18–39.

Quine, W. O. 1948. On what there is. *Review of Metaphysics* 2 (1): 21–38.

Riccio, G. E., and T. A. Stoffregen. 1988. Affordances as constraints on the control of stance. *Human Movement Science* 7 (2–4): 265–300.

Rietveld, E. 2008. Situated normativity: The normative aspect of embodied cognition in unreflective action. *Mind* 117 (468): 973–1001.

Rietveld, E. 2012. Bodily intentionality and social affordances in context. In *Consciousness in Interaction: The Role of the Natural and Social Context in Shaping Consciousness*, ed. F. Paglieri, 207–226. John Benjamins.

Rietveld, E., and J. Kiverstein. 2014. A rich landscape of affordances. *Ecological Psychology* 26 (4): 325–352.

Sanders, J. T. 1997. An ontology of affordances. *Ecological Psychology* 9 (1): 97–112.

Searle, J. R. 1980. The intentionality of intention and action. *Cognitive Science* 4 (1): 47–70.

Sebanz, N., H. Bekkering, and G. Knoblich. 2006. Joint action: Bodies and minds moving together. *Trends in Cognitive Sciences* 10 (2): 70–76.

Spivey, M. J., and R. Dale. 2006. Continuous dynamics in real-time cognition. *Current Directions in Psychological Science* 15 (5): 207–211.

Stoffregen, A. 2003. Affordances as properties of the animal-environment system. *Ecological Psychology* 15 (2): 115–134.

Surtees, A. D., and I. A. Apperly. 2012. Egocentrism and automatic perspective taking in children and adults. *Child Development* 83 (2): 452–460.

Turvey, M. T. 1992. Affordances and prospective control: An outline of the ontology. *Ecological Psychology* 4 (3): 173–187.

Zukow-Goldring, P. 2012. Assisted imitation: First steps in the seed model of language development. *Language Sciences* 34 (5): 569–582.

14 The Role of Affordances in Pretend Play

Zuzanna Rucińska

Pretending is often conceptualized as an imaginative or symbolic capacity, positing mental representations in its explanation. Traditional explanations hold that pretending is achieved by adding new *meaning* to the object pretended with. There is no denying that in object-substitution pretense (such as the banana-phone game), the agent uses the object differently from what the object usually designates. For example, when a banana is played as if it were a phone, in the present context it means "phone." As Vygotsky ([1934] 1987) notices, in play, children step away from what objects usually mean and make them into something else. In the intellectualist framework, meanings are understood as given by mental contents, and they are imposed on (rather than found in) the reality. The intellectualist assumption is that without representing the meaning of what is to be acted out, one could not get engaged in pretense in the first place. Such a change of meaning is said to be done by manipulation of mental contents, whether these contents are belief-like (Leslie 1987; Nichols and Stich 2000, 2003) or imagining-like (Currie 2004, 2006; Van Leeuwen 2011).[1] That pretense requires mental representations in its explanation (such as mental plans or models) is thought to be justified by considering that pretense is itself representational, as McCune and Agayoff

---

1. For example, Leslie (1987) proposes a decoupling mechanism that decouples primary representations of objects ("this is a banana") into copies ("this is a banana*") and then manipulates them into pretense representations ("this banana is a phone*"). Nichols and Stich (2000, 2003) propose that the content of the initial premise specifies the impending pretense play episode by entering a task-specific cognitive mechanism called the Possible Worlds Box. The PWB contains tokens of primary representations, whose function is "not to represent the world as it is or as we'd like it to be, but rather to represent what the world would be like given some set of assumptions that we may neither believe to be true nor want to be true" (2000, 122). On the imagistic spectrum, Van Leeuwen proposes that "(nonveridical) mental imagery can be *integrated* into the perceptual field and that *this* form of imagining delivers the objects we relate to in constructing pretence action, such as make-believe" (2011, 56); "the imaginings that are most immediate to the production of pretence are spatially rich; they are perceptually formatted or structured as *representations of bodily movement*" (67; italics added).

exemplify: "The capacity to utilize such internal models of previous experience is considered to be the foundation of the capacity to engage in mental representation, and hence pretending" (2002, 44). If we were to borrow the jargon of Searle (1983), the *direction of fit* is supposed to be meaning (how you think about or imagine the banana) to environment (how you act on the banana), adding the new meaning "phone" to the banana.

Embodied and enactive cognition theorists avoid recourse to mental representations to explain cognitive phenomena, looking for mentality in the interactions and not in encapsulated mental representations (see, e.g., Varela, Thompson, and Rosch 1991; Hutto and Myin 2013). Embodied and enactive cognition theorists also seek alternative explanatory tools for cognition; inspired by ecological psychology, they look at affordances to play such roles (e.g., Chemero 2003, 2009; Rietveld and Kiverstein 2014). Any enactive account that proposes affordances as explanatory tools would propose, contra Searle, to get rid of the jargon of "direction of fit" altogether, emphasizing the mutuality of the environment and the animal in creating meaningful interactions. In this chapter, I propose an account of pretense in line with this mutuality.

Moreover, Gibson, who coined the term "affordances" in modern Western ecological psychology literature, claimed that the environment is already *meaningful* in the sense of providing opportunities for particular kinds of behavior: "The meaning or value of a thing consists of what it affords" (Gibson 1982, 407). For example, to say that the tree is meaningful to an animal that seeks shelter is just to say that the tree affords using it as a shelter to that animal. This equates with saying that whatever affordances afford can already be seen as meaningful.[2] In seeing the banana, we do not impose the meaning "banana" on it but directly see it as affording: eating, grabbing, playing phone with. As such, the banana "means" all these things: it is a food, an object, a toy, with respect to the possibilities of the animal. Thus Gibson's notion of meaning should not be conflated with a conception of meaning as being anything like mental content. There is no need to add representational content to specify what the object "means" in use. Therefore affordances could be considered as alternative explanatory tools to mental representations in making sense of cognitive capacities.

Whether affordances can be applied to explaining basic forms of pretense is the topic of this chapter. I propose an alternative way to explain pretending with the use of affordances, instead of mental representations, as explanatory tools. The alternative account is based on enactivism; it proposes to explain pretending through dynamic interactions of environmental affordances and animal effectivities in context. It shows that a specific notion of affordance has to be appropriated for affordances to play the relevant explanatory roles in pretense. This analysis opens up a discussion of the nature of affordances, clarifying how the

---

2. As Noë clarifies: "Gibson took this feature of his theory to be quite radical, for it suggested that we directly perceive meaning and value in the world; we do not impose meaning and value on the world" (2004, 105).

environment and the animal play a role in shaping affordances (sec. 1). It then clarifies which conception of affordances is most compatible to explain pretending; I suggest that a particular conception of affordances as dispositional properties of the environment, which also introduces animal effectivities that complement the affordances (à la Turvey 1992), can make affordances explanatorily useful (sec. 2). I then show how the environmental affordances with animal effectivities, placed in the right context formed by social factors, could form an explanation of basic kinds of pretend play (sec. 3). I propose that the explanatory posits of the alternative account (affordances, effectivities, social factors) mutually interact to provide a coherent explanation of pretense, showing that the roles cultural and social factors play in actualizing pretense in a specific context are nontrivial.

1   Affordances According to Gibson and His Followers

The purpose of this section is to clarify how to think of affordances with respect to explaining pretense. There is an ongoing debate in the affordance literature about how to conceive of affordances.[3] The only consensus with respect to affordances is that they are possibilities for action. What they are (properties or relations), where they are located (in the environment or cutting across the environment-animal dichotomy), and how they work (whether they invite actions or not) are a matter of great debate.

How should we think of affordances so that they are the best explanatory posits of pretense? What is the most fruitful way of thinking about affordances for the purposes of explaining pretense? This section will clarify some possible ways of thinking about affordances and, on that basis, suggest the most fruitful way to think about affordances to provide a coherent affordance-based explanation of basic pretense.[4] Showcasing the many conceptions of affordances serves to show which one in the end might do the best job to explain basic pretense. I argue that conceiving of affordances as dispositional properties of the environment (inspired by Turvey 1992) might best serve the explanatory function of an affordance-based account of pretense, as it divides the explanatory burden equally between the animal and the environment.[5]

3. The term "affordance" has been used in multiple ways in the literature on ecological psychology (including Gibson [1979] 1986, 1982; Turvey 1992; Reed 1996; Chemero 2003, 2009; Noë 2004; Withagen et al. 2012; Rietveld and Kiverstein 2014); the concept of an affordance features in Gestalt psychology and phenomenology, as well.

4. This chapter does not aim to settle the issue of what affordances really are. Rather, it sets the stage for an analysis of how different conceptions of affordances impact the ways they can explain pretense.

5. It should be clarified that my account is only inspired by Turvey's and may not be attributed to Turvey as such. Turvey conceives of affordances merely in biological terms; for example, the nutrients of the environment can be said to have dispositional properties of being "edible." However, it is not likely that Turvey would apply the notion of affordance beyond basic biological functions, as I do here (e.g., to phoneness of a banana).

The coining of the notion of affordance has been attributed to the ecological psychologist James Gibson. Gibson emphasized the relational nature of affordances, as existing in the environment for the animals. As he famously writes: "The affordances of the environment are what it offers the animal, what it provides or furnishes, either for good or ill" ([1979] 1986, 127). On the one hand, affordances are to be placed in the environment. They might be thought of as properties of the environment. As Gibson writes:

The affordances of the environment are permanent, although they do refer to animals and are species-specific. ... The perception of what something affords should not be confused with the "coloring" of experience by needs and motives. Tastes and preferences fluctuate. Something that looks good today may look bad tomorrow but what it actually offers the observer will be the same. (1982, 410)

Hence, on the one hand, affordances could be thought of as mind-independent ecological phenomena; they do not change as the need of the observer changes.[6] On the other hand, affordances could be seen as resources that the environment offers only to animals that have the capabilities to perceive and use those resources; in that sense, it is mind dependent (see Chemero 2003, 182). For example, in Gibson's words:

An affordance is neither an objective property nor a subjective property; or it is both if you like. ... It is equally a fact of the environment and a fact of behaviour. It is both physical and psychical, yet neither. An affordance points both ways, to the environment and to the observer. ([1979] 1986, 129)

According to this description, affordances seem to go beyond mere animal-environment coupling; they occupy an ontological space of their own; perhaps they can be considered as a relation. Hence, if it is a "fact of behavior" of the animal, it is difficult to understand how they are "mind-independent," as the first quote suggests.

The supposed inconsistency has sparked a great discussion, and many interpretations of the nature of Gibsonian affordances, and how best to think of them, have been proposed since (including the interpretations that I discuss here, e.g., by Turvey [1992], Reed [1996], Chemero [2003, 2009], or Rietveld and Kiverstein [2014]). What is most unclear is the relationship between the animal and the environment in shaping affordances. Are affordances (possibilities for action) to be found in the environment independently of the animals, or are they already shaped by the animals; finally, are they dependent on both the environment and animals equally?

One could hold that affordances are completely independent of the animals and totally dependent on the world. Reed (1996) is considered to be the most dedicated realist about affordances. He claims that affordances are properties of the environment that are resources of that environment. As resources, affordances play the evolutionary role of selecting the

---

6. This is the interpretation of Withagen et al. (2012). They claim that affordances "are not properties of the phenomenological world that depend upon the state of the observer; rather, they are ecological phenomena that exist in the environment. ... Hence, according to Gibson, affordances are opportunities for action that exist in the environment and do not depend on the animal's mind" (251).

animals and their skills. It could be said that they exist in the world before animals. For example, a banana tree's affording nutrition or shelter selects for animals that can act on such affordances, such as monkeys.

One could hold the opposing view that affordances are totally dependent on the animals. Arguably, Varela, Thompson, and Rosch (1991) come close to considering affordances in such a way. They claim that "certain properties are found in the environment that are not found in the physical world per se. The most significant properties consist in what the environment affords to the animal, which Gibson calls affordances" (1991, 203). However, they disagree with Gibson that affordances do not depend in any way on the perceptually guided activity of the animal (203). If we follow Varela et al., affordances should be understood as dependent on how the animal perceives the world.[7] Mutual interaction, or necessary coupling, between the environment and the animal is crucial.

Yet although they speak of mutuality, Varela et al. seem to focus more on the capacities of the animal than on the properties of the environment in securing affordances. Several things they say come close to suggesting this; for example, "The meaning of this or that interaction for a living system is not prescribed from outside but is the result of the organization and history of the system itself" (1991, 157).[8] Hence Varela et al.'s enactivism promotes a view of affordances that exist or come into existence only when a self-sustaining organism is present. Perhaps in Varela et al.'s view we can only speak of the world affording action when it shows up for creatures. For example, the affordance of the tree to be sheltered under or climbed may only show up to subjects who can do these things (e.g., monkeys); the animal is the one who turns what the world offers into meaningful invitations for action. Thus, while affordances may still be considered dispositional in Varela et al.'s account in the sense that affordances are not just dependent on the animal, what can be said certainly is that in their account, the organism/animal would play a much more important role than the environment in securing meaningful interactions. While Varela et al. stress the partnership between the animal and the environment, the partnership seems not to be equal.

7. This is in line with their notion of enactivism, according to which there are no properties of the world independent of the interaction or mutual enforcement. "[Enactivism] questions the centrality of the notion that cognition is fundamentally representation. Behind this notion stand three fundamental assumptions. The first is that we inhabit a world with particular properties, such as length, color, movement, sound, etc. The second is that we pick up or recover these properties by internally representing them. The third is that there is a separate subjective 'we' who does these things" (Varela, Thompson, and Rosch 1991, 9).

8. This reflects Varela's earlier stand on the origin of meaning of cellular organisms. Consider how Varela, first and foremost a biologist, discusses this origin: "Meaning can only arise for those systems, which assert their own identity vis-à-vis their environment, that is, for systems with a degree of autonomy. ... In general, however, a living system brings forth its own world of relevance, and is not given in advance. The meaning of this or that interaction is not given by an outside designer, but is the result of the organisation of the system itself and its history" (1988, 152).

Then one could hold the view that affordances are equally dependent on the animals and their environments. Such a view would claim that affordances are to be found in the world as real possibilities for action, but which action is afforded in particular circumstance depends on how an animal interacts with it. Turvey could be considered to hold such a view about affordances. He thinks that "an affordance is an invariant combination of properties of substance and surface taken with reference to an animal" (1992, 174). What Turvey suggests is that we consider affordances as dispositional properties of the environment. As such, they dispose the environment to be in a certain way, relative to an animal. He considers a disposition to be a property of a thing that is latent or possible. The disposition to act is prior to the action, and it is actualized when "conjoined in suitable circumstances" (178). Hence the dispositions of the animals pair up with, and are actualized by, dispositions of the environment, just as the dispositions of the environment need the dispositions of the animal to be actualized.

To match the affordances in the environment, we need the counterpart dispositions of the animal, or *effectivities*. They complement the affordances.[9] Effectivities can be biological dispositions, like digestion; for example, a banana can afford food only to animals that have the disposition to digest it. However, Turvey's notion of effectivities can be extended to involve various capacities and capabilities of animals shaped by histories of interactions. Introducing effectivities as animal-relative counterparts of affordances shows the dynamic relationship between the environment and the animal. It is not a view about the emergence of affordances in the interaction; the animal does not have to be actually interacting with the object for it to have an affordance. Rather, the idea is that the possibility for action resides in the banana only insofar as it is matched with an animal who has latent dispositions (before the interaction) to manipulate the object in relevant ways.

Thus, for each affordance in the environment, there must be a complementary effectivity in the animal. The animal does not have to act on the affordance, but the affordance is there in virtue of there being an animal that can interact with it. Affordances only get to work when animal effectivities are involved. That is because effectivities allow affordances to become "manifest." As Turvey claims:

7.1.   The circumstances actualizing a disposition or causal propensity of a thing Z involve some thing X, other than Z, forming part of Z's environment.

7.2.   This X, the complement of Z, must have a disposition matching (in the mathematical sense of "dual to") Z, for Z's disposition to actualize (i.e., if Z is refractible—has a disposition to become refracted—then X must have a disposition to refract). (1992, 178–179)

Hence objects in the environment (Z) and animals (X) have certain dispositions. Affordances are the dispositions of the environment; effectivities are the dispositions of the animals.

9. "Whereas an affordance is a disposition of a particular surface layout, an effectivity is the complementing disposition of a particular animal. An effectivity, as the term suggests, is the causal propensity for an animal to effect or bring about a particular action, to manifest what is needed for (an action) to be realized" (Turvey 1992, 179).

Affordances need to be complemented, or paired with actualizing circumstances, where the dispositions can become manifest. To give an example, an object (a banana) has an affordance of "edibility" only if there is an animal that has the effectivity of "digesting."

Similarly, to stress the mutuality of animals and environments, one could view affordances not as properties of environments (dependent or independent of the animals) but as relations (à la Chemero 2003, 2009).[10] What this means is that affordances are features of whole situations that involve animals and environments; it is not the object that affords but the situation.[11] It is the environment (or situation as a whole), not the individual object in the environment, that affords certain behaviors. The environmental relata of affordances (what we called "affordances") are features or aspects of the environment, whereas organismal relata of affordances (what we called "effectivities") are animal capacities or body scales. With regard to whether affordances exist without animals, Chemero (2003) clarifies that affordances have the quality of *being lovely* (pace Dennett 1998).[12] This means that affordances depend on there being a potential observer, not an actual act of observation. This feature can exist even if there are no animals (as proposed by Turvey). For example, a banana is edible insofar as there exists an animal that can eat it. The edibility affordance does not emerge with the presence of the monkey; on the contrary, it is there all the time, waiting to be actualized by the monkey (an animal with the relevant dispositions, or effectivities).

While Chemero speaks of individual affordances as relations between "features" of the environment and the abilities of an organism, Rietveld and Kiverstein (2014) take the relationship further, from a potential animal into a potential *form of life*.[13] In their view,

10. As Chemero (2003) proposes: "I argue that affordances are not properties of the environment; indeed, they are not even properties. Affordances, I argue, are relations between particular aspects of animals and particular aspects of situations. ... Affordances, which are the glue that holds the animal and environment together, exist only in virtue of selection pressure exerted on animals by the normal physical environment. They arise along with the abilities of animals to perceive and take advantage of them" (184, 190).

11. To operationalize the view that affordances are relations between particular aspects of animals and particular aspects of situations, Chemero explains that for a relation "affords Q" to hold between a subject and the environment is to say that the environment affords Q to the subject, where Q is a behavior (2003, 186–187).

12. "Affordances do not disappear when there is no local animal to perceive and take advantage of them. They are perfectly real entities that can be objectively studied and are in no way figments of the imagination of the animal that perceives them. So, ecological psychology is not a form of idealism. However, affordances do depend on the existence of some animal that could perceive them, if the right conditions were met" (Chemero 2003, 193).

13. "Affordances are possibilities for action the environment offers to a form of life, and an ecological niche is a network of interrelated affordances available in a particular form of life on the basis of the abilities manifested in its practices; its stable ways of doing things. An individual affordance is an aspect of such niche" (Rietveld and Kiverstein 2014, 334).

affordances are relational to animal species. Rietveld and Kiverstein claim that one should focus on what the relations are between the kind of animal (a "form of life") and its ways of living (i.e., its practices) in the environment the animals find themselves in. Moreover, affordances, in their view, show up in two forms: forming the *landscape* of affordances and the *field* of affordances. The *landscape* encompasses all the possibilities for animal's action, even if they are not perceptible or if they cannot be acted on by the animal at a moment. The *field* of affordances is a subpart of the landscape that is available to a particular animal at a particular time. Even though they need not be acted on, the affordances in the *field* of affordances entail the relevant affordances to a particular animal.

Rietveld and Kiverstein call the field affordances "solicitations" or "invitations," as the affordances in the field can *solicit* or *invite* particular behavior to an animal.[14] Solicitations are the affordances that are relevant to the animal's concerns (Rietveld and Kiverstein 2014, 342). For example, a banana affords eating to all monkey-forms-of-life (forming the landscape of affordances) but invites eating to a particular monkey only when it is hungry (forming a field of affordances). Thus, as affordances are considered to be mere possibilities for action, solicitations have been introduced into our theoretical vocabulary as those affordances that invite action.

To summarize this section, there are different conceptions of affordances, where the animal plays either no role in what is afforded, or the crucial role in bringing about affordances, or an *even* role, as the affordances emerge in dynamical interaction between the animal and the environment. Importantly, all the proposed affordance-based stories actively deny mental representations. Accepting any of them secures the possibility that a nonmental representational explanation of pretense that uses affordances could be provided. Even the accounts that emphasize the role of the animal need not posit mental representations; what plays a relevant part on the "animal" side of the relationship is either the animal dispositions, concerns, capacities, or practices found in the form of animal life.[15] The question remains about

14. Withagen et al. (2012) also endorse the notion of affordances as opportunities for action, but they elaborate their story by giving affordances the quality of inviting (or soliciting) behavior. They say: "[We] think of affordances as action-relevant properties of the environment that are defined with respect to the animals' action capabilities but exist independently of their needs and intentions. However, by suggesting that affordances can also invite behavior, we move beyond Gibson's original conception of affordances as mere action possibilities. ... [We] conceive of affordances ... as action possibilities that can invite" (255).

15. One may worry that if the animal is doing more work, then affordances are tacitly mental representations, and there is room for mental representational structures to be doing the relevant work. This worry, however, is not entirely justified, as we can conceptualize the mind in action in terms different from mental representations. By contrast, if one were to focus only on the environment because one fears mental representations, one would be motivated by an extremely problematic view of the mind. Thanks to Martin Weichold for this point.

how best to characterize affordances with respect to the animal and the environment so as to explain pretense. Some accounts work better than others with respect to providing such explanation of basic pretense. In the next section, I consider the following questions: What follows about pretense from these conceptions of affordances? How well are they suited to explain basic forms of pretense, like the eighteen-month-old's playing "phone" with a banana—if they can explain it at all? Answering these questions allows us to choose which conception of affordance may best explain pretense.

2 Which Conception of Affordance Best Serves to Explain Pretense?

The issue is whether we can extend the resources of the affordances to deal with pretense. To do that, we must understand what plays the relevant role for allowing pretend interactions. The best notion of affordance for explaining pretense should focus on how the environment, as well as the animal, partakes in explaining a pretend activity, without recourse to mental representations. In this section, I explain why the dynamical conception of affordance that assigns equal importance to the role of the environment and the animal is most fruitful for explaining pretense.

If the environment is all that is important, such a position should have to commit to the view that the "phoneness" affordance of the banana exists in the banana as its property, completely independent of there being any subjects. For example (as long as affordances can be used to explain pretense), even the "phoneness" affordance would have to have influenced the evolution of the right types of animals. However, for "phoneness" to exist first, independently of the animal (and its sociocultural practices of using phones), is just not likely. Hence if affordances are resources of the environment that select for the animals, this notion of affordance would not make affordances good explanatory tools of pretense.[16]

If the animal is most important, for things to afford "phone-play," the animal must possess a relevant capacity; the capacity in question would be "graspability," "pick-up-ability," and other capacities that amount to the capacity "playing phone with." It would have to be said that the subject *could* use this capacity in different contexts, in the presence of different objects. Hence the object would not be as important as the capacity of the subject, or what the subject can do with the object. With this notion of affordance, perhaps the "phoneness" affordance could be considered as most independent of the actual object at hand (the banana); another object would seem to do just as well. However, one might worry that bringing the "playing phone with" capacity to any object can be mistaken, or wrongly applied, as

---

16. Reed could say that the conception of affordance does not extend to the sociocultural realm but belongs solely to the biological realm. Yet a priori exclusion of affordances (possibilities for action) as social possibilities, applicable to cultural contexts, is also unjustified.

there would not be an object with its own properties that would regulate what is afforded.[17] For example, if only the animal concerns are important, a hungry monkey should see all sorts of objects as "food." Similarly, pretenders wanting to play "phone" would use any means achievable to play "phone," even if bananas are not available. In this view, a banana would afford "phone" just as much as, presumably, a table. Again, it is not likely that this conception of affordance would best explain our experience with how children pretend, which is by often using bananas or phone-shaped objects, and not other objects, to play "phone" with (Piaget 1962; Leslie 1987; Nichols and Stich 2000, 2003).

Another position is that the animal capacities could be found in a whole set of animals or forms of life. Hence the "phoneness" of a banana would be found in the landscape of affordances. This means that we do not look at dispositions of a particular animal but look at practices of types of animals. In the case of pretense, there would be a practice of playing phone that is part of our human form of life that shapes the landscape of affordances, where cultural niches that practice phone play exist. In this story, the particular animal is not as important as the animal's sociocultural context that forms the landscape of affordances. The way of using bananas as phones is independent of the individual, but not independent of the society the individual is part of. However, what may be problematic in this account is that, to explain pretense, one would have to commit to the idea that as long as there are people in the form of life who happen to play with the banana as a phone, the banana can afford "phone" to any individual sharing the same form of life, even if that individual does not have a personal history of interactions with bananas as phones. For example, just as someone who has never engaged in parkour would have to see the wall as climbable (because such a practice exists in one's culture or form of life), so a child would have to see the banana as affording "phone" even if the child was never engaged in a context involving phones. This is also unlikely. At best, this account could explain how the child can see the banana as affording "phone" to other people in their form of life. Yet if the question is how to explain an individual child's pretense, this story is not the most promising.

The position of equal mutuality would suggest that the banana affords "phoneness" only if there is a subject with the effectivity to play "phone" with a banana (dispositions that could be shaped by, e.g., the history of interactions with phones) and if there are objects with right dispositions to be like phones (bananas, due to their shape and size). This story depicts a dynamic relationship between the objects in the environment and animals, where the object is just as important as the animal that it interacts with. Only in suitable circumstances can the affordance-effectivity pair be actualized. This is a good candidate for the best conception of affordance to explain pretense.

What kinds of things are effectivities? The term "effectivities" used in this chapter is broad, in the sense that apart from "dispositions" narrowly construed (such as mere bodily

17. In cognitivist literature, this could be considered as a problem of misrepresentation.

dispositions), they also encompass all sorts of moods, capacities, biological setups, and cultural and individual history (including training) of the animal that shape the animal's dispositions. I discuss these factors briefly later in the chapter.

First, evolutionary or biological setups determine what the animal is naturally disposed to do. Consider the case of a fly. It affords "mate" to another fly of the opposite sex, but it affords "food" for a frog. The frog is likely to be fixed in its repertoires, or at least not as flexible as other animals; frogs are likely to be responsive to the "food" affordance of a fly in ways selected for by evolution. A subset of dispositions will be naturally selected for, fixing what the frog should do. Natural selection can put processes into play that make creatures initially responsive only to certain affordances.[18]

Moreover, through *training*, an animal can develop new dispositions and so expand its initial habitual repertoire. Training is just a special case of a history of interactions that shapes current dispositions. Consider the following case of monkeys using "reverse pliers" (Umiltà et al. 2008). Experimenters succeeded in teaching the monkeys to use a counterintuitive pliers (one where the monkey needs to grip the tool to open the pliers and let go of the grip to close the pliers) to pick food. After six to eight months of training, the monkeys were able to do it. How did the monkeys learn to use the new pliers? They responded to the only possibility the new pliers allowed relevant to picking food up, which was to work with them in a counterintuitive way. Umiltà et al. concluded that "the capacity to learn tool use appears, therefore, to be based on two elements: the goal-centered organization of primate motor cortex and an appropriate interaction with the external world" (2211–2212). The organization of the monkeys' motor cortex can be considered as one of their effectivities, the reverse pliers as the environmental affordances, and the two in pair allowed the monkeys to develop the capacity to use the reverse pliers. This explanation allows us to understand the monkeys' behavior without having to attribute to the monkeys the capacity to manipulate mental contents.[19]

18. With regard to evolutionary setup, Withagen et al. (2012) claim that certain affordances were selected for. Consider their example of the solicitation to "flight" when encountering a lion on a safari: "[Affordances] that are crucial for survival and reproduction (e.g., objects or animals that afford danger, shelter, or nutrition) are likely to attract or repel the agent. For example, a human that encounters a lion on a safari is likely to flee into his car as soon as possible. Such an affordance will be acted upon immediately, irrespective of the intentions of the actor at that moment in time (e.g., to take a picture, to pee)" (256).

19. This example does not yet conclusively show that mental representations are not in play. One could propose that the monkey forms a mental representation of the functional structure of the tool to use it "in the correct way"; this explanation is also more likely than an explanation asking for the learning procedure to be accomplished at the subpersonal level of explanation (organization of the motor cortex) in only six to eight months. However, positing mental representations is not necessary, either. The reference to training focuses on training an embodied skill or new capacity through exploration

What is really different between human children and monkeys is the speed with which the child can shift between affordances; that can be due to humans' inherent *flexibility*. The human child maintains the capacity to learn quickly and switch quickly between contexts thanks to a more flexible makeup.[20] However, often, mere flexibility is not enough to account for more expert pretense; some training is required, as well. That pretend play also needs to be practiced is exemplified by professional stand-up comedians, who can act with the same object as if it is routinely something else.[21] Their innovative pretending is an on-the-spot, embodied activity of switching between many affordances. The object solicits to the actors many strange behaviors, and bringing them forth needs to be practiced. Through *repeated exposures* to and *manipulations* of these objects, such creative responsiveness to an object's affordances can be developed by the actors.

With regard to the *cultural history* of the animal in explaining how affordances invite actions, engaging in culturally shaped routines influences how one responds to the environment. For example, a chair is often perceived as something to sit on, though it affords many other types of behavior too. In that sense, cultural history is a factor in delimiting and shaping which affordances we respond to. Culture is part of the animal's history of interactions; what participation in patterned practices shapes are one's abilities and concerns (what one cares about). What follows from the history of cultural interactions is that the objects can present themselves as explicitly and exclusively good for a specific, culturally established purpose, at the same time appearing to be unsuitable for other purposes.[22]

Finally, *personal history* affects what one is attracted to; clearly, an object may solicit different actions to individuals who are of the same culture, or may solicit different actions to the same individual at different times. Consider Withagen et al.'s examples:

Indeed, members of the same culture are often attracted to different objects or are invited by the same object to do different things. As an example, although chocolate may afford eating for the vast majority of people, there is substantial variation in whether and how people are attracted to it. Some people are

---

and trial and error. It should be understood as an acquisition of a new bodily skill, not just a novel organization of the motor cortex. While making a functional representation cannot be ruled out, it seems unlikely that it would take as long as six to eight months to represent the new goal of the object functionally.

20. Perhaps the plasticity of the brain can account for this. In addition, the speed with which new behaviors can be learned can be accounted for by, for example, action readiness potentials in the neural networks; see Bruineberg and Rietveld 2014.

21. See, for example, the comedians of the show *Whose Line Is It Anyway?*

22. Consider a mailbox; it canonically affords posting letters. That is its purpose, and that is what we have learned to use it for. Although, technically, as a box with a slit it affords putting trash in it, not many would act on that affordance, as we are shaped by society to believe that putting anything other than letters in a mailbox is inappropriate.

almost addicted and cannot wait to eat it; others might not like it and prefer to eat something else. It is important to note that the invitation can also vary over time and might change on a moment-to-moment basis. For example, a person who initially liked chocolate but had suffered from gastroenteritis after eating it is likely to be repelled by its affordance for some time. (2012, 256)

To summarize, the animal dispositions gained from the history of interactions, either biological, cultural, or from training, explain why certain affordances are seen as inviting. Importantly to the account of nonrepresentational pretense, there is no reason to think that these factors or the effect of these factors needs to be *mentally represented*. But as Withagen et al. claim, that an affordance stands out by result of culture (like a chair's affordance to sit on it)

[does] not mean that the inviting character is a mental product. ... Instead, cultural variations are better thought of as variations in perceptual-motor skills ... giving rise to a particular responsiveness to certain affordances in the environment. ... Like cultural variation, individual differences in perception can also be explained in terms of variation in what information is exploited or in the bodily responsiveness to such information. (2012, 256)

To conclude this section, the affordance-based alternative explanation necessarily involves the concept of effectivity for action, which explains how certain possibilities for action can come about to invite or solicit specific animal behaviors. Affordances of the objects do not, on their own, invite action; animal effectivities have to be matched, as they do important work in explaining why certain behaviors can take place. The story is not complete without understanding what it is about the animal that needs to coexist with affordances to explain the animal's responsiveness to affordances in the environment. Importantly, the animal does no choosing of which affordances it will respond to. Such choosing is usually associated with deliberation and is explained by positing mental contents. There is simply a mutual attraction of affordances to effectivities. Paired together, the effectivities allow the affordances to be inviting. Thus the story can be cast in terms of animal-environment mutuality: just as the objects we shape have a pull on us with respect to how to engage with them (Malafouris 2008), so the affordances of the environment can invite specific behaviors when the right effectivities are in place. Yet this is not the whole story, because solicitations do not necessarily bring about action, and there are still many relevant solicitations that can encourage different behaviors in the same situations. So what actually drives the animal to respond to one solicitation rather than another? What makes specific behaviors (like pretend behaviors) inviting? I target these questions in the final section.

3   How Do Effectivities Invite Specific Pretense Actions?

From the last section, it should be clear that affordances have no soliciting power on their own; it is our histories of interacting that form our dispositions, or effectivities, that allow affordances to invite actions. Thus our dispositions make it such that in some or even many situations, certain affordances are dominant. However, affordance-effectivity pairs alone are not enough to explain the special kind of responsiveness needed to explain specific behaviors, such as pretend play. There are still many affordance-effectivity pairs that can invite various behaviors. Consider Turvey's claim: "X and Z have multiple dispositions—m and n, respectively. To actualize Wpq, the juxtaposition function j must be such as to filter p and q from the array of m x n dispositions possessed by X and Z" (1992, 179).[23]

Applied to a situation involving pretense, the agent (X) and the banana (Z) have multiple dispositions, such as "softness," "foodness," or "phoneness" of the banana (m collection of affordances) and "disposition to grab," "eat," or "capacity to play phone with" of the agent (n collection of effectivities). To actualize the pretense act, a juxtaposition must occur specifically of the "phoneness" of the banana (p affordance from the m collection) and the "capacity to play phone with" of the agent (q effectivity from the n collection). Only this juxtaposition can filter other affordances and effectivities out. However, saying that the "phoneness" affordance of bananas requires the "playing phone with" effectivity of the animal begs the question.[24]

So far it has not been explained how particular pretending occurs. After all, there are many affordances and just as many effectivities. So what determines that the right affordance-effectivity pair comes forth? What explains which affordance-effectivity pair is actualized that would lead to the pretend action? I discuss two answers to this question. The first is that some affordances naturally strongly invite; these are, for example, canonical affordances (Costall 2012). Thus there are special canonical affordance-effectivity pairs that can explain some pretense. The second is that it is the immediate context, shaped by other people (either participants or mere spectators), that specifies which affordance-effectivity pairs come up. While these are not the only possibilities for what can actualize pretense affordances, both canonical affordances and other people are clear examples of factors that do not require positing mental representations.

What makes specific behaviors inviting can be a set of special affordance-effectivity pairs. The idea is that perhaps there are more basic affordance-effectivity pairs, which strongly

23. While Turvey does not expand further on the idea of how the filtering is done, it could be thought of as a probability function that explains the outcome of the juxtaposition as the probability distribution of affordance-effectivity pairs. Thanks to Christian Tewes for this point.

24. It may look as if the matching effectivities have to be "phone-like as well." After all, how can a child have the requisite effectivities, without positing that the child has a "phone-like" history of engagement with the banana? The worry is that positing "phone-like" effectivities would be presupposing what we are trying to explain. To address this worry, a solution is that early forms of interaction with the banana do not have to be "phone-like" per se. It is enough that the child has a history of playing with the banana and not using it in a typical "banana-like way" (such as eating it). With other contextual factors in place (to be discussed in the next section), "phone-like" behavior with the banana can occur.

invite. For example, there might be affordances that stand out because of cultural engagements. These can be thought of as canonical affordances (Costall 2012). Such affordances are the first to strongly invite or solicit action. For example, an animal's feeding practices set up bananas to canonically afford "eating," but in winter, the animal's practices can set them up for "storing." Canonical affordances are shaped by wider sociocultural context, such as past practices.[25]

Canonical affordances can explain some pretense. One way to think about pretense being canonical is to think of it as stepping from one practice into another, such as from the "banana-eating" practice to the socially established "banana-phone game" practice. Being part of such a practice would require being enculturated in a "phone" practice, which would amount to having a certain *know-how* to use the banana as a phone, acquired through, for example, past interaction with and exploration of the banana, or imitating others who have used the banana as a phone. The "phone play" is afforded because the "phone" practice (stemming from available exposure to uses of phones and routine engagement with phones) is an established practice that forms the relevant play context.

However, the canonical "banana-phone" practice cannot be the whole story; after all, children play in creative ways, such as pretending that a banana is a hat or a gun, without entering into an established practice. It is unlikely that all these pretend-play scenarios with a banana amount to canonical practices. Thus, while it is likely that with respect to the banana-phone itself there is a canon of placing the banana to the ear (as this is the most-quoted example of pretense in philosophical and psychological literature), referring to the canonical affordance of "phone" for bananas in play context does not explain how children pretend something else with a banana just as easily. Moreover, acting on pretense affordances often requires stepping away from what one is often or normally solicited to do. Pretense acts are not likely to be canonical acts, because, as mentioned in the introduction, in pretense, one treats an object differently from what it usually (or canonically) affords (banana as phone).[26] The pretense situations will be less common or unorthodox. For example, a banana is more often encountered in a "breakfast" context than in a "play" context, and so it should first and foremost afford "food," not "phone." This requires explaining how the child

---

25. As Costall explains: "The concept of 'canonical affordances' itself alerts us to those important cases where the affordances of some thing are not simply shared between people but also normatively predefined. Yet the affordance of any artefact is not confined to that object in isolation, but depends on a 'constellation' … of not only other objects but also events. The affordances of artefacts are not usually self-contained but depend upon a wider context of other artefacts (as in the case of a toolkit) but also upon the encompassing practices in which they go together" (2012, 91).

26. That is, unless they are toys or objects that primarily afford the pretense play (e.g., toy phone for "phone" pretense, or bear costume for "bear" pretense). However, having specified toys is not a requirement for typical object-substitution play found in early pretense engagements of young children, and it does not explain the banana-phone game.

inhibits the primary or canonical solicitations of objects when the child does something unusual with the objects in pretense. As nonpretense behaviors are typically solicited, this may require a bypassing mechanism to be in place for the child to step away from the "eating" affordance of the banana.[27]

However, even if some form of inhibition of affordances were needed, it is not settled that the inhibiting must be achieved by a procedure that involves mental representations. The "stepping away" does not require intellectual choice in the way cognitivists presuppose, whereby one represents the banana as "banana" and then has to create a new representation of the banana as "phone" (Leslie 1987; Nichols and Stich 2000, 2003). It can be the context (including other people in it) that an individual has found himself or herself in, which shapes which affordances are inviting.[28] One can say that there is no decontextualized situation where a banana affords "first and foremost" eating; in the right play context, the banana does not solicit eating at all, but only phoning. It is the context that influences and sets the stage for our various ways of acting on things. To make a comparison, consider an example of being in a situation that involves chairs. On their own, chairs canonically afford sitting on; however, in the context of play, that may not be the action they invite; they will instead invite jumping on or piling up to make a fort. So in the right play context, the chair will not invite what it culturally specifies in the first place. The present context one finds oneself in with objects will dynamically bring forth new possibilities for action with those objects.

Other people can also influence which affordance-effectivity pairs gets acted on. They can also enable the "stepping away" from canonical practices, forming the immediate context where pretense affordances can be inviting. When directly engaged in the pretend play, others influence the play; in acting together, people thus create new forms of play. The mutuality effect of individual player and other players on shaping actions when they are attuned to one another is clear (see, e.g., Reddy and Morris 2004; De Jaegher and Di Paolo 2007). It could be said that other agents form a "self-world-other" structure, or an animal-environment-other dynamic (Froese and Di Paolo 2009), where in creating an intersubjective context, they directly affect the actions of the individual through immediate interaction. However, what I

27. In the standard story, the "stepping away" from original meanings is explained by a mental representational mechanism like decoupling (Leslie 1987; Nichols and Stich 2000, 2003). Similarly, one would have to explain how stepping away from canonical affordances occurs so that alternative affordances can be acted on. The worry for the present account is that only mental representational structures have been proposed so far to be able to allow the stepping away (I will call this the "stepping away" worry).

28. While in some pretense (especially more the complex pretense of older children) the manifestation of affordance-effectivity pairs may be triggered by the choice of the agent that is based on imaginative or symbolic capacities, my intent here is to show that the context could often be enough to play this determining role.

have not yet addressed is other people who are *not* directly engaging in interactions, but are simply in the background, as making an important contribution to the individual's actions on objects in their environments. Other people can form the actualizing circumstances that explain why some affordances, like the pretense ones, are found inviting. For example, the "phone" action is invited from the matching of the banana's affordances and the player's effectivities; the matching then needs to be done in the right context. Something in the immediate context, such as other people, completes an explanation of how pretense actions come about. For example, one is strongly disposed to play phone with the banana when the circumstances are inviting, and such a disposition is brought forth by the mere presence of other people who create playful contexts.

To clarify the role of others in shaping the animal-environment dynamic, consider an analogy to being vulnerable. McKitrick (2003) discusses how objects like sensors can make a city less vulnerable to an attack, or how the presence of bodyguards can make someone who walks in a park alone (let's call her Joan) feel less vulnerable after dark.[29] It is the dynamic between the city and its attackers (not between the city and the sensors) that is changed by the sensors; similarly, it is the dynamic between Joan and the park that is affected by the presence of the bodyguards, not the dynamic between Joan and the bodyguards. Joan is less vulnerable because an entourage of bodyguards surrounds her. She is not less vulnerable with respect to the bodyguards themselves (Joan might be scared of, or shy with respect to, the bodyguards) but less vulnerable with respect to her environment, the dark park (where someone could attack her). The mere presence of bodyguards "turns on" Joan's disposition to be less vulnerable in the park. The point is simply that other people (just as objects) can influence the dynamic between the environmental affordances and animal effectivities without being the focus of that dynamic themselves. Hence the mere presence of others can drive pretend phone play with a banana. Others influence the context in which children play, making the context playful. For example, the parents of a child interacting with a banana could make the child's acting on "phone" affordances more likely by smiling and approving of the various ways the child explores the banana. Thanks to this playful context, the pretense of children can be more creative, going beyond the canon.

29. "A military target, a city, is protected by a Star Wars–like defense system. The system has sensors that bring out defenses when there is a threat, rendering the city invulnerable. However, the sensors and anti-aircraft weapons are all located outside the borders of the city and are built, maintained, and staffed by a foreign country. Should the defense system be disabled, or should the foreign power withdraw its protection, the city would change from being invulnerable to being vulnerable. However, the city may remain intrinsically the same, or internally the same in all ways that are relevant to its vulnerability. … Changing a thing's environment can make it vulnerable. By adding the defense system, the city changes with respect to its vulnerability. Walking alone in Central Park at night, Joan would be vulnerable. Accompanied by an entourage of bodyguards, she would be less vulnerable" (McKitrick 2003, 161).

This proposal is in line with Turvey's position. He also stresses the need for the right actualizing context to the affordance-effectivity pairs:

Let *Wpq* (e.g., a person-climbing-stairs system) = *j(Xp, Zq)* be composed of different things *Z* (person) and *X* (stairs). Let *p* be a property of *X* and *q* be a property of *Z*. Then *p* is said to be an affordance of *X* and *q* the effectivity of *Z* (i.e., the complement of *p*), if and only if there is a third property *r* such that

(i)   *Wpq* = *j(Xp, Zq)* possesses *r*
(ii) *Wpq* = *j(Xp, Zq)* possesses neither *p* nor *q*
(iii) Neither *Z* nor *X* possesses *r*.

Thus, a person cannot execute locomotion in the highly particular manner of stair climbing unless a sloped surface is underfoot composed of adjacent steps with suitable dimensions (of rise and horizontal extent). When it is, then the disposition to locomote in this highly particular way is actualized. (Turvey 1992, 180)

To clarify, something is an affordance and an effectivity if and only if there is a third property given by the context. One such property can be a sloped surface. Applying Turvey's conception of affordances to the pretense context, we can explain banana-phone pretend play by using a similar structure. Let *Wpq* (a child pretend playing that a banana is a phone) be made of different things *Z* (the child) and *X* (the banana). Let *p* (phoneness) be a property of *X* (banana). Let *q* ("capacity to play phone with") be a property of *Z* (the child). Then *p* (phoneness) is said to be an affordance of *X* (banana) and *q* ("capacity to play phone") the effectivity of *Z* (the child). The effectivity is a complement of the affordance; hence we can conceive of the effectivity as the flexibility to play "phone," the history of interactions with the phone, know-how about using phones, and so on. Then what actualizes this affordance-effectivity pair is the right circumstance, such as being in a playful context, which is shaped by the presence of toys, other people playing, or narrative contexts. The third property can be another agent acting as an extrinsic disposition to action. In short, once the right context is in play, it solicits action.

To conclude this section, calling on special canonical affordance-effectivity pairs and other people in the immediate context of play is just one possible way to explain how affordances can be used to explain basic forms of pretense.[30] Yet factors like other people, and not just the individual's concerns, moods, intentions, or explicit choices, can also affect what affordances come forth. Intersubjective engagements play a big role, among other contextual factors, in shaping pretense. Once we recognize the need for affordance-effectivity mutuality

---

30. This is not the only way to step away from one practice to another; we respond to affordances all the time, and so we could switch about even without the presence of others, but due to some new occurrence in our contexts. Also, we may change our moods or attitudes about what practice we take ourselves to be in, and that change could bring about different affordances as inviting, as well. On how moods or "current concerns" could play a role in shaping the playful context, see Bruineberg and Rietveld 2014.

in dynamically bringing forth or inviting responses, we can accept that the relevant explanatory work is done by multiple agential, environmental, and social factors. What filters the affordance-effectivity pairs is the context. To explain why a specific affordance-effectivity pair is acted on, we can refer not only to the canonical practices with objects or right moods of the subjects but also to the intersubjective context, where other people can influence the way certain actions strongly invite, and potentially drive, behaviors. Elaborating on when exactly these factors do the relevant explanatory work, and to what extent they explain pretense alone or together, requires studying specific examples of acts of pretense of specific individuals.

Conclusion

This chapter has shown that we can provide an affordance-based story to explain basic cognitive engagements, such as object-substitution pretend play. The story involves environmental affordances, animal effectivities, and context (where we find other people) that make the explanation dynamic. Affordances on their own do not explain why certain actions are taken; we are not responsive to all the possibilities for action that the world affords. Only some affordances can invite behaviors to action. Explaining this requires further specification of what makes affordances inviting. Animal effectivities, dispositions and capacities shaped by the history of interactions, and context, which includes other people, bring about the pretense affordances for action. The animal both creates and responds to the invitations of affordances; in that sense, the explanation involving affordances is dynamic. A specific affordance-effectivity pair comes about either through sociocultural norms and practices that make certain acts of pretense possible via "canonical affordances," or by others who form the intersubjective contexts that shape and influence what and how we pretend.

Thus what can explain actions like pretending involves object affordances, subject effectivities, and the social context, which does not posit mental representations. Although the account of nonmental representational pretense is not complete (more work needs to be done to ensure that the enactive explanation is the best explanation of pretense and can scale up to other more complex forms of pretense), this chapter lends support to the idea that an explanation of pretense that does not posit mental representations can be achieved.

References

Bruineberg, J., and E. Rietveld. 2014. Self-organization, free energy minimization, and optimal grip on a field of affordances. *Frontiers in Human Neuroscience*. doi:10.3389/fnhum.2014.00599.

Chemero, A. 2003. An outline of a theory of affordances. *Ecological Psychology* 15 (2): 181–195.

Chemero, A. 2009. *Radical Embodied Cognitive Science*. MIT Press.

Costall, A. 2012. Canonical affordances in context. *Avant* 3 (2): 85–93.

Currie, G. 2004. *Arts and Minds*. Oxford University Press.

Currie, G. 2006. Rationality, decentring, and the evidence for pretense in non-human animals. In *Rational Animals?* ed. S. Hurley and M. Nudds. Oxford University Press.

De Jaegher, H., and E. Di Paolo. 2007. Participatory sense-making: An enactive approach to social cognition. *Phenomenology and the Cognitive Sciences* 6 (4): 485–507.

Dennett, D. 1998. *Brainchildren*. MIT Press.

Froese, T., and E. Di Paolo. 2009. Sociality and the life-mind continuity thesis. *Phenomenology and the Cognitive Sciences* 8 (4): 439–463.

Gibson, J. J. [1979] 1986. *The Ecological Approach to Visual Perception*. Houghton Mifflin.

Gibson, J. J. 1982. Notes on affordances. In *Reasons for Realism: The Selected Essays of James J. Gibson*, ed. E. Reed and R. Jones, 401–418. Erlbaum.

Hutto, D. D., and E. Myin. 2013. *Radicalizing Enactivism: Basic Minds without Content*. MIT Press.

Leslie, A. 1987. Pretense and representation: The origins of "theory of mind." *Psychological Review* 94:412–426.

Malafouris, L. 2008. At the potter's wheel: An argument for material agency. In *Material Agency: Towards a Non-anthropocentric Approach*, ed. C. Knappett and L. Malafouris, 19–36. Springer.

McCune, L., and J. Agayoff. 2002. Pretending as representation: A developmental and comparative view. In *Pretending and Imagination in Animals and Children*, ed. R. W. Mitchell, 43–55. Cambridge University Press.

McKitrick, J. 2003. A case for extrinsic dispositions. *Australasian Journal of Philosophy* 81 (2): 155–174.

Nichols, S., and S. Stich. 2000. A cognitive theory of pretense. *Cognition* 74:115–147.

Nichols, S., and S. Stich. 2003. *Mindreading: An Integrated Account of Pretense, Self-Awareness, and Understanding of Other Minds*. Oxford University Press.

Noë, A. 2004. *Action in Perception*. MIT Press.

Piaget, J. 1962. *Play, Dreams, and Imitation in Childhood*. Trans. C. Gattegno and F. M. Hodgson. Norton.

Reddy, V., and P. Morris. 2004. Participants don't need theories: Knowing minds in engagement. *Theory and Psychology* 14 (5): 647–665.

Reed, E. S. 1996. *Encountering the World*. Oxford University Press.

Rietveld, E., and J. Kiverstein. 2014. A rich landscape of affordances. *Ecological Psychology* 26 (4): 325–352.

Searle, J. 1983. *Intentionality: An Essay in the Philosophy of Mind*. Cambridge University Press.

Turvey, M. T. 1992. Affordances and prospective control: An outline of the ontology. *Ecological Psychology* 4 (3): 173–187.

Umiltà, M. A., L. Escola, I. Intskirveli, F. Grammont, M. Rochat, F. Caruana, A. Jezzini, V. Gallese, and G. Rizzolatti. 2008. When pliers become fingers in the monkey motor system. *Proceedings of the National Academy of Sciences of the United States of America* 105 (6): 2209–2213.

Van Leeuwen, N. 2011. Imagination is where the action is. *Journal of Philosophy* 108 (2): 55–77.

Varela, F. J. 1988. Structural coupling and the origin of meaning in a simple cellular automaton. In *The Semiotics of Cellular Communication in the Immune System, NATO ASI Series H23*, ed. E. E. Sercarz et al., 151–161. Springer.

Varela, F. J., E. Thompson, and E. Rosch. 1991. *The Embodied Mind: Cognitive Science and Human Experience*. MIT Press.

Vygotsky, L. S. [1934] 1987. Thinking and speech. Trans. N. Minick. In *The Collected Works of L. S. Vygotsky*, vol. 1: *Problems of General Psychology*, ed. R. W. Rieber and A. S. Carton, 39–285. Plenum Press.

Withagen, R., H. J. de Poel, D. Arauja, and G. Pepping. 2012. Affordances can invite behaviour: Reconsidering the relationship between affordances and agency. *New Ideas in Psychology* 30:250–258.

15   Ornamental Feathers without Mentalism: A Radical Enactive View on Neanderthal Body Adornment

Duilio Garofoli

Neanderthals were a species of the genus *Homo* closely related to *Homo sapiens*, living between circa 250 and 38 ka. They evolved along relatively separate evolutionary paths from modern humans for several hundred thousand years, Neanderthals inhabiting a cold area in Europe, though also reaching regions in the Near East, whereas modern humans lived in a warmer African environment (Harvati 2015). Neanderthals were thus European aborigines, likely evolved from a local population of the more archaic *Homo heidelbergensis* species (Stringer 2002).

The discovery of Neanderthals occurred within a scientific context deeply entrenched in ideas of modern human perfection and colonial dogmas according to which primitives, including nonindustrialized historical populations, were located at an inferior cognitive and cultural level (Bednarik 2013, chap. 3). However, Neanderthals challenged the position of modern humans as located on the highest rank of such a *scala naturae* of living beings. Indeed, despite a first attempt to depict them as feral humans (Zilhão 2012, 36), Neanderthals appeared too similar to modern humans to justify the idea of a modern superiority based on merely anatomical considerations. Most importantly, Neanderthals were provided with big brains, with a range of 1,250 to 1,700 cubic centimeters, which is larger in absolute size than the modern human brain (Schoenemann 2006). It is unclear whether this larger absolute size was also associated with an increased encephalization quotient, namely, the size of their brain scaled against their body size, due to the Neanderthals' higher muscularity (A. Gallagher 2014; Ruff, Trinkaus, and Holliday 1997). In any case, we have no reason to believe that their encephalization quotient was different from that of modern humans, who have both smaller brain and body sizes. Many scholars therefore shifted the focus from anatomy to cognition, attempting to show that the specialness of *Homo sapiens* lies in a set of behavioral features uniquely ascribed to the "moderns." In this way, scholars devoted great efforts to defining criteria for the identification of "behavioral modernity" within the archaeological record (Henshilwood and Marean 2003; McBrearty and Brooks 2000; Wadley 2001). Over time, the notion of behavioral modernity has gradually been approximated to cognitive modernity by implicitly assuming that "modern behavior" necessarily requires the existence of a "modern mind" shared by all human beings (e.g., Conard 2010; Henshilwood, d'Errico, and Watts 2009; Klein and Steele 2013).

According to Nowell (2010), the archaeological community today has reached an agreement about the nature of quintessentially modern behavior, which has been associated with the rise of symbolically mediated lives, as reflected by linguistic abilities and the symbolic storage of information within material culture typical of many ethnographic contexts (Henshilwood 2007; Henshilwood and Marean 2003; Noble and Davidson 1991; Wadley 2001). The presence of artifacts considered to be symbols was both consistent and qualitatively various in the European Upper Paleolithic. The most relevant examples are artifacts used as body ornaments (Vanhaeren and d'Errico 2006), ivory figurines (Conard 2009), depictions of therianthropy (Wynn, Coolidge, and Bright 2009), and parietal art (Fritz and Tosello 2007). This evidence initially led scholars to argue that the European Upper Paleolithic represented a revolutionary event in the technological, social, and cultural complexity of modern humans (e.g., Mellars and Stringer 1989). The human revolution model is synergic with the idea that some mutational event occurred in the brain of some Middle Stone Age African populations that were facing a bottleneck at circa 60 ka, causing enhanced cognitive phenotypes, armed with a modern set of abilities, to quickly replace the unenhanced ones and migrate out of Africa (Klein 2000; Mellars 2005).

However, evidence of symbolic behavior, represented by engraved objects (H. Anderson 2012; Texier et al. 2010), ochre fragments possibly used for body painting (Barham 2002; Henshilwood, d'Errico, and Watts 2009), and, most of all, perforated shells interpreted as body ornaments (e.g., Bar-Yosef Mayer, Vandermeersch, and Bar-Yosef 2009; Bouzouggar et al. 2007; d'Errico et al. 2005; Vanhaeren et al. 2006), was registered in several early modern human African sites. This evidence was used to reject the idea that modern humans incurred a discrete event of revolution and enhancement, arguing for an incremental acquisition of behavioral modernity.

During the last years, proponents of the multiregional model rejected the idea that behavioral modernity was a prerogative of modern humans only (d'Errico 2003; Villa and Roebroeks 2014; Zilhão 2007). In contrast, they claimed that this condition was reached at different times and locations, following diverse and not necessarily incremental trajectories of human development (Hovers and Belfer-Cohen 2006). Most crucially, partisans of the multiregional model argue that late Neanderthal populations in Europe were also following their own idiosyncratic trajectory for acquiring behavioral modernity, which was either influenced by the arrival of modern humans in Europe or possibly related to reasons intrinsic to the Neanderthal world (d'Errico 2003; Zilhão 2007). These arguments are mostly grounded in the recent findings of "symbolic" artifacts in some Neanderthal populations in Europe, comparable with those appearing in the modern human Middle Stone Age. According to some lines of argument, such ornaments were developed before the incursion of *Homo sapiens* in Europe (Zilhão 2013), reflecting either independent development or the importing of ideas and concepts within the Neanderthal world (d'Errico 2003; Zilhão

2007). Consequently, advocates of the multiregional model support the thesis that the same cognitive and biological bases were present in humans since the Middle Pleistocene (Speth 2004; Zilhão 2007, 2011a, 2011b), and hence modern and archaic populations were cognitively equivalent (for reviews, see d'Errico and Stringer 2011; Harrold 2009; Nowell 2010).

In cognitive archaeology, the conditional approach aims at identifying the minimal cognitive requirements that are necessary and sufficient to explain the existence of certain practices within the archaeological record of past human populations (e.g., Abramiuk 2012, 31–33; Wynn and Coolidge 2009). Following this agenda, some authors have considered a set of cognitive abilities as sufficient conditions to support the cognitive equivalence between archaic and contemporary populations (e.g., Henshilwood and Dubreuil 2011). They have further elaborated that such abilities are necessarily bound to the presence of body ornaments in the archaeological record of early modern humans in Africa (Middle Stone Age, from circa 100 ka onward). This analysis could be applied also to analogous phenomena of body adornment in late Neanderthal cultures, therefore apparently supporting the case for Neanderthal cognitive equivalence with modern humans (Zilhão 2007; see below). Nevertheless we have several reasons to believe that the very connection between body ornaments and cognitive equivalence at the core of this position is deeply problematic. After examining the relationship between symbolism and cognitive equivalence in the modern human Middle Stone Age, I will illustrate these problems by referring to the evidence of Neanderthal extraction of feathers for presumably ornamental reasons as a case study.

1 The Material Symbolism Package

Early body ornaments in the Middle Stone Age are primarily represented by the finding of perforated shells at various African locations, with particular emphasis on those related to the Still Bay and Howiesons Poort in South Africa, respectively dated at circa 77–72 ka and 65–59 ka (d'Errico et al. 2005; Henshilwood et al. 2004). At the site of Blombos, South Africa, *Nassarius kraussianus* shells were intentionally perforated by applying pressure-based puncturing methods with a tool through their aperture (d'Errico et al. 2005; Henshilwood et al. 2004). These artifacts have been interpreted as body ornaments due to the presence of some use-wear facets on their surface, produced by the friction of the beads against one another, suggesting that they were strung and worn as pendants (e.g., Henshilwood and Dubreuil 2011, 374). Furthermore, these objects were found in locations distant from the natural environment where they are usually situated (d'Errico et al. 2005), thus indicating long-distance transportation of the shells from coastal to inland areas, plausibly related to their social value, given the scarce importance of these mollusks for nourishing (Roberts 2015). The first attempts to conclude for cognitive equivalence of archaic and contemporary populations in archaeological literature have been moved along the idea that such shell bead ornaments implied the storage of information outside the human brain (d'Errico et al. 2003; Henshilwood and Marean 2003). The construction and sharing of such a symbolic meaning for these items required the use of a "modern" syntactical language (d'Errico et al. 2003; d'Errico et al. 2005; Henshilwood et al. 2004; Mellars 2005; Stiner 2014; Watts 2009). Furthermore, modern language is considered to tap a set of abilities, such as working memory capacity, which would strengthen the case for cognitive equivalence with current human populations (Wynn and Coolidge 2004, 2007). However, these initial proposals led some scholars, in particular the linguist Rudolph Botha (2008, 2010), to invite caution. In fact, the semantic ascription of symbolic meaning to a shell bead ornament does not necessarily warrant the existence of structures such as grammar and syntax that characterize contemporary languages.

Critiques of the previous kind have superseded simplistic associations between artifacts and ill-defined cognitive abilities, such as "modern" language, with more specific theories. Henshilwood and Dubreuil (2011), in particular, have proposed a connection between body ornaments, symbolism, and cognitive equivalence by arguing that body adornment requires a "cognitive package" of highly demanding cognitive functions, which underlies the ability to create material symbols. According to their view, symbolism implies that individuals understand that artifacts are imbued with meanings, and these meanings are collectively constructed and shared. Shell beads are thus to be considered symbols, because they require an individual to understand how others look at him or her and to acknowledge that the contents of other people's beliefs are crucial to his or her own position within the game of reputation.

These ideas are reinforced by the fact that shell beads showed consistency in their distribution through time and space. At Blombos Cave, for example, pierced shells were found throughout different archaeological layers within the same location (Vanhaeren et al. 2013, 514), showing a durable tradition of caring about one's own aspect within the social space. This evidence was used to argue that body ornaments were not the result of idiosyncratic and transient behaviors but were adopted within a set of socially shared and well-established canons (d'Errico et al. 2003; Kuhn and Stiner 2007a). Redundancy and standardization are allegedly considered by themselves evidence of the symbolic character of these ornaments, because they reflect the presence of shared and well-defined referential meaning to be communicated. Further confirmation of their symbolic use, according to Kuhn and Stiner (2007a, 2007b), lies in the fact that shell bead ornaments, such as necklaces or bracelets, were subject to the incremental addition of units, due to the small size and durability of the shells. Therefore, beyond visual display and aesthetics, shell beads could have been used to create composite jewels with the aim of communicating values such as "status," "wealth," or "identity."

At the cognitive level, these behaviors require one to inhibit one's own perspective, focusing on how others see one's aspect, an ability known as level-2 perspective taking (Henshilwood and Dubreuil 2009; Apperly 2011). Furthermore, they require associating an abstraction to a material sign, possibly by adopting a "stand for" semiotic function (such as in the case of linguistic semantic constructions) or, at the minimum, a "linked to" function (Henshilwood and Dubreuil 2011, 391). Ultimately, an individual has to understand that this relationship is conventionally shared with other social agents, and accordingly regulate his or her own action. Most crucially, the meaning of an ornament is processed by sharing the relationship between a mental representation (e.g., "coolness," "status," or "wealth") and an artifact (a shell) through full-blown mindreading (fig. 15.1).

Following the argument raised by Hutto (2008), and Bermúdez (2003, 175), mindreading appears to depend on language, and in particular on the possibility of building metarepresentations of the kind "I know that you know that the ornament means 'coolness'"

Figure 15.1

The protagonist (left) understands what the ornament means for the target (right) in a mentalistic way. She first builds up a connection between an abstract concept ("coolness") and an item (a shell)—not shown. Then she looks at the action of the target (i.e., projected arms, emotional bodily features) and seeks to explain it through a process of full-blown mindreading. In this way, the protagonist makes an inference to the mental contents of the target (a) and realizes that *the target knows that the ornament means* "coolness" (b). Hence, she concludes that the target is showing interest toward the shell because he shares with her the same meaningful artifact-concept relationship. (The proportions of the shell have been altered for illustrative reasons.)

(Henshilwood and Dubreuil 2009, 59; d'Errico et al. 2005, 19). Such a philosophical stance is reinforced by empirical evidence that shows a high degree of correlation between language development and theory of mind in children (Milligan, Astington, and Dack 2007). The concept of "coolness" appears to further rely on language, since it seems to be represented by a theory, namely, a Bayesian set of causally connected events (Rehder 2007), or a definition (Piccinini 2011), which is based on establishing normative conditions for belonging to a conceptual category. Although these requirements cannot confirm that modern syntactic structures, such as language recursion, are necessary to produce ornaments, the complexity of language and cognitive processes required to build metarepresentations or abstractions, contra Botha (2008, 2010), would prove substantial equivalence with contemporary human cognition. In the rest of the chapter, I refer to this body of abilities as the "material symbolism package" (MSP).

2   Neanderthal Feathers as Analogous Ornaments

Evidence of body adornment in late Neanderthal archaeological sites is mostly represented by shell beads (Zilhão et al. 2010; Zilhão 2011b), perforated bone pendants (Zilhão 2012, 38), and raptor talons presumably adopted for ornamental reasons (Radovčić et al. 2015; Romandini et al. 2014). These artifacts share physical properties that make them suitable for visual display, and the creation of compositional jewels. In consequence, they can be considered analogous to the Middle Stone Age perforated shells (Johansson 2013). However, this analogy has been often contested because these ornaments, although spatially distributed across a range of sites (Zilhão 2012), appear only seldom in the Neanderthal archaeological record, favoring idiosyncratic interpretations over actual forms of symbolism based on shared customs and cultural constructs (Klein 2008; Kuhn and Stiner 2007b; Wynn and Coolidge 2012, 119–122).

Quite recently, though, evidence of the intentional removal of flight feathers (remiges) from various species of birds with scavenging behaviors (i.e., raptors, falcons, and corvids), across a broad series of Neanderthal sites, has been adopted to defend the reality of Neanderthal symbolism (Finlayson et al. 2012; Morin and Laroulandie 2012; Peresani et al. 2011). The ornamental reasons behind the extraction of feathers have been inferred from the fact that the feathers were collected from species that are usually not highly valued for nutrition in modern cultures (Finlayson et al. 2012, 7). In addition, the bones associated with parts of the animals richer in flesh appear unprocessed, whereas the vast majority of cut marks are situated on the wing bones (mostly humeri and ulnae), low in meat but anchors for the flight feathers (fig. 15.2).

In line with symbolic use, feathers can be aesthetically relevant, due to their glossy properties, and suitable to be adopted within a system of aesthetic canons. The preference for dark colors of the selected feathers could have been part of these shared values. Furthermore, feathers are rare and difficult to provide, requiring a great investment of resources and time,

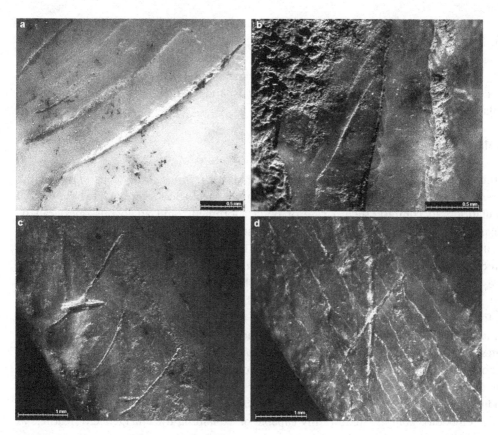

Figure 15.2
Evidence of cut marks on scavenging bird wing bones: (a) distal diaphysis of *Pyrrhocorax pyrrhocorax* (red-billed chough) humerus; (b) proximal diaphysis of *Pyrrhocorax pyrrhocorax* humerus; (c) proximal diaphysis of *Pyrrhocorax pyrrhocorax* humerus; (d) distal diaphysis of *Milvus milvus* (red kite) radius. From Finlayson et al. 2012. Open source.

thus becoming an effective way to communicate wealth and status of the wearer (Johnstone 1995). Ethnographic and historical analogies show that ornamental feathers have received great consideration in human history, spanning most of the cultures that have been studied, including contemporary contexts (Amkreutz and Corbey 2008; Tella 2011). Most importantly, the study by Finlayson and colleagues has shown that the extraction of feathers was spatially distributed across several locations and vertically transmitted for several millennia at least at Gibraltar (Gorham Cave, ca. 57–28 ka), suggesting that this phenomenon was not idiosyncratic but based on long-lasting traditions. In this way, feathers would satisfy the condition of "reference to durable shared values," which some scholars have considered as the basis of true symbolism.

Overall, ornamental feathers represent an optimal case study for Neanderthal symbolism. If we accept that these artifacts were symbolic ornaments analogous to the Middle Stone Age shell beads, then it follows that the same cognitive requirements, namely, the MSP, would equally be necessary to process these ornaments, allowing us to conclude for cognitive equivalence between Neanderthals and contemporary modern humans (Finlayson et al. 2012).

3   Aims and Methods

In the remainder of the chapter, I argue that the MSP is not necessarily constrained by the presence of presumably ornamental feathers in the archaeological record of late Neanderthal populations. In contrast, such a case of body adornment (and analogous ones) can be explained by more minimalistic abilities, at least for what concerns the construction and social maintenance of the ornaments' meaning. To make my argument, I adopt a combination of the most relevant strands of theory in radical embodied and enactive cognitive science (henceforth REC). In particular, I follow the agenda proposed by the philosopher Anthony Chemero (2009) and known today as the epistemological version of REC. According to this proposal, the essence of REC consists in understanding the extent to which it is possible to explain sophisticated behaviors without relying on computations over mental representations as key components of cognitive tasks. In the same fashion, I argue that radical embodied strategies based on direct social perception, material scaffolding, and narrative practice can obviate the need for sharing abstract social standards through mindreading to produce ornamental feathers. Adding to my previous analysis (Garofoli 2015a), I argue that radical embodied abilities are not only capable of explaining the emergence of feathers as incidental phenomena limited to short time frames. In fact, they suffice also to address the long-term maintenance of "ornamental systems" and their communicative scope without the risk of reintroducing mentalistic positions.

In addition to minimalism, radical enactivism also provides the possibility of avoiding important metaphysical issues that plague evolutionary psychology approaches to the emergence of the MSP. I dedicate particular attention to the neurocentric and disembodied assumptions underlying the idea that the MSP evolved as a prior, innately specified adaptation to "enable" the production of ornaments, adopting material engagement theory as an alternative account.

I will show that the material engagement with ornaments represents a necessary condition for the emergence of symbolism, whereas the contrary cannot be assumed without incurring great metaphysical costs. Artifact-mediated cognitive transformations are critical conditions that need to be identified within the archaeological record, extending the scope of the conditional approach from a proximal relationship between artifacts and cognitive properties to long-term transformations.

4 Ornaments REC-onsidered

The radical enactive view of early adornment is based on the idea that some objects might have special features, such as, for example, reflecting light in a way that "ordinary" objects do not. Within a material world made of, for example, stone tools, bones, and wood, the shiny properties of shell nacre or the glistening aspect of a raptor feather might have attracted the interest of local hominids more than ordinary items could. An innate preference for glossy versus dull objects could have been further responsible for these behaviors, as assessed within several empirical studies of modern human children (Coss, Ruff, and Simms 2003; Danko-McGhee 2006) and adults (for a review, see Meert, Pandelaere, and Patrick 2014). The possibility that Neanderthals could have ascribed an aesthetic and emotional value to these objects is also supported by the presence of special items in their culture and that of their ancestors, such as collected crystals or fossils with no utilitarian use (Hayden 1993, 123). At the same time, Neanderthals were also highly complex social beings, as supported by the transmission of sophisticated behavioral practices, such as hafting (Wragg-Sykes 2015) and coordinated hunting maneuvers (Churchill 2014, 242). These practices suggest the existence of social niches for the conveyance of information in Neanderthal cultures, wherein the behavior of social agents in relation to other individuals was given great consideration (Sterelny 2012, chap. 3). These elements support the conception that the initial emotional experience of a Neanderthal for a feather might have motivated her to remove it from the bird just laboriously captured. In this way, a Neanderthal agent (henceforth "the protagonist," left character in fig. 15.3) might have carried around a feather, attracting the interest of her band members.

The feather activates an emotional experience in the individual looking at it (henceforth "the target," right character in fig. 15.3), which is displayed by an embodied reaction toward the item itself (Garofoli 2015a). The protagonist, in this way, can directly perceive such an embodied reaction toward the feather and understand that the object is special for her companion, being thus motivated to keep it. Experimental evidence shows that embodied expressions, in particular the orientation of gaze, can reveal a target's emotional commitment toward an object in both infants (Hoehl, Wiese, and Striano 2008) and adults (Ulloa et al. 2014). Indeed, gaze orientation, which is strongly associated with emotions (Adams et al. 2003; Hadjikhani et al. 2008), can transfer to an object the intentionality of a person looking at it, lading the item with affective meaning (Becchio, Bertone, and Castiello 2008). This explains why embodied expressions of people looking at objects can durably alter the preference of infants for these items, leading them to favor (Flom and Johnson 2011) or avoid (Hoehl, Wiese, and Striano 2008) these objects, in relation to the kind of emotion conveyed by the others' facial expressions and gaze.

At the core of the REC approach adopted here lies the idea that the embodied actions and emotions perceived by the two agents within this dynamic, relational system do *by themselves* express the meaning of the feather (Garofoli 2015a). Thus, when the protagonist leaves the

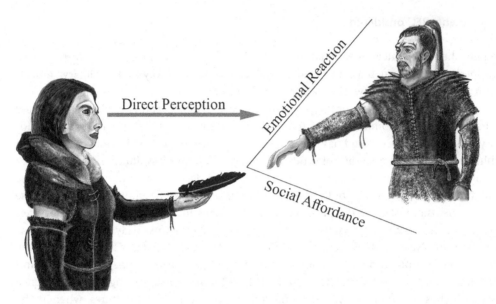

Figure 15.3
In the radical enactive setting, a Neanderthal protagonist (left) shows a feather to a target (right). Due to its glossy material properties, the feather enacts an embodied emotional reaction of interest within the target, thus bringing forth a social affordance for the protagonist. The target's emotional reaction embodies by itself the meaning of the feather, which can be directly perceived in action by the protagonist. This obviates the need for identifying connections between objects and abstract concepts within the mind of the target through a mentalistic process (see fig. 15.1). In contrast, the protagonist can rely on her past perceptual experience when turning the item into a body ornament.

scene and proceeds to apply holes to her clothes, prepare a string, or entwine her own hair to fix the feather to her body, what she has in mind, and what motivates her conduct, is the embodied emotional reaction of the target toward the feather. In this way, the protagonist needs to add no additional meaning to the feather in her head, no abstract, theoretical, or even definitional representation that must be mentalized and imposed on the object to make sense of it. The conceptual meaning of the feather appears therefore to be kept in memory as a form of "situated action," namely, a concept that is grounded in perception.[1]

Furthermore, since the meaning of the feather is directly perceived in embodied activity, no inference to the mental contents of the target is required to explain why he is reacting in

1. This concept format is close to the idea of situated concepts defended by Barsalou (1999) and Prinz (2002), although, unlike their approach (Prinz and Barsalou 2000), not formulated in representational terms. Concepts such as "cool feather" are reconstructive processes entering in resonance with the invariant structure characterizing, for example, the scene in figure 15.3 (see Robbins 2006 for a theory of "direct memory").

that way (e.g., I know that the subject will display the following body form if he believes the feather means "coolness," such as I do). Therefore no form of metarepresentational mind-reading is necessary to assess the meaning of the feather.

Conversely, the same cognitive abilities are adopted by the target to understand the meaning of the feather. When the target sees the feather turned into a body ornament, he does not need to understand that the ornament *stands for* "coolness," nor does he need to guess this relationship in the mind of the protagonist. The target just sees an individual adorned with an aesthetic item (a feather) and directly perceives it as cool. Meaning, therefore, is not a fact independently built up by the two agents, who use internal rules, divorced from the world, to interpret other minds, nor does it mysteriously emerge from the mere interaction of the two agents. In fact, meaning is created within the relationship between two individuals who are not naive but have memories and past experiences (De Jaegher and Di Paolo 2012).

A deeper phenomenological and cognitive analysis might help clarify how meaning emerges within this relational artifact-hominid system. When generating an embodied emotional reaction in the target, the feather brings forth a structure, which is invariant with those characterizing other emotional contexts that Neanderthal agents have previously experienced. Having engaged with this invariant structure during the past, the protagonist is aware of the sensorimotor contingencies that specify "a positive emotion toward an object."[2] In accord with past experience, these sensorimotor contingencies are recognized within the body features of the target. However, the protagonist can now perceive something new, namely, the connection between the emotional structure and the properties of the feather. In this way, through its material features, the feather actively provides a social affordance. It exposes a structure that resonates with past memories and allows creating new meaningful relationships to be picked up by the Neanderthal agents. Thus the feather acts as a keystone within a dynamic system created by the reciprocal engagement of the two Neanderthal agents, therefore exerting material agency (Malafouris 2008a) within a participatory sense-making system (De Jaegher and Di Paolo 2007; Fuchs and De Jaegher 2009).

In sum, following the epistemological version of REC as a minimalistic tool in conditional cognitive archaeology, this cognitive and phenomenological analysis shows that the adoption of ornamental feathers by Neanderthals, at least for what concerns the construction of meaning within the short term, does not necessarily constrain the presence of the MSP.

5   Mentalism and the Origins of Meaning

The conditional approach does not simply focus on identifying *what* kind of cognitive abilities are minimally required to explain the existence of artifacts within fixed,

2. The awareness of sensorimotor contingency invariance defines a "situated concept." Situated concepts resonate with the invariant structure of reality, allowing agents to reconstruct meaning, rather than standing for it, such as in more conservative embodied models (e.g., Barsalou 1999).

decontextualized scenarios, which are abstracted from time and devoid of any historical and developmental dimension. In contrast, this approach needs to focus also on providing the most plausible explanations for the *origins* of the cognitive mechanisms through which artifacts come into existence (Haidle et al. 2015). Applied to the current case study, this means that we need to focus on where the meaning of feathers and the cognitive abilities required to construct that meaning originate from.

The cognitive abilities comprised by the MSP, namely, abstract concepts such as "coolness" or "status," a "stand for" semiotic function, and a full-blown theory of mind, exist before and are necessary for the formation of the meaning of ornaments (d'Errico et al. 2005, 19; Henshilwood and Dubreuil 2009, 61). Accepting this position, as Kuhn and Stiner (2007b, 48) seem to propose, raises the question about the derivation of these abilities. A possible solution to this riddle lies in assuming a hard-core evolutionary psychology argument (e.g., Cosmides and Tooby 2013), according to which the MSP represents the result of a set of cognitive adaptations, possibly stemming from both domain-specific and domain-general alterations of the mental architecture (Henshilwood and Dubreuil 2011, 364). Within this neo-Darwinian approach, all the components that represent the MSP are innately specified functions that emerge as a solution to an adaptive issue within Paleolithic societies (e.g., maximizing social stability) and generate body ornaments as a result of this evolutionary design (for similar arguments, see Buss 2005). In a more extreme fashion, one could argue that the brain hosts content detectors (e.g., Lieberman, Tooby, and Cosmides 2007; New, Cosmides, and Tooby 2007), which are capable of connecting specific aspects of the world, such as a shell, to an innate body of knowledge, represented by the concept of "coolness," as the outcome of innately specified Darwinian algorithms (for a contemporary application, see Mithen 1994, 2014).

Along this line of argument, partisans of cognitive equivalence could maintain that the presence of body ornaments in the early modern human and late Neanderthal archaeological record reflects the existence of a complete MSP in both species, comprising a series of fixed meanings, symbolic functions, and metarepresentational abilities. This package could have been either created by mutational enhancement or promptly triggered, when requested by adaptive needs, from a "silent" state that evolved millennia before (for critique, see Nash 2014).

Nevertheless, these positions are affected by deep metaphysical issues. Structures of the previous sort appear as homuncular entities that are capable of making judgments about aspects of the world on the basis of their innately specified contents. This conception inevitably leads to a series of dualistic stances implying that neural homunculi are conscious agents, rather than simple biological systems capable of responding or not under certain conditions (e.g., S. Gallagher 2008). Such neurocentric and neo-Darwinian explanations associating the emergence of ornaments with complete cognitive packages therefore incur theoretical costs that reduce their plausibility (Tallis 2011). In line with the tenets of material engagement theory (Iliopoulos and Garofoli 2016; Malafouris 2013), the aforementioned

REC approach could offer an alternative view to the emergence of early body ornaments and human cognition capable of avoiding these issues with hard-core evolutionary psychology arguments.

6   Feathers as Transformative Devices

In section 4, I discussed the idea that an individual wearing feathers could be considered cool on the basis of positive emotions elicited in group members by the feathers. A situated concept of "cool individual" can therefore emerge from this specific perceptual situation concerning embodied emotions and feathers. However, within the long term, the perceptual system learns to recognize the structure of a cool individual within different situations, unrelated to feathers. This invariance of deep perceptual structure represents the ground for the emergence of an abstract concept of "coolness" out of a concrete perceptual level. An analogous argument can be supported in relation to the adoption of feathers as indexes of social values, such as "status," from an initial level of iconic body alteration for aesthetic purposes (Iliopoulos 2016). Given that scavenging birds are difficult to capture or find, feathers can represent a resource of quite limited availability. Thus access to feathers can be granted only to particular individuals who have enhanced social and material assets. Building on an initial situated concept based on aesthetic values, the feathers' rarity could contribute to creating a series of more complex situated concepts, now specifying the perceptual structure of "status." These concepts can be applied to a broad gamut of social situations in the long term, becoming increasingly more abstract in defining social hierarchy values.

At the same time, changes in tonal inflection (i.e., prosody) can be adopted to emphasize particular aspects of action and perception (e.g., "aha!" or "mmmmmhhhh"; Mithen 2005) within intersubjective systems, thereby assisting the construction of situated concepts. This can gradually lead to forming durable and reliable words, on the basis of the mutual engagement between hominids, things, and invariant sounds (e.g., Brown, Collins, and Duguid 1989). A rise in the complexity of embodied social concepts can further bring forth the need for creating words to define them, and on the other hand, language itself can scaffold the construction of more abstract social concepts (Clark 1997, chap. 10). For example, an abstract theory of "coolness" connecting a set of causally related events grounded in perception, or a definition of "status" that specifies normative conditions for what being a chief means, can effectively be constructed when language provides a vehicle to bind events or conditions within a unitary concept. In the long term, narratives can emerge from this context and be used as mnemonic artifacts to ground the meaning of this intricate network of social relationships. Likewise, mentalistic mindreading, which supporters of the MSP consider to be necessary for body ornamentation, emerges as a consequence of the linguistic awareness that is produced through these narratives (Hutto 2008).

Within this perspective of material engagement, the conclusions about the origins of cognitive functions are diametric to those reached by the MSP model. Feathers, acting as transformative devices on some basic cognitive processes, bring forth and allow the construction of highly demanding cognitive abilities, rather than presupposing them for being brought into existence (e.g., Henshilwood and Dubreuil 2009, 61). According to this approach, concepts like "coolness" and "status," semiotic functions, and metarepresentations do not constitute a priori conditions for the production of body ornaments. On the contrary, material engagement with ornaments represents a condition for the emergence of such sophisticated meanings and cognitive abilities (albeit not the only possible one).

In line with these principles, the appearance of ornaments within the archaeological record of past human populations cannot be considered as evidence for symbolism and the MSP. In contrast, the material engagement theory invites us to think that early ornaments are nonsymbolic transformative devices that hominids can use to enact symbolic abilities over the long term (Malafouris 2008b). Thus, these artifacts should be interpreted as evidence *against* symbolism and a "complete" MSP (Garofoli 2016).

7   Scaffolding Maintenance: A Threat for Minimalism?

At this point, skeptics might concede that the REC abilities introduced in the previous section can obviate the need for the MSP in the initiation and use of ornaments within the short term and that material engagement is at the same time necessary for the emergence of the MSP. However, they might rebut by claiming that mentalism would be reintroduced by the long-term maintenance of ornaments, which would be manifest in the formation of abstract shared standards, constructed and maintained through narrative practice.

The mentalistic argument for language, symbolism, and cognitive equivalence would thus be based on the following propositions:

(a) Neanderthals were provided with language before the emergence of abstractions such as "coolness" or "status," which were brought forth by the ornaments.
(b) Language likely took part in the long-term stabilization of the feathers as indexical markers, by supporting the creation of social narratives of shared norms and values.
(c) These social narratives are necessarily grounded in abstractions shared through metarepresentational mindreading.

Concerning (a), it is implausible that language in Neanderthals was fully brought forth by ornaments alone, following a linear trajectory from prosody to concrete and ultimately abstract words. The hypothesis that an articulated language was already present in the Neanderthal ornament makers is supported by theoretical arguments against the possibility of effectively coordinating group hunting behaviors, such as ambushing, without a productive language with syntactical structures (Zilhão 2007). Furthermore, current experimental contexts show the existence of a correlation between the transmission of lithic technologies and the effectivity of communicative strategies, with a maximum enhancement associated with the use of detailed linguistic explanations (Morgan et al. 2015). As a consequence, it is difficult, albeit perhaps not impossible, to imagine that highly demanding technologies, such as birch bark pitch hafting, were transmitted by Neanderthals without relying at least on a basic level of language articulation. Nevertheless, this language form did not necessarily have to include all the features of contemporary languages (Johansson 2013; Wynn and Coolidge 2004).

In relation to (b), namely, the actual involvement of language in maintaining the ornamental use of feathers, it is crucial to ask if language is a necessary condition for the maintenance of signification with feathers. The radical enactive scenario I have illustrated shows that the primacy of language in maintaining a set of social meanings might have been exaggerated. In contrast, we should turn to the importance of artifacts in actively constituting contexts of meanings grounded in embodied relationships (Coward 2016). In the example discussed in section 6, the relationship between body adornment and "status" is made available by the persistence of the feathers on the body of the band leader, which can be used to reenact what is meant to be a chief at any time. In this way, the feathers provide a structure that can persist over the long term, instantiating a custom, namely, part of a cultural niche that is vertically transmitted. Thus, in grounding a network of embodied meanings and social relationships, feathers can be maintained without the necessary involvement of language.

However, critics might contend that a maintenance strategy grounded in the mere crystallization of perceptual structures mediated by artifacts is too extreme. A reliable transmission of highly complex social meanings, such as those related to social hierarchy, would be implausible without assuming that such concepts were stabilized through a series of norms expressed within narratives (Henshilwood and Dubreuil 2009). In this way, if language-based abilities were already applied within other Neanderthal contexts, such as the transmission of spear-making techniques, it is reasonable that similar verbal artifacts were quickly created and adapted to ornamentation to support its maintenance over the long term. Indeed, a selective impenetrability of language to the emergence of signification through ornaments would appear as merely ad hoc.

The argument with (c) implies that the construction of such narratives of norms requires that Neanderthals were capable of metarepresenting the world or building up theoretical knowledge shared in a mentalistic way. Thus mentalism is displaced from building ornament semantics to shaping a context of norms. To clarify these aspects, it is worth referring to Martin Byers's (1994, 1999) position on material signification. According to this theory, feathers, for example, do not mean "status" in a wordlike fashion. Rather, when these artifacts, with their material properties, are situated within a context of norms, they come to "warrant" a particular meaning. A notorious application of Byers's theory is the case of a passport, which warrants the possibility of crossing a frontier given that the passport's physical properties, namely, the style in which it is made, match the standard expected by the frontier guards. In this case, the passport's features by themselves allow an individual to pass, because they are part of an agreement that binds both the guards and the traveler. One might argue that a passport warrants free passage at the frontier because, in phenomenological terms, the frontier guards interpret the meaning of the passport as a warranting tool by relying on a series of mentalistic norms they share with the traveler. These norms are of the kind "I know that the other knows that the passport allows him to pass." In this way, what the passport warrants is still determined through a mentalistic approach, even if now there is no direct ascription of a meaning to a sign (as in language semantics) and no understanding of this relationship through mindreading.

Even this minimal level, therefore, would require that the feathers are associated at least with a core set of abilities present in the MSP, which is now evident in the creation of norms, rather than in the construction of the ornament's meaning. In any case, the presence of the MSP would give credit to the thesis that Neanderthals and modern humans were, at least in a qualitative sense, cognitively equivalent.

8   Maintenance without Mentalism: Behavioral-Contextual Narratives

Although the argument just introduced might appear sound, we have reasons to be skeptical about the validity of (c). The construction of "warranting contexts" through narratives does not necessarily require linguistic metarepresentations. Fiebich (2016) has recently introduced an important distinction between mentalistic and behavioral-contextual narratives. While the former are based on explaining the action of social agents by focusing on mental contents "as such," the latter exclusively target behavioral features and chains of actions. The structure of language underlying the production of behavioral-contextual narratives could thus be represented by a series of propositional chunks (possibly limited to monoclausal or coordinative propositions), which refer to sensorimotor aspects of behavior and perceptual concepts. Such narratives could, for instance, take the following shape:

The hunter killed the kite and is now wearing its bones and feathers. Thus the adorned man is strong, and everyone follows him.

This example shows that it is possible to construct contexts of norms, wherein ornaments exert a significative role, based on the mastery of such behavioral rules, which are grounded in the correlation between chunks of words and situated concepts.[3] Most crucially, these narratives do not include linguistic metarepresentations of the kind "I know that you know." In line with a series of proposals in enactive cognitive science (Gallagher and Hutto 2008; Hutto

---

3. In this chapter, I focus on the possibility of building metarepresentations through language. A parallel problem concerns the nature of Neanderthal language. The relationship between metarepresentational abilities and specific properties of language is not well known. It could be said here that the minimal language properties necessary to construct behavioral contextual narratives admit the presence of a form of protolanguage, composed by hierarchy, grammar, and monoclausal structure. This is compatible with the idea that Neanderthals were provided with several features of most contemporary languages, although not all of them (Johansson 2013).

2011; Hutto and Kirchhoff 2015), the exposition of behavioral-contextual narratives can deeply alter hominid perception, enhancing the salience of relevant aspects of the environment. Indeed, these narratives are based on words tokening sensorimotor aspects of reality and are thus considered to directly resonate with the perceptual concepts acquired by individuals. That is, behavioral-contextual narratives can be integrated within the perceptual system of young Neanderthals, allowing them to notice salient sensorimotor contingencies (O'Regan and Noë 2001) and affordances for action in the world (Chemero 2003). Radicalizing Kim Sterelny's (2003, 203; 2012) position, a cognitive niche engineered through narrative practice can be transmitted downstream and maintained as a scaffold for the cognitive development of further generations. In this way, the example of behavioral-contextual norms presented here could have allowed new generations to directly perceive the concept of "status" embodied by the feathers worn by a band member, without the need to enact this meaning *ex novo* through the long-term process described in figure 15.3. In sum, by engineering the cognitive niche, behavioral-contextual narratives can allow the construction and maintenance of the ornaments' meaning without necessarily tapping metarepresentational language and theory of mind (contra d'Errico et al. 2005, 19; and Henshilwood and Dubreuil 2009). This is a radical enactive way of saying that ornaments can "warrant" their meaning without mentalism.

Conclusion

The argument discussed in this chapter has deep implications for several domains of knowledge. Concerning cognitive archaeology, the radical enactive approach can explain the adoption and maintenance of allegedly ornamental feathers without tapping into a qualitative level of meaning that finds evidence in contemporary humans, namely, metarepresentation.

Preliminary analyses concerning the relationship between metarepresentation and cognitive architecture theories seem to support the idea that a metarepresentational system has a different organization from an architecture capable of processing visuospatial and propositional meanings at the basis of direct social perception, situated concepts, and behavioral-contextual narratives (Garofoli 2015b). Indeed, representing other representations seems to instantiate a different quality of invariance from that constituted by sensorimotor and propositional meanings (Barnard 2010). At the level of neurological realization, there is no a priori reason to believe that a neural architecture sufficient to process these latter levels of invariance can plastically acquire metarepresentational abilities through cultural and material engagement and neuronal recycling (M. Anderson 2014; Dehaene and Cohen 2007) without a biological alteration in its structure (Garofoli 2013, 2016). Thus, in contrast with the tenets of the multiregional model in cognitive archaeology, the presence of body ornaments in the Neanderthal record, even in the case of genuine innovation (Zilhão 2012), does not warrant cognitive equivalence with modern humans. Additional research focused on mapping the radical enactive abilities introduced here on a theory of mental organization is required to more thoroughly assess the validity of such preliminary conclusions.

Further conditional reasons to believe that ornamental feathers do not warrant the presence of metarepresentations lie in the origins of meaning and cognitive abilities. If we wish to avoid neurocentric tenets about the evolutionary selection of a complete MSP in Neanderthal populations, we have to consider feathers as nonsymbolic transformative devices that might scaffold the emergence of symbolism (Malafouris 2008b). However, there is no warranty that Neanderthals could also have reached a form of symbolic cognition that requires the abilities included in the MSP, for the simple fact that they had started this process of transformation. Such a connection would indeed have teleological implications. At the same time, assuming that Neanderthals could perform their cognitive transformation to symbolism because (1) modern humans succeeded in this task and (2) Neanderthals and modern humans shared part of their transformative trajectory is logically circular. Indeed, this approach assumes as a premise that Neanderthals, like modern humans, can enact a metarepresentational level, which in fact is the point to be argued for (see Garofoli 2016). In contrast, the conditions constraining a metarepresentational level ought to be identified within the archaeological record of Neanderthals (Mithen 2014).

The conditional approach developed here, however, refers only to a single aspect of Neanderthal material culture. The method for theoretical validation that I have repeatedly defended with my work (Garofoli 2016; Garofoli and Haidle 2014) maintains that every theory proposed in cognitive archaeology can be evaluated only by contextualizing it within a set of other similar analyses produced on other aspects of Neanderthal material culture (for a similar contextual approach, see Johansson 2013; Mithen 2014; Zilhão 2007). This allows one to constrain some logical possibilities on the basis of the consistency with the context of late Neanderthal material culture as a "whole." In other words, if other sophisticated practices identified in the archaeological record necessarily require mentalistic explanations that overlap with the MSP to be justified, then there would be at least the possibility that mentalism would be reintroduced also within body adornment.

Ultimately, from a phenomenological perspective, the reliance on mentalism and metarepresentation in human cognitive life has often been overstated by modernist intuitions. In contrast, critical arguments from enactive cognitive science contend that these abilities, albeit possible, are likely not even primary in contemporary human cultures, which rely extensively on more direct forms of social perception (de Bruin and de Haan 2009; Hutto 2011; S. Gallagher 2008). By these lights, one of the differences that made the difference for modern humans, at least partly explaining our evolutionary success within the long timescale, could have been just a mentalistic drop of ink in a radically embodied glass of water.

Acknowledgments

I wish to thank the editors, Christoph Durt, Thomas Fuchs, and Christian Tewes, for the invitation to contribute to this book. Special thanks to Christian Tewes and Christoph Durt for providing assistance and valuable comments during the preparation of this chapter.

References

Abramiuk, M. A. 2012. *The Foundations of Cognitive Archaeology*. MIT Press.

Adams, R. B., H. L. Gordon, A. A. Baird, N. Ambady, and R. E. Kleck. 2003. Effects of gaze on amygdala sensitivity to anger and fear faces. *Science* 300 (5625): 1536

Amkreutz, L., and R. Corbey. 2008. An eagle-eyed perspective: *Haliaeetus albicilla* in the Mesolithic and Neolithic of the Lower Rhine Area. In *Between Foraging and Farming: An Extended Broad Spectrum of Papers Presented to Leendert Louwe Kooijmans, Vol. Analecta Praehistorica Leidensia*, ed. H. Fokkens, B. J. Coles, A. L. van Gijn, J. P. Kleijne, H. H. Ponjee, and C. G. Slappendel, 167–180. Leiden University.

Anderson, H. 2012. Crossing the line: The early expression of pattern in Middle Stone Age Africa. *Journal of World Prehistory* 25 (3–4): 183–204.

Anderson, M. L. 2014. *After Phrenology: Neural Reuse and the Interactive Brain*. MIT Press.

Apperly, I. 2011. *Mindreaders: The Cognitive Basis of "Theory of Mind."* Psychology Press.

Bar-Yosef Mayer, D. E., B. Vandermeersch, and O. Bar-Yosef. 2009. Shells and ochre in Middle Paleolithic Qafzeh Cave, Israel: Indications for modern behavior. *Journal of Human Evolution* 56 (3): 307–314.

Barham, L. 2002. Systematic pigment use in the Middle Pleistocene of South-Central Africa. *Current Anthropology* 43 (1): 181–190.

Barnard, P. J. 2010. From executive mechanisms underlying perception and action to the parallel processing of meaning. *Current Anthropology* 51 (1): 39–54.

Barsalou, L. W. 1999. Perceptual symbol systems. *Behavioral and Brain Sciences* 22 (4): 577–660.

Becchio, C., C. Bertone, and U. Castiello. 2008. How the gaze of others influences object processing. *Trends in Cognitive Sciences* 12 (7): 254–258.

Bednarik, R. G. 2013. *Creating the Human Past: An Epistemology of Pleistocene Archaeology*. Archaeopress.

Bermúdez, J. L. 2003. *Thinking without Words*. Oxford University Press.

Botha, R. 2008. Prehistoric shell beads as a window on language evolution. *Language and Communication* 28 (3): 197–212.

Botha, R. 2010. On the soundness of inferring modern language from symbolic behaviour. *Cambridge Archaeological Journal* 20 (3): 345–356.

Bouzouggar, A., N. Barton, M. Vanhaeren, F. d'Errico, S. Collcutt, T. Higham, E. Hodge, et al. 2007. 82,000-year-old shell beads from North Africa and implications for the origins of modern human behavior. *Proceedings of the National Academy of Sciences of the United States of America* 104 (24): 9964–9969.

Brown, J. S., A. Collins, and P. Duguid. 1989. Situated cognition and the culture of learning. *Educational Researcher* 18 (1): 32–42.

Buss, D. M. 2005. *The Handbook of Evolutionary Psychology*. John Wiley.

Byers, A. M. 1994. Symboling and the Middle-Upper Palaeolithic transition: A theoretical and methodological critique. *Current Anthropology* 35 (4): 369–399.

Byers, A. M. 1999. Communication and material culture: Pleistocene tools as action cues. *Cambridge Archaeological Journal* 9 (1): 23–41.

Chemero, A. 2003. An outline of a theory of affordances. *Ecological Psychology* 15 (2): 181–195.

Chemero, A. 2009. *Radical Embodied Cognitive Science*. MIT Press.

Churchill, S. E. 2014. *Thin on the Ground: Neandertal Biology, Archeology, and Ecology*. John Wiley.

Clark, A. 1997. *Being There: Putting Brain, Body, and World Together Again*. MIT Press.

Conard, N. J. 2009. A female figurine from the basal Aurignacian of Hohle Fels Cave in southwestern Germany. *Nature* 459 (7244): 248–252.

Conard, N. J. 2010. Cultural modernity: Consensus or conundrum? *Proceedings of the National Academy of Sciences of the United States of America* 107 (17): 7621–7622.

Cosmides, L., and J. Tooby. 2013. Evolutionary psychology: New perspectives on cognition and motivation. *Annual Review of Psychology* 64:201–229.

Coss, R. G., S. Ruff, and T. Simms. 2003. All that glistens: II. The effects of reflective surface finishes on the mouthing activity of infants and toddlers. *Ecological Psychology* 15 (3): 197–213.

Coward, F. 2016. Scaling up: Material culture as scaffold for the social brain. *Quaternary International* 405:78–90.

Danko-McGhee, K. 2006. Favourite artworks chosen by young children in a museum setting. *International Journal of Education through Art* 2 (3): 223–235.

de Bruin, L., and S. de Haan. 2009. Enactivism and social cognition: In search of the whole story. *Cognitive Semiotics* 4 (1): 225–250.

Dehaene, S., and L. Cohen. 2007. Cultural recycling of cortical maps. *Neuron* 56 (2): 384–398.

De Jaegher, H., and E. Di Paolo. 2007. Participatory sense-making. *Phenomenology and the Cognitive Sciences* 6 (4): 485–507.

De Jaegher, H., and E. Di Paolo. 2012. Enactivism is not interactionism. *Frontiers in Human Neuroscience* 6.

d'Errico, F. 2003. The invisible frontier: A multiple species model for the origin of behavioral modernity. *Evolutionary Anthropology: Issues, News, and Reviews* 12 (4): 188–202.

d'Errico, F., C. Henshilwood, G. Lawson, M. Vanhaeren, A.-M. Tillier, M. Soressi, F. Bresson, et al. 2003. Archaeological evidence for the emergence of language, symbolism, and music: An alternative multidisciplinary perspective. *Journal of World Prehistory* 17 (1): 1–70.

d'Errico, F., C. Henshilwood, M. Vanhaeren, and K. Van Niekerk. 2005. *Nassarius kraussianus* shell beads from Blombos Cave: Evidence for symbolic behaviour in the Middle Stone Age. *Journal of Human Evolution* 48 (1): 3–24.

d'Errico, F., and C. B. Stringer. 2011. Evolution, revolution or saltation scenario for the emergence of modern cultures? *Philosophical Transactions of the Royal Society B: Biological Sciences* 366 (1567): 1060–1069.

Fiebich, A. 2016. Narratives, culture, and folk psychology. *Phenomenology and the Cognitive Sciences* 15(1): 135–149.

Finlayson, C., K. Brown, R. Blasco, J. Rosell, J. J. Negro, G. R. Bortolotti, G. Finlayson, et al. 2012. Birds of a feather: Neanderthal exploitation of raptors and corvids. *PLoS One* 7 (9): e45927.

Flom, R., and S. Johnson. 2011. The effects of adults' affective expression and direction of visual gaze on 12-month-olds' visual preferences for an object following a 5-minute, 1-day, or 1-month delay. *British Journal of Developmental Psychology* 29 (1): 64–85.

Fritz, C., and G. Tosello. 2007. The hidden meaning of forms: Methods of recording Paleolithic parietal art. *Journal of Archaeological Method and Theory* 14 (1): 48–80.

Fuchs, T., and H. De Jaegher. 2009. Enactive intersubjectivity: Participatory sense-making and mutual incorporation. *Phenomenology and the Cognitive Sciences* 8 (4): 465–486.

Gallagher, A. 2014. Absolute and relative endocranial size in Neandertals and later Pleistocene Homo. *HOMO: Journal of Comparative Human Biology* 65 (5): 349–375.

Gallagher, S. 2008. Direct perception in the intersubjective context. *Consciousness and Cognition* 17 (2): 535–543.

Gallagher, S., and D. D. Hutto. 2008. Understanding others through primary interaction and narrative practice. In *The Shared Mind: Perspectives on Intersubjectivity*, ed. J. Zlatev, T. Racine, C. Sinha, and E. Itkonen, 17–38. John Benjamins.

Garofoli, D. 2013. Critique of "How Things Shape the Mind: A Theory of Material Engagement," by Lambros Malafouris. *Journal of Mind and Behavior* 34 (3–4): 299–310.

Garofoli, D. 2015a. Do early body ornaments prove cognitive modernity? A critical analysis from situated cognition. *Phenomenology and the Cognitive Sciences* 14:803–825.

Garofoli, D. 2015b. Neanderthal cognitive equivalence: Epistemological problems and a critical analysis from radical embodiment. PhD thesis, University of Tübingen. http://hdl.handle.net/10900/64291.

Garofoli, D. 2016. Cognitive archaeology without behavioral modernity: An eliminativist attempt. *Quaternary International* 405:125–135.

Garofoli, D., and M. N. Haidle. 2014. Epistemological problems in cognitive archaeology: An anti-relativistic proposal towards methodological uniformity. *Journal of Anthropological Sciences* 92:7–41.

Hadjikhani, N., R. Hoge, J. Snyder, and B. de Gelder. 2008. Pointing with the eyes: The role of gaze in communicating danger. *Brain and Cognition* 68 (1): 1–8.

Haidle, M. N., M. Bolus, M. Collard, N. J. Conard, D. Garofoli, M. Lombard, A. Nowell, C. Tennie, and A. Whiten. 2015. The nature of culture: An eight-grade model for the evolution and expansion of cultural capacities in hominins and other animals. *Journal of Anthropological Sciences* 93:1–28.

Harrold, F. B. 2009. Historical perspectives on the European transition from Middle to Upper Paleolithic. In *Sourcebook of Paleolithic Transitions*, ed. M. Camps and P. Chauhan, 283–299. Springer.

Harvati, K. 2015. Neanderthals and their contemporaries. In *Handbook of Paleoanthropology*, ed. W. Henke and I. Tattersall, 2243–2279. Springer.

Hayden, B. 1993. The cultural capacities of Neandertals: A review and re-evaluation. *Journal of Human Evolution* 24 (2): 113–146.

Henshilwood, C. S. 2007. Fully symbolic Sapiens behaviour: Innovation in the Middle Stone Age at Blombos Cave, South Africa. In *Rethinking the Human Revolution: New Behavioural and Biological Perspectives on the Origins and Dispersal of Modern Humans*, ed. C. Stringer and P. Mellars, 123–132. Cambridge University Press.

Henshilwood, C. S., F. d'Errico, M. Vanhaeren, K. Van Niekerk, and Z. Jacobs. 2004. Middle Stone Age shell beads from South Africa. *Science* 304 (5669): 404.

Henshilwood, C. S., F. d'Errico, and I. Watts. 2009. Engraved ochres from the Middle Stone Age levels at Blombos Cave, South Africa. *Journal of Human Evolution* 57 (1): 27–47.

Henshilwood, C. S., and B. Dubreuil. 2009. Reading the artifacts: Gleaning language skills from the Middle Stone Age in southern Africa. In *The Cradle of Language*, vol. 2, ed. R. Botha and C. Knight, 61–92. Oxford University Press.

Henshilwood, C. S., and B. Dubreuil. 2011. The Still Bay and Howiesons Poort, 77–59 ka. *Current Anthropology* 52 (3): 361–400.

Henshilwood, C. S., and C. W. Marean. 2003. The origin of modern human behavior. *Current Anthropology* 44 (5): 627–651.

Hoehl, S., L. Wiese, and T. Striano. 2008. Young infants' neural processing of objects is affected by eye gaze direction and emotional expression. *PLoS One* 3 (6): e2389.

Hovers, E., and A. Belfer-Cohen. 2006. "Now you see it, now you don't": Modern human behavior in the Middle Paleolithic. In *Transitions before the Transition: Evolution and Stability in the Middle Paleolithic and Middle Stone Age*, ed. E. Hovers and S. L. Kuhn, 295–304. Springer.

Hutto, D. D. 2008. *Folk Psychological Narratives: The Sociocultural Basis of Understanding Reasons.* MIT Press.

Hutto, D. D. 2011. Elementary mind minding, enactivist-style. In *Joint Attention: New Developments in Psychology*, ed. A. Seeman, 307–341. MIT Press.

Hutto, D. D., and M. D. Kirchhoff. 2015. Looking beyond the brain: Social neuroscience meets narrative practice. *Cognitive Systems Research* 34:5–17.

Iliopoulos, A. 2016. The material dimensions of signification: Rethinking the nature and emergence of semiosis in the debate on human origins. *Quaternary International* 405:111–124.

Iliopoulos, A., and D. Garofoli, eds. 2016. The material dimensions of cognition: Reconsidering the nature and emergence of the human mind. *Quaternary International* 405:1–136.

Johansson, S. 2013. The talking Neanderthals: What do fossils, genetics, and archeology say? *Biolinguistics* 7:35–74.

Johnstone, R. A. 1995. Sexual selection, honest advertisement, and the handicap principle: Reviewing the evidence. *Biological Reviews of the Cambridge Philosophical Society* 70 (1): 1–65.

Klein, R. G. 2000. Archeology and the evolution of human behavior. *Evolutionary Anthropology: Issues, News, and Reviews* 9 (1): 17–36.

Klein, R. G. 2008. Out of Africa and the evolution of human behavior. *Evolutionary Anthropology* 17 (6): 267–281.

Klein, R. G., and T. E. Steele. 2013. Archaeological shellfish size and later human evolution in Africa. *Proceedings of the National Academy of Sciences of the United States of America* 110 (27): 10910–10915.

Kuhn, S. L., and M. C. Stiner. 2007a. Body ornamentation as information technology: Towards an understanding of the significance of early beads. In *Rethinking the Human Revolution*, ed. P. Mellars, K. Boyle, O. Bar-Yosef, and C. Stringer, 45–54. McDonald Institute for Archaeological Research.

Kuhn, S. L., and M. C. Stiner. 2007b. Paleolithic ornaments: Implications for cognition, demography and identity. *Diogenes* 54 (2): 40–48.

Lieberman, D., J. Tooby, and L. Cosmides. 2007. The architecture of human kin detection. *Nature* 445 (7129): 727–731.

Malafouris, L. 2008a. At the potter's wheel: An argument for material agency. In *Material Agency: Towards a Non-anthropocentric Approach*, ed. C. Knappett and L. Malafouris, 19–36. Springer.

Malafouris, L. 2008b. Beads for a plastic mind: The "blind man's stick" (BMS) hypothesis and the active nature of material culture. *Cambridge Archaeological Journal* 18 (3): 401–414.

Malafouris, L. 2013. *How Things Shape the Mind: A Theory of Material Engagement.* MIT Press.

McBrearty, S., and A. S. Brooks. 2000. The revolution that wasn't: A new interpretation of the origin of modern human behavior. *Journal of Human Evolution* 39 (5): 453–563.

Meert, K., M. Pandelaere, and V. M. Patrick. 2014. Taking a shine to it: How the preference for glossy stems from an innate need for water. *Journal of Consumer Psychology* 24 (2): 195–206.

Mellars, P. 2005. The impossible coincidence: A single-species model for the origins of modern human behavior in Europe . *Evolutionary Anthropology* 14 (1): 12–27.

Mellars, P., and C. Stringer. 1989. *The Human Revolution*. Edinburgh University Press.

Milligan, K., J. W. Astington, and L. A. Dack. 2007. Language and theory of mind: Meta-analysis of the relation between language ability and false-belief understanding. *Child Development* 78 (2): 622–646.

Mithen, S. J. 1994. From domain specific to generalized intelligence: A cognitive interpretation of the Middle/Upper Paleolithic transition. In *The Ancient Mind: Elements of Cognitive Archaeology*, ed. C. Renfrew and E. B. W. Zubrow, 137–156. Cambridge University Press.

Mithen, S. J. 2005. *The Singing Neanderthals: The Origins of Music, Language, Mind, and Body*. Weidenfeld & Nicolson.

Mithen, S. J. 2014. The cognition of *Homo neanderthalensis* and *H. sapiens*: Does the use of pigment necessarily imply symbolic thought? In *Dynamics of Learning in Neanderthals and Modern Humans*, vol. 2, ed. T. Akazawa, N. Ogihara, H. C. Tanabe, and H. Terashima, 7–16. Springer.

Morgan, T., N. Uomini, L. Rendell, L. Chouinard-Thuly, S. Street, H. Lewis, C. P. Cross, et al. 2015. Experimental evidence for the co-evolution of hominin tool-making teaching and language. *Nature Communications* 6 (6029). doi:10.1038/ncomms7029.

Morin, E., and V. Laroulandie. 2012. Presumed symbolic use of diurnal raptors by Neanderthals. *PLoS One* 7 (3): e32856.

Nash, A. 2014. Are Stone-Age genes created out of whole cloth? Evaluating claims about the evolution of behavior. *Dialectical Anthropology* 38 (3): 305–332.

New, J., L. Cosmides, and J. Tooby. 2007. Category-specific attention for animals reflects ancestral priorities, not expertise. *Proceedings of the National Academy of Sciences of the United States of America* 104 (42): 16598–16603.

Noble, W., and I. Davidson. 1991. The evolutionary emergence of modern human behaviour: Language and its archaeology. *Man* 26 (2): 223–253.

Nowell, A. 2010. Defining behavioral modernity in the context of Neandertal and anatomically modern human populations. *Annual Review of Anthropology* 39:437–452.

O'Regan, J. K., and A. Noë. 2001. A sensorimotor account of vision and visual consciousness. *Behavioral and Brain Sciences* 24 (5): 939–973.

Peresani, M., I. Fiore, M. Gala, M. Romandini, and A. Tagliacozzo. 2011. Late Neandertals and the intentional removal of feathers as evidenced from bird bone taphonomy at Fumane Cave 44 ky BP, Italy. *Proceedings of the National Academy of Sciences of the United States of America* 108 (10): 3888–3893.

Piccinini, G. 2011. Two kinds of concept: Implicit and explicit. *Dialogue* 50 (1): 179–193.

Prinz, J. J. 2002. *Furnishing the Mind: Concepts and Their Perceptual Basis*. MIT Press.

Prinz, J. J., and L. W. Barsalou. 2000. Steering a course for embodied representation. In *Cognitive Dynamics: Conceptual and Representational Change in Humans and Machines*, ed. E. Dietrich and A. Markman, 51–77. MIT Press.

Radovčić, D., A. O. Sršen, J. Radovčić, and D. W. Frayer. 2015. Evidence for Neandertal jewelry: Modified white-tailed eagle claws at Krapina. *PLoS One* 10 (3): e0119802.

Rehder, B. 2007. Property generalization as causal reasoning. In *Inductive Reasoning: Experimental, Developmental, and Computational Approaches*, ed. A. Feeney and E. Heit, 81–113. Cambridge University Press.

Robbins, S. E. 2006. Bergson and the holographic theory of mind. *Phenomenology and the Cognitive Sciences* 5 (3–4): 365–394.

Roberts, P. 2016. "We have never been behaviourally modern": The implications of material engagement theory and metaplasticity for understanding the Late Pleistocene record of human behaviour. *Quaternary International* 405:8–20.

Romandini, M., M. Peresani, V. Laroulandie, L. Metz, A. Pastoors, M. Vaquero, and L. Slimak. 2014. Convergent evidence of eagle talons used by late Neanderthals in Europe: A further assessment on symbolism. *PLoS One* 9 (7): e101278.

Ruff, C. B., E. Trinkaus, and T. W. Holliday. 1997. Body mass and encephalization in Pleistocene Homo. *Nature* 387:173–176.

Schoenemann, P. T. 2006. Evolution of the size and functional areas of the human brain. *Annual Review of Anthropology* 35:379–406.

Speth, J. 2004. News flash: Negative evidence convicts Neanderthals of gross mental incompetence. *World Archaeology* 36 (4): 519–526.

Sterelny, K. 2003. *Thought in a Hostile World: The Evolution of Human Cognition*. Blackwell.

Sterelny, K. 2012. *The Evolved Apprentice: How Evolution Made Humans Unique*. MIT Press.

Stiner, M. C. 2014. Finding a common bandwidth: Causes of convergence and diversity in Paleolithic beads. *Biological Theory* 9 (1): 51–64.

Stringer, C. 2002. Modern human origins: Progress and prospects. *Philosophical Transactions of the Royal Society B: Biological Sciences* 357 (1420): 563–579.

Tallis, R. 2011. *Aping Mankind: Neuromania, Darwinitis, and the Misrepresentation of Humanity*. Acumen.

Tella, J. L. 2011. The unknown extent of ancient bird introductions. *Ardeola* 58 (2): 399–404.

Texier, P.-J., G. Porraz, J. Parkington, J.-P. Rigaud, C. Poggenpoel, C. Miller, C. Tribolo, et al. 2010. A Howiesons Poort tradition of engraving ostrich eggshell containers dated to 60,000 years ago at Diepkloof Rock Shelter, South Africa. *Proceedings of the National Academy of Sciences of the United States of America* 107 (14): 6180–6185.

Ulloa, J. L., C. Marchetti, M. Taffou, and N. George. 2014. Only your eyes tell me what you like: Exploring the liking effect induced by other's gaze. *Cognition and Emotion* 29 (3): 460–470.

Vanhaeren, M., and F. d'Errico. 2006. Aurignacian ethno-linguistic geography of Europe revealed by personal ornaments. *Journal of Archaeological Science* 33 (8): 1105–1128.

Vanhaeren, M., F. d'Errico, C. Stringer, S. L. James, J. A. Todd, and H. K. Mienis. 2006. Middle Paleolithic shell beads in Israel and Algeria. *Science* 312 (5781): 1785–1788.

Vanhaeren, M., F. d'Errico, K. L. Van Niekerk, C. S. Henshilwood, and R. M. Erasmus. 2013. Thinking strings: Additional evidence for personal ornament use in the Middle Stone Age at Blombos Cave, South Africa. *Journal of Human Evolution* 64 (6): 500–517.

Villa, P., and W. Roebroeks. 2014. Neandertal demise: An archaeological analysis of the modern human superiority complex. *PLoS One* 9 (4): e96424.

Wadley, L. 2001. What is cultural modernity? A general view and a South African perspective from Rose Cottage Cave. *Cambridge Archaeological Journal* 11 (2): 201–221.

Watts, I. 2009. Red ochre, body painting, and language: Interpreting the Blombos ochre. In *The Cradle of Language*, vol. 2, ed. R. Botha and C. Knight, 93–129. Oxford University Press.

Wragg Sykes, R. 2015. To see a world in a hafted tool: Birch pitch composite technology, cognition and memory in Neanderthals. In *Settlement, Society and Cognition in Human Evolution: Landscapes in the Mind*, ed. F. Coward, R. Hosfield, M. Pope, and F. Wenban-Smith, 117–137. Cambridge University Press.

Wynn, T., and F. L. Coolidge. 2004. The expert Neandertal mind. *Journal of Human Evolution* 46 (4): 467–487.

Wynn, T., and F. L. Coolidge. 2007. Did a small but significant enhancement in working memory capacity power the evolution of modern thinking? In *Rethinking the Human Revolution*, ed. P. Mellars, K. Boyle, O. Bar-Yosef, and C. Stringer, 79–90. McDonald Institute for Archaeological Research.

Wynn, T., and F. L. Coolidge. 2009. Implications of a strict standard for recognizing modern cognition in prehistory. In *Cognitive Archaeology and Human Evolution*, ed. S. de Beaune, F. L. Coolidge, and T. Wynn, 117–127. Cambridge University Press.

Wynn, T., and F. L. Coolidge. 2012. *How to Think Like a Neandertal*. Oxford University Press.

Wynn, T., F. Coolidge, and M. Bright. 2009. Hohlenstein-Stadel and the evolution of human conceptual thought. *Cambridge Archaeological Journal* 19 (1): 73–84.

Zilhão, J. 2007. The emergence of ornaments and art: An archaeological perspective on the origins of "behavioral modernity." *Journal of Archaeological Research* 15 (1): 1–54.

Zilhão, J. 2011a. Aliens from outer time? Why the "human revolution" is wrong, and where do we go from here? In *Continuity and Discontinuity in the Peopling of Europe*, ed. S. Condemi and G. C. Weniger, 331–366. Springer.

Zilhão, J. 2011b. The emergence of language, art and symbolic thinking. In *Homo symbolicus: The Dawn of Language, Imagination and Spirituality*, ed. C. S. Henshilwood and F. d'Errico, 111–131. John Benjamins.

Zilhão, J. 2012. Personal ornaments and symbolism among the Neanderthals. *Developments in Quaternary Science* 16:35–49.

Zilhão, J. 2013. Neandertal-modern human contact in western Eurasia: Issues of dating, taxonomy, and cultural associations. In *Dynamics of Learning in Neanderthals and Modern Humans*, vol. 1, ed. T. Akazawa, Y. Nishiaki, and K. Aoki, 21–57. Springer.

Zilhão, J., D. E. Angelucci, E. Badal-García, F. d'Errico, F. Daniel, L. Dayet, K. Douka, et al. 2010. Symbolic use of marine shells and mineral pigments by Iberian Neandertals. *Proceedings of the National Academy of Sciences of the United States of America* 107 (3): 1023–1028.

IV Embodiment and Its Cultural Significance

16   Neoteny and Social Cognition: A Neuroscientific Perspective on Embodiment

Vittorio Gallese

The evolution of consciousness can scarcely be matched as a momentous event in the history of life; yet I doubt that its efficient cause required much more than a heterochronic extension of fetal growth rates and patterns of cell proliferation. There may be nothing new under the sun, but permutation of the old within complex systems can do wonders.
—Stephen Jay Gould, *Ontogeny and Phylogeny*

In the present chapter I address the notion of embodiment from a neuroscientific perspective, by emphasizing the crucial role played by bodily relations and sociality for the evolution and development of distinctive features of human cognition. To do so, I critically frame the neuroscientific approach and discuss it against the background of the Evo-Devo paradigm, emphasizing the relationship between ontogenesis and evolution. As recently argued by the philosopher Marco Mazzeo (2015), the paleontologist Stephen Jay Gould pointed out that humans are neotenic creatures: they retain in adulthood formerly juvenile features, produced by the retardation of somatic development (Gould 1977, 483).

By adopting this perspective, some distinctive features of human social cognition can be explained without resorting to prominent *deus-ex-machina*-like explanations of the apparent human cognitive discontinuity with respect to other nonhuman primates and mammals, such as putative gene mutations triggering the dawning of language (see Pinker 1994, 1997). I argue that neoteny further corroborates the crucial role played by embodiment, here spelled out by adopting the explanatory framework of embodied simulation, in allowing humans to engage in social relations, and make sense of others' behaviors. I briefly sketch embodied simulation as a new model of social perception and cognition, discussing its supposed neural underpinnings. The neotenic account of human social cognition also helps clarify the different levels of description involved in the relationship between human cognitive traits and their supposed neural underpinnings. I also explain why the neurophysiological level of description can be accounted for in terms of bodily formatted representations and reply to some recent criticisms of this notion. I propose that this approach can fruitfully be used to shed new light onto nonpropositional forms of communication and social understanding and onto distinctive human forms of meaning making. I conclude by showing how neoteny and embodiment can be usefully applied to the study of an important aspect of human social cognition: the experience of man-made fictional worlds.

1   Critical Neuroscience

Is the level of description offered by neuroscience necessary and sufficient to provide a thorough and biologically plausible account of the human mind and, for what mostly concerns us here, of intersubjectivity and human social cognition? Let's face it: the contemporary hype in popular media on the heuristic value of cognitive neuroscience mostly rests on the results of brain imaging techniques, and particularly of functional magnetic resonance imaging (fMRI). The fMRI technique, though, only indirectly "sees" brain activity, by measuring neurons' oxygen consumption. Such a measure is also indirect, as it depends on the local difference between oxygenated and deoxygenated hemoglobin. Oxygenated and deoxygenated hemoglobin behave differently within the strong magnetic field created by the big coil inside which participants' heads are placed. By measuring this functional parameter, local neural activity in terms of different MRI signals can be estimated only indirectly.

The fMRI temporal resolution is even worse, since it is in the order of a few seconds, whereas neurons' action potentials last less than one millisecond. Clearly, fMRI cannot match such temporal resolution because, as we have seen, it measures the delayed and prolonged local hemodynamic response. Furthermore, and most importantly, human brain imaging can provide correlations only between particular brain patterns of activation and particular behaviors or mental states; it is neutral about the possible causal role played by those same brain states to determine behaviors and mental states.

All of these limitations have made some researchers adopt a critical stance toward the heuristic value of cognitive neuroscience.[1] Such criticisms in their most radical form include claims against the possibility to naturalize consciousness and the mind. Peter Reynert recently wrote:

Neuroscience is a variant of naturalism which reduces aspects of consciousness and human existence to brain processes. ... The absurdity of naturalism implies that human consciousness and existence are conceptualized with notions and theories that cannot be applied to them, because they do not belong to the ontological region called nature. (Reynert 2016, 54–75)

It is beyond the scope and limits of this chapter to offer a defense of naturalism when applied to cognitive neuroscience. Nevertheless, I would like to clarify a few points. There cannot be any mental life without the brain. Thus, cognitive neuroscience is *necessary* to shed new light on the human mind and existence. Is this level of description also *sufficient*? Probably not.

---

1. See De Vos and Pluth 2016.

Certainly, it isn't sufficient if we study the brain by isolating it from the body, from the world, and from other individuals (Gallese 2000, 2003). In other words, cognitive neuroscience should resist the solipsistic "brain-in-a-vat"-like attitude purported by classic cognitivism and study how situated brain-bodies map their mutual relations and their interactions with the physical world.

A common misconception of cognitive neuroscience consists in perceiving its reductionism not as a necessary methodological strategy but as an ontological statement, according to which a necessary and totalitarian identity theory between brain and behavior, brain and psychology, or brain and cognition should be established. Perhaps neuroscience occasionally endorses such identity theories. However, this is not necessarily the case, and certainly to the best of my knowledge it is not the case here. Cognitive neuroscience should investigate human nature first and foremost by clarifying what human *experience* consists in.

Already fifteen years ago I proposed to study social cognition and intersubjectivity empirically by carefully distinguishing three levels of description: the phenomenological, the functional, and the subpersonal (Gallese 2001). However, it should be added that knowing whether or not some specific neurons and neural networks are active when we experience ourselves in relation to objects and to others gives us a totally new perspective on human nature, because it enables us to deconstruct many of the words and concepts we normally use when referring to human nature and mental life.

The experience we make of the world, as humans, indeed always exceeds the words and concepts we normally use to describe it. Action, perception, cognition, subject, object, intersubjectivity, and language may be very differently conceived when we begin with a subpersonal neurocognitive level of investigation (Ammaniti and Gallese 2014) rather than our typical everyday view of the world.

Furthermore, the issue of intersubjectivity, like all other issues related to human nature, should be framed within both phylogenetic and ontogenetic perspectives, by studying how human intersubjectivity relates to the social relations of other animal species and to their underpinning neural mechanisms. We must study if and how the way human beings develop their intersubjective skills relates to the way other animals develop theirs.

If there is one lesson I have learned from neuroscience, it is that even at its innermost and subpersonal level of description, intersubjectivity speaks of the quintessential nature of human beings as situated, feeling, and interacting bodies. Being, feeling, acting, and knowing are different modalities of how our bodies relate to the world. These modalities all share a constitutive underpinning bodily root, mapping onto distinct and specific ways of functioning of brain circuits and neural mechanisms. Such specific ways of functioning of the human brain cannot be fully understood if the brain is considered and studied separately from the sensing and acting body to which it is connected and of which it is integral part. Most importantly, I have learned this lesson mainly from the neurofunctional comparative study of nonhuman primates' brains and behaviors. This is the main reason why I think that the scientific study of intersubjectivity and the human self requires a comparative approach—the only approach capable of connecting distinctive traits of human nature to their likely phylogenetic precursors. In so doing, neurophysiological mechanisms and the cortical networks expressing them can be related to several aspects of primates' social cognitive behavior and thus be thoroughly investigated. We must study the biological roots of conceptual notions like intersubjectivity and the self to better understand their nature, genesis, structure, and properties. Such deflationary notions of the same concepts and a detailed study of their evolutionary origin are most successful if driven by a meticulous investigation of the underpinning neurophysiological mechanisms, most often available at the level of the single neuron only from the study of nonhuman primates' brains.

This comparative approach has enabled enormous progress of our knowledge of the functions of the human brain. A telling example comes from the study of vision. Researchers made discoveries in the functional organization of nonhuman primates' cortical visual system by correlating the discharge activity of single neurons in macaques' visual cortices with different parameters of the visual stimuli the macaques were looking at.[2] Interestingly enough, these results later led to similar investigations on the human visual brain by means of fMRI, which enabled the detection of a similar functional architecture, in spite of the species difference and, most importantly, of the different scale at which these investigations were carried out: a few hundred recorded neurons at best, in the case of experiments carried out on macaques' brains, and hundreds of thousands if not millions of activated neurons detected by local increase of blood flow, in the case of the human brain. In spite of the large-scale magnification implied when confronting single-neuron data from macaques and fMRI results from humans, some important functional features are nevertheless manifest across these different levels of description.

This remarkable functional similarity across different species and scales of investigation clearly isn't a mere coincidence. Rather, it is likely that some functional properties of the primate brain exhibit a sort of "fractal quality" so that they can be detected no matter what the level of investigation is and the species in which it is carried out. For these reasons, fMRI cannot be the sole neurocognitive approach to human social cognition, but must be complemented by other approaches compensating for some of its deficiencies. For example, transcranial magnetic stimulation (TMS), when used to reversibly inactivate restricted cortical regions, can provide some causal insight into their functional role, particularly if complemented by the study of brain-damaged patients. High-density electroencephalography (EEG), by means of its high temporal resolution, can compensate for the temporal limitations of fMRI.

The evidence from single-neuron studies in macaques contradicts some widely held assumptions about the functional organization of the human brain and about the best empirical strategy to investigate it. According to these assumptions, given the complexity of the human brain, the contents it processes would be exclusively mapped at the level of large

2. For a comprehensive review of this literature, see Zeki 1993.

populations of broadly tuned neurons. In spite of the almost astronomical figures characterizing the human brain (about 100 billion neurons, each of which connects with thousands of other neurons), its complexity perhaps does not parallel such astronomical figures, or at least not in such a way to deny any heuristic value to the single-neuron approach. Indeed, when trying to interpret human brain imaging results, when results from analogous brain areas of macaques are lacking, quite often fMRI results are interpreted in a rather ad hoc way: for example, when mindreading tasks lead to the activation of a set of human brain areas, the "explanation" given consists of the tautological statement that mindreading is "implemented" there.

Brain size may have less to do with the complexity of behavioral repertoires and cognitive capacity than has been generally assumed. Larger brains are, at least partly, a consequence of larger neurons that are necessary in large animals because of basic biophysical constraints. Larger brains also contain greater replication of neuronal circuits, adding precision to sensory processes, detail to perception, more parallel processing, enlarged storage capacity, and greater plasticity. But these advantages most likely are not sufficient to produce the qualitative shifts in behavior that are often assumed to accompany increased brain size, or at least not in a one-to-one manner (Chittka and Niven 2009). In this respect, as I will argue in the next section, neoteny can offer alternative explanations.

2   Neoteny and Social Cognition

There is no doubt that humans, despite sharing almost their entire genetic endowment with chimpanzees, are cognitively very different from them. Human language and human symbolic capacities can perhaps be explained without resorting to the discontinuities founded on theories of "cognitive catastrophes"—genetic big bangs, as in the case of the so-called grammar gene invoked by Steven Pinker (1994, 1997). The distinctive character of the human mind proves more comprehensible, or at least more easily approachable, if we investigate it empirically after having set aside the self-consoling recipes of some quarters of classic cognitive science and evolutionary psychology.

Neoteny may be a key issue to better understand the distinctive qualities of human social cognition. The term "neoteny" comes from the Greek νέος τείνειν, which means literally "tending to youth," and it refers to the slowed or delayed physiological and somatic development of forms of life. In chapter 10 of his *Ontogeny and Phylogeny*, Stephen Jay Gould writes:

In practically all human systems, postnatal growth either continues long past the age of cessation in other primates, or the onset of characteristic forms and phenomena is delayed to later times. The brain of a human baby continues to grow along the fetal curve; the eruption of teeth is delayed; maturation is postponed; body growth continues longer than in any other primate. (Gould 1977, 371)

And, as Marco Mazzeo recently pointed out:

The key sentence here is: "maturation is postponed." The deferment of sexual maturity in humans allows what detractors of neoteny take on as an objection, i.e. slower growth rates and longer developmental times. ... Neoteny as intended by Gould provides an evolutionary justification to the fact that *Homo sapiens* has cognitive and sensomotoric skills that enable him to speak a language. (Mazzeo 2015, 119)

Premature birth in humans often happens because of the impossibility for the fetus's head to pass through the female uterus. As emphasized by Mazzeo, human birth is indeed a paradox: while it is premature if compared to our developmental times, it is just its being premature that enables a peculiar quality and quantity of exposure to social relations, to sensory stimulation, and to language.

   The characterization of human beings as fragile, defective, and underdeveloped creatures has surfaced many times in the history of human thought, likely starting with Plato's *Protagoras*.[3] Gould's thesis about neoteny echoes arguments proposed earlier by Helmuth Plessner and Arnold Gehlen, as they were both strongly influenced by the theories of the Dutch anatomist Louis Bolk (1866–1930). Indeed, Gould writes:

Bolk emphasized that his list of [neotenic] features reflected two different phenomena: the physiological retardation of development ("the retardation hypothesis of anthropogenesis"—1926a, 470) and the somatic retention of ancestral juvenile proportions ("the fetalization theory of anthropogeny," 1926a, 469). The two phenomena are joined in "the narrowest causal connection" ("in engstem kausalen Zusammenhang," 1926b, 12) because delayed development prolongs fetal growth rates and conserves fetal proportions: "The essential in his form is the result of a fetalization, that of his life's course is the result of retardation. These two facts are closely related, for, after all the fetalization is the necessary consequence of the retardation of morphogenesis" (1926a, 470–471). If fetalized form is a consequence of retarded development, then the key to human evolution lies in the cause of this retardation: "There is no mammal that grows so slowly as man, and not one in which the full development is attained at such a long interval after birth. ... What is the essential in Man as an organism? The obvious answer is: The slow progress of his life's course. ... This slow tempo is the result of a retardation that has gradually come about in the course of ages" (1926a, 470). "Human life progresses like a retarded film" (1929, 1). (Gould 1979, 359–360)

According to Gould, neoteny is a parsimonious mechanism through which nature accomplishes macroevolutionary changes by means of the modification of the genetic mechanisms presiding over the timing and rhythm of ontogenetic development. King and Wilson (1975) proposed that the difference between humans and chimps might be due to genetic changes in a few regulatory systems.

   Recent research in genetics on the family of Hox genes, or homeotic genes, which are in charge of controlling the body plan and specify how anatomical structures develop in different segments of the body, appears to confirm King and Wilson's hypothesis. The French neurobiologist Alain Prochiantz (1989, 2012) has argued that the apparently small genetic

---

3. For a thorough discussion of this point, see Gualandi 2013; see also Mazzeo 2003.

difference separating humans from chimps acquires great importance if one considers that most of this difference lies in the genes that regulate the ontogenetic development of humans. Indeed, there is a significant excess of genes related to the development of the prefrontal cortex that show neotenic expression in humans relative to chimps and macaques (Somel et al. 2009; see also Sakai et al. 2011).

The implications for the phylogeny and ontogeny of human social cognition should be obvious. Human postnatal brain growth and maturation occur within a world of social bodily relations, thus making human development more dependent on parental cares and relations than that of nonhuman primates. On average, for the first year of their life humans can barely move around; thus they spend a considerable amount of their time looking at what others do. Neoteny thus amplifies and prolongs learning processes and potentiates family relationships. Stronger family relationships potentiate the possibility to transfer knowledge across generations, helping to build oral traditions, which, in turn, allow for the development of cultural practices.

Epigenetics clearly further boosts this processes, because it shows that DNA, the material constituent of our genes, does not code in a rigid and deterministic way the synthesis of proteins, but is deeply influenced by our relationship with the world. The physical, but also material, historical, social, and cultural environment in which we live, the type of human relations characterizing our life—in essence our *Umwelt*—influence the way genes work. DNA does not change, but its expression does, and, even more interestingly, these changes are passed on to descendants. Given these facts, it is not by chance that in recent decades we have witnessed a "biocultural" turning point within the human sciences. This turning point has questioned the rigid and obsolete distinction between nature and culture.

I posit that neoteny and epigenetics cannot help but boost humans' bodily simulative proclivity, contributing to the development and shaping of more sophisticated forms of social engagement and interaction. The relative bodily immobility caused by delayed development also promotes enhanced attention paid to the world of interacting others that surrounds human newborns and infants. The social world hence becomes akin to a pervasive cinematic screen (see Gallese and Guerra 2015) whose content, made of the actions, emotions, and sensations of others, far from being merely passively registered by the infant, becomes a cogent solicitation not only to develop ever growing forms of social relations and interactions, but also to actively model the way others' actions, emotions, and sensations are mapped, thus literally shaping the functionality of developing brain circuits and the activity of the neurons they are made of. In this way, human cognitive development and social learning are far less dependent on predetermined "genetic instructions" than for all other primates, as they rely so heavily on individuals' idiosyncratic historical experiences within a given sociocultural environment characterized by active interpersonal interactions.

Neoteny reveals that the genetic body plan specified by regulatory genes affects the developmental morphology of the body, linking it to cognition. This last aspect leads to a further capital role of neoteny for embodied cognition: its "exapted" quality. "Exaptation" (Gould and Lewontin 1979) refers to the shift in the course of evolution of a given trait or mechanism, which is subsequently reused to serve new purposes and functions. Neoteny favors and facilitates the development of cognitive exapted qualities, such as the development of more sophisticated forms of communication and symbolic expression, by reusing neural resources originally selected for other purposes (see Gallese 2000, 2014).

One of the earliest material indications of human symbolic expression, the engraved pieces of ochre found in Blombos Cave near Cape Town, South Africa, is characterized by geometric incisions, clearly the outcome of intentional actions by a human hand. Researchers have proposed that these specimens, dating to about 70,000 years BC, might be the earliest forms of abstract representation and conventional symbolic material tradition recorded thus far (Henshilwood et al. 2002; d'Errico et al. 2003). It is difficult not to speculate that these abstract engravings could be the outcome of the exaptation in "ritualized" form of everyday utilitarian hand gestures normally used to build tools, cut meat, or flense an animal to manufacture clothes. The human brain-body system developed over millions of years to interact with a physical world inhabited by inanimate three-dimensional objects and other animate bodies. The human hand—which is oddly represented next to the mouth in the sensorimotor cortex, despite the fact that hand and mouth are not anatomically contiguous, almost as if betraying their common social and communicative vocation—at some particular time freed itself from the mere utilitarian role of tool-maker, thanks to which man could hunt, feed, and clothe himself, and build dwellings in which to live. The human hand became a builder of symbols: at first, probably, body ornamentations of which no traces are left, then jewels, and finally representations of the world on rock walls and caves. Symbolic expression, which since modernity we have also learned to recognize as artistic production, is the ability to transform material objects by giving them a meaning that they would not have in nature by themselves, through the action of the hand. The hand creating symbols turns pragmatics into po(i)etics, thus enabling us to give presence to meaning, to produce truth starting from a knowledge that up to that moment had been exclusively put to the service of utilitarian goals.

Following this logic, the process of symbol making, despite its articulation as progressive abstraction and externalization from the body, seems to maintain its ties to the body, not only because the body is the instrument of symbol production, but also because, as I will propose later on, the body is the main instrument of symbols' reception.

3   Bodily Selves, Embodied Simulation, and the "Body-Snatchers"

Strictly speaking, as argued elsewhere, it is incorrect to state that the mind and social cognition are embodied *in* the body (see Gallese 2000, 2014; Gallese and Cuccio 2015). This notion entails a still dualistic conception of human nature. Despite an almost unanimous agreement on the fact that human cognition is somehow related to the material nature of the human brain, many philosophers and cognitive scientists still hold that we must draw a clear-cut distinction between cognitive-linguistic processes and sensorimotor processes. Furthermore, according to this mainstream view, the brain is divorced from the body, the former being modeled as an information-processing device, "a sort a magic box" housing algorithmic wonders, while the body is considered unable to shed light on the above-mentioned information-processing algorithms supposedly presiding over the cognitive aspects of human existence.

Organisms, however, are not information-processing machines, and the meaning they attribute to their life in the world does not consist of information processing: meaning is first and foremost *experienced* meaning. Embodied simulation and its relation to language and cognition cast severe doubts on this hypothesis and support instead the thesis that human cognition is tightly and necessarily dependent on the kind of body we have. Humans, as all other organisms, evolved by adapting to a physical world that obeys a series of physical laws, which in turn shaped their brain-body and the way their brain-body interacts with that physical world.

The body—the acting and sensing body—plays a constitutive role in what we call the *minimal self*. The concept of minimal, prereflective, or "core self" (Rochat 2004) as a minimalist level of subjective experience highlights the potential contribution of bodily experience to its constitution. Some aspects of the minimal self proposed by contemporary philosophical and empirical research include the notion of perspective and first-person perspective, the "mineness" of the phenomenal field, the question of transparency, the embodiment of point of view, and issues of agency and ownership (see Cermolacce, Naudin, and Parnas 2007). The philosophical tradition of phenomenology emphasizes the necessity of embodiment of the self for all the above cited aspects of self experience. As proposed by Cermolacce, Naudin, and Parnas, in phenomenology,

the field of experience is not yet considered to be *subjective* because this predicate already implies that there is a subject. For phenomenology, the very idea of the *subject* articulates itself in experience. In this sense, the manifestation and appearing of experience are the *conditions for* the experience of the subject in question. (Cermolacce, Naudin, and Parnas 2007, 704)

This perspective has important implications for empirical studies of social cognition: instead of searching for the neural correlates of a predefined, explicit, and reflective self-knowledge, empirical research can start investigating which kind of experience allows an implicit and prereflective self-knowledge to emerge and how it does so. Recent evidence shows that (1) the multisensory integration leading to the experience of our body as our own is not merely due to perceptual association of the visual and proprioceptive inputs, but is conditioned by whether we are able to perform actions with a given body part; (2) we recognize ourselves as agents because of the congruence between self-generated movements and their expected consequences; and (3) the first-person perspective can be derived from the phenomenological idea that the world *appears* as constrained by a mobile bodily self, that is, by the spatially

located point of view, and the orientation and the attitudes relative to the subject's sensorimotor background capacities (for a review, see Gallese and Sinigaglia 2011a, 2011b).

This evidence prompted the hypothesis that there is a sense of body that is enactive in
nature—that the most primitive sense of self is bodily self. According to this hypothesis,
the body is primarily given to us as a source of action, that is, as the variety of motor potentialities defining the horizon of our interaction with the world (Gallese 2000; Gallese and
Sinigaglia 2010, 2011a, 2011b). Our body, as a self-propelled, mobile bodily self, is experienced as specifying the variety of motor potentialities defining the horizon of the world we
interact with. This is what Erwin Straus's (1960) notion of the human body plan (*Bauplan*)
refers to.

Researchers have recently demonstrated the relationship among the minimal sense of self,
self-body recognition, and the cortical motor system. Ferri et al. (2011) showed that the
motor experience of one's own body, even at a covert level, allows an implicit and prereflective bodily self-knowledge to emerge, leading to a self/other distinction ("self-advantage").
Ferri et al. (2012) corroborated this hypothesis by demonstrating that the implicit self-
advantage correlates with a strong activation of the cortical motor system: the contralateral
ventral premotor cortex is uniquely and specifically activated during mental rotation of the
participants' own dominant hand. Furthermore, processing participants' hands activates a
bilateral cortical network formed by the supplementary and presupplementary motor areas,
the anterior insula and the occipital cortex.

These results show a tight relationship between the bodily self-related multimodal integration carried out by the cortical motor areas specifying the motor potentialities of one's
body and guiding its motor behavior and the implicit awareness one entertains of one's body
as one's own body and of one's behavior as one's own behavior. The notion I am proposing
of the minimal self as bodily self allows us to link the self's openness to the world to the
motor potentialities entailed by its bodily nature.

Even more interesting is that the discovery of mirror neurons (MNs) in the brain of
macaques (Gallese et al. 1996; Rizzolatti et al. 1996; Rizzolatti, Fogassi, and Gallese 2001) and
the subsequent discovery of mirror mechanisms (MMs) in the human brain (see Gallese,
Keysers, and Rizzolatti 2004; Rizzolatti and Sinigaglia 2010) demonstrate that there is a direct
modality of access to other bodily selves' behavior and—possibly—to their meaning. This
access modality is different from the explicit attribution of propositional attitudes. MNs are
motor neurons that respond not only to the execution of movements and actions, but also
during their perception when executed by others. The relational character of behavior as
mapped by the cortical motor system enables the appreciation of purpose without relying on
explicit propositional inference.

If one considers not only the properties of MNs but also those displayed by neighboring
canonical neurons and F4 neurons in macaques or by their human analogues (for review, see
Gallese 2014), it turns out that the functionality of the cortical motor system literally carves
out a pragmatic *Umwelt*, dynamically surrounding our body. The profile of our peripersonal
space is not arbitrary: it maps and delimits a perceptual space that expresses—and is

constituted by—the motor potentialities of the body parts it surrounds. Manipulable objects, such as the coffee mug sitting on my desk, are not only 3D shapes with a given size, orientation, color, texture, and contrast. The coffee mug I am now looking at is the potential target of my intentional action, and it is mapped as such by my cortical motor system through canonical neurons (Murata et al. 1997; Rizzolatti, Fogassi, and Gallese 2000; Raos et al. 2006; Umiltà et al. 2007, 2008). I submit that an important component of my perceptual experience of the coffee mug is determined, constrained, and ultimately constituted by the limits posed by what my body can potentially do with it. The body thus constrains how its motor potentialities can be mapped by the brain circuits controlling its motor behavior.

This evidence reveals how the brain can map intentional actions. Such mapping appears to be more "basic" with respect to the standard propositional account of the representation of action. The intentional character, the "aboutness" of our mind is deeply rooted in the intrinsic relational character of the bodily format of bodily actions as they are mapped by distinct populations of cortical motor neurons. This, in turn, shows how intrinsically intertwined action, perception, and cognition are (Hurley 1998, 2008; Gallese 2000).

The same motor brain circuits that control the ongoing behavior of individuals within their environment also map distances, locations, and objects in that very same environment, thus defining and shaping in bodily constrained motor terms their representational content. The way the visual world is represented by the motor system incorporates agents' idiosyncratic ways of interacting with it, and such idiosyncratic ways are, in turn, shaped and determined by the body and by the way it functions. To put it simply, the producer and repository of representational content is not the brain per se, but the brain-body system, by means of its interactions with the world of which it is a part.

Action constitutes only one dimension of the rich set of experiences involved in interpersonal relations. Indeed, every interpersonal relation implies a multiplicity of states such as, for instance, the experience of emotions and sensations. As originally hypothesized by Goldman and Gallese (2000), empirical research demonstrated that the very same nervous structures involved in the subjective experience of emotions and sensations are also active when such emotions and sensations are recognized in others. Thus, a multiplicity of "mirroring" mechanisms is present in our brain. I have proposed that these mechanisms, thanks to the "intentional attunement" they generate (Gallese 2006), allow us to recognize others as our fellow beings, likely making intersubjective communication and mutual implicit understanding possible.

All of these aspects can be more thoroughly understood when defined by a unifying functional process, neither confined to mindreading nor committed to a propositional representational format. According to my interpretation, embodied simulation is just this sort of functional mechanism.[4] The functional architecture of embodied simulation seems

---

4. For the sake of concision, I cannot address here other relevant aspects of embodied simulation in relation to mental imagery and language. On these topics, see Gallese and Lakoff 2005; Gallese 2008, 2014; Gallese and Sinigaglia 2011a; Gallese and Cuccio 2015.

to constitute a basic characteristic of our brain, making possible our rich and diverse inter-subjective experiences, as it is at the basis of our capacity to empathize with others. Intentionality, the aboutness of our mental content is—in the first place—an exapted property of the action models instantiated by the cortical motor system to guide the body's behavior (see Gallese 2000, 34) and by the somatosensory and visceromotor cortical networks mapping our sensations and emotions. These cortical networks instantiate not only causative properties but also content properties, mapped in bodily format. The reuse of neural resources is a constitutive part of our observing and making sense of others. Observing others acting or experiencing emotions and sensations elicits our sensorimotor and emotion-related resources as if we were acting, or experiencing similar emotions and sensations, and *not* as if we were *explicitly representing* ourselves acting or experiencing those very same emotions and sensations. This is not a stepwise process: the activation of MMs entails the functional attribution to others of the very same actions, emotions, and sensations they map.

The notion of bodily formatted representations has triggered several criticisms. For sake of concision, I will briefly focus here on the objections raised by Shaun Gallagher (2015a, 2015b).[5] Gallagher argued that in recent years there has been a

reactionary move by some theorists, who I'll call the *body snatchers* because in some real sense they devise a version of embodied cognition that leaves the body out of it. They still retain the term "embodied," but in fact, for them, the body, per se, is not necessarily involved in the real action of cognition. Rather, the real action, all the essential action, occurs in the brain. (Gallagher 2015a, 97, italics added)

The main target of Gallagher's recent work is the notion of bodily formatted representations, as proposed by Goldman (Goldman and de Vignemont 2009; Goldman 2012). However, in a subsequent paper Gallagher puts me in the same ballpark, stating that

we can specify in precise terms what embodiment means in Gallese's theory since his concept of reused neural mechanisms is closely tied to the recently introduced concept of body- or B-formatted representations. This concept clearly steers the interpretation of reuse back to more classic representational conceptions of simulation and away from a more enactive interpretation. (Gallagher 2015b, 40)

According to Gallagher, this

reduces embodiment to neural representations fully located within an individual's brain. It's a version of embodied cognition that does not involve the body per se: it's a version of social cognition that does not rely on the social dimension of interaction to explain anything of importance; it's a theory of inter-subjectivity that does not depend on intersubjective processes except in the most minimal causal sense. … Social cognition, on these theories, seems a lonely, solipsistic activity. (Gallagher 2015b, 41)

5. For a reply to Gallagher's arguments against embodied simulation, see Gallese and Sinigaglia 2011a; Gallese 2014.

I find it rather odd to be listed among the supporters of a solipsistic and disembodied view of intersubjectivity, since my empirical research during the last twenty years or so, and the theoretical speculations it has stimulated, has moved exactly in the opposite direction, contributing to give to embodied cognition a renewed and, at last, empirically based thrust.[6] I am afraid that Gallagher, while misconstruing my position, does not pay enough attention to the link between the body and the functional properties characterizing the multimodal motor neurons controlling its behavior. Gallagher properly acknowledges that the brain cannot be understood without linking it to the body, but he fails to use this notion when interpreting the functional relevance of neurons and what their activity can account for. When arguing against B-formatted representations as a proper way to describe how embodied cognition works and as evidence of an embodied subpersonal level of description of social cognition, Gallagher paradoxically conceives of neurons and what they map in the very same way as classic cognitive science does, that is, as the instantiation of information-processing algorithms. In other words, Gallagher seems to ignore the fact that cortical motor, somatosensory, and visceromotor neurons are functioning the way they do *precisely because* they are connected to the body.

When planning and executing a motor act such as grasping an object with my hand, bodily factors (e.g., biomechanical, dynamical, and postural) constrain what the neurons controlling such grasping can map or represent. This neural representational format in turn constrains the B-formatted representation of a single motor outcome (e.g., grasping something) or of a hierarchy of motor outcomes (e.g., grasping something for eating it), making it different from a propositional representation of that outcome or of that hierarchy of outcomes (Gallese and Sinigaglia 2011a). As I argued before, the body is not only the transcendental, irreducible condition of possibility for us to experience the world and others; it is also the source of the peculiar quality of the neural functions presiding over its motor potentialities (see Gallese 2000; Gallese and Sinigaglia 2010, 2011a, 2011b).

How does Gallagher explain that we experience our body as a dynamic binding principle operating at the level of motor intentionality? How does Gallagher explain that motor intentionality underlies action-related bodily self-awareness, determining the way bodily movements are experienced and bound in the temporal domain together with their sensory effects? Can all these aspects be explained, as Gallagher does, by referring to "full bodied interactions with others in complexly contextualized situations ... irreducible to processes found within one individual" (Gallagher 2015b, 41)? Do human social cognition and intersubjectivity just really boil down to Tango dance steps?

6. As aptly put by Marc Slors: "The question is what the chances are that the hypothetical early enactivist theory of social perception would predict neural resonance to be a part of the sensorimotor contingencies underlying social perception. What are the chances that the motor side of these contingencies would have been predicted to consist partly in the neural mimicking of the observed action? The chances would be very slim, I think, and that may be an understatement" (Slors 2010, 443).

To rightly emphasize the crucial role played by social interactions in shaping the human mind and in allowing social cognition is not enough, as it begs the question of how such interactions are made possible, controlled, mapped, and "directly perceived." The appeal to direct perception of others' mental states fails to account for the fact that abilities to make sense of others' actions, emotions, and sensations can be modulated by perceivers' abilities to act and to experience emotions and sensations. Furthermore, how is it that the actions of others are perceived as something we can respond to? The main problem I have with this form of enactivism heralded, among others, by Gallagher is that it seems to confound the explanations with the *explananda*. Finally, given that Gallagher seems to acknowledge the plausibility of neural reuse, I do not see why the fact that our motor systems were "built primarily for action" (Gallagher 2015b, 42) should in principle contradict the fact that they can be reused to understand others' motor outcomes and basic motor intentions, through their mapping in B-format representations.

Can we really give up the notion of representation? I don't think so. Let's briefly see why, with the help of Aristotle. For Aristotle animals are essentially characterized by two aspects: feeling/perception (*aisthesis*) and desire (*orexis*). These two aspects are closely intertwined. In fact, perception cannot be thought of as merely recording what is in the world. Computational machines do that. Perception in animals, by contrast, is always accompanied by the experience of pleasure or pain, and it is from this intermingling of perception, pleasure, and pain that we can explain most of our behaviors. For Aristotle, as for many others after him, like Freud, it is precisely the pursuit of pleasure and the avoidance of pain that drive us to act. Following this argument, the desire that moves us to act implies the notion of mental representation. No animal could want something that has no representation, unless that something is, for example, always directly present in the immediate perceptual context. However, desires are not always directed at what is present. Often animals have to search for what they desire, be it food, water, or a mate. The search for something that is not present implies its mental representation. Absence, what is not here or is not yet here, is the paradigm of mental representation. According to Aristotle (2008, 2012), this level of representation, which he described as *phantasia aistetiché* or sensitive imagination, is deeply rooted in our senses. The Aristotelian notion of *phantasia aistetiché* recalls the phenomenon of mental imagery. The imagination or representation of something, in fact, correlates with the activation of the brain areas linked to this type of experience. Man clearly shares this type of B-format representations with other animals.

The advantage of this position with respect to radical enactivism consists in the fact that it can be also applied to distinctive forms of human social cognition, such as the experience of symbolic fictional worlds, or when beholding man-made images—and hence in the absence of "full-bodied interactions with others in complexly contextualized situations" (Gallagher 2015b, 41).

4 The Neotenic Gaze: Liberated Embodied Simulation and Aesthetic Experience

Being a human self not only means experiencing physical reality, but also being able con- ceive possible worlds, to surrender to imagination and to fictional worlds. An interesting topic for cognitive neuroscience is how our brain-body system enables us to navigate in real and fictional worlds, constantly switching among them. Embodied simulation, as a new model of perception and cognition, also reveals the constitutive relationship between the body and symbolic expression, for example by showing that the human experience of man- made images—broadly speaking—should always be understood as a natural form of rela- tional experience. We live in relation with other people and objects that are present in our real world, but we also live in relation with people and objects that are part of the imagi- nary fictional worlds created by human symbolic expression, which in the course of our cultural history we have come to identify as art. Both kinds of relationship are rooted in our brain-body system. To grasp the basis of the complex multimodality these relationships imply, we have to understand our own brain and body. The very same forms of sociality that enable symbolic and artistic expressions and their reception are, at their basis, a further exemplification of intersubjectivity conceived of as intercorporeality. Neuroscience allows us to understand how the line between what we call reality and the imaginary is much less sharp and clear than one might think. Indeed, experiencing an emotion and imagining it are both underpinned by the activation of partly identical brain circuits, although differ- ently connected in these two different cognitive and phenomenal situations. Similarly, to see something and to imagine it, to act and to imagine acting, share the activation of partly common brain circuits. A recent high-density EEG study showed that the brain cir- cuits that inhibit action execution are partly the same that allow us to imagine acting (Angelini et al. 2015). All these examples of dual activation pattern of the same brain circuits represent a further expression of embodied simulation and the related notion of neural reuse.

A further advantage of embodied simulation consists in the possibility of addressing human forms of creative symbolic expression in terms of social performativity. Indeed, the biocultural approach to the naturalization of art and aesthetics, heavily influenced by cultural anthropology, emphasizes the performative character of human creativity. In *The Perception of the Environment*, Tim Ingold writes:

Hunters and gatherers of the past were painting and carving, but they were not "producing art." ... We must cease thinking of painting and carving as modalities of the production of art, and view art instead as one rather peculiar, and historically very specific objectification of the activities of painting and carving. (Ingold 2000, 131)

In a similar vein, Ellen Dissanayake writes: "Art is not an ornamental and dispensable luxury, but intrinsic to our species. ... Art as a behavioral complex is an inherited tendency to act in a certain way, given appropriate circumstances" (1992, 224).

I posit that embodied simulation is congruent with this approach and can shed new light on aesthetic experience in at least two ways: first, because of the bodily feelings triggered by the outcomes of human symbolic expression, and by means of the embodied simulation they evoke. In such a way, embodied simulation generates the peculiar *seeing-as* characterizing our aesthetic experience of the images we look at. Second, because of the intimate relationship between the symbol-making gesture and its reception by beholders, in virtue of the motor representation that produces the image and, by means of simulation, enables its experience (Freedberg and Gallese 2007; see also Gallese and Di Dio 2012; Gallese 2012, 2014; Gallese and Gattara 2015). When looking at a graphic sign we unconsciously simulate the gesture that has produced it (Heimann, Umiltà, and Gallese 2013). Embodied simulation, through its plasticity and modulation, might be also the vehicle of the projective qualities of our aesthetic experience, where our personality, cultural identity, the context, our mood and disposition, shape the way we relate to a given perceptual object. Embodied simulation, if conceived of as the dynamic instantiation of our implicit memories, can relate perceptual object and beholder with a specific, unique, and historically determined quality. This projective quality of embodied simulation complements its receptive features.

However, there is a clear distinction between our experience of the real world and our experience of the worlds of fiction. Our relationship with fictional worlds is double-edged: on the one hand, we pretend them to be true, while, on the other, we are fully aware they are not. When beholding a painting at an art museum, for example, several powerful framing effects take place. First, we find ourselves in a context where the images hanging on the wall are taken to be artworks. Second, once we let the image capture our attention, the frame surrounding it almost disappears, as we are fully absorbed by the image. As the Italian philosopher Alfonso Iacono put it, we experience a sort of "tail of the eye" vision, which, according to him, characterizes all of our relationships with the "intermediate worlds" of fiction (Iacono 2005, 2010). Our appreciation of man-made images implies intermediate worlds where territory and map do overlap. As Iacono wrote, one enters the picture through the frame forgetting about having entered. This is at the origin of the process of naturalization—that is, the process that makes artificial, historical, and changeable events appear natural, eternal, and unmodifiable (Iacono 2010, 84).

Despite the fact that the body is at the core of our perceptions, our understanding, and our imagination, our relationship with fictional worlds is still mainly explained in purely cognitive terms, that is, following Coleridge, in terms of "suspension of disbelief." This explanation, however, is partial at best. It has been proposed (Wojciehowski and Gallese 2011) that embodied simulation can be relevant to our experience of fictional worlds because of the feeling of body they evoke by means of activating the potentiation of the mirroring mechanisms. In such a way, embodied simulation generates the specific attitude informing our aesthetic experience. Such potentiation supposedly boosts the bodily memories and imaginative associations fictional content can awaken in our minds, thus providing the idiosyncratic character of its appreciation.

How is such potentiation achieved? One important context-dependent aspect characterizing our relationship to fictional worlds deals with our distancing from the unrelated external world, which remains at the periphery of our attentional focus, very much like the frame surrounding the image we are beholding. According to my hypothesis, such distancing, this temporary suspension of the active grip on our daily occupations, liberates new simulative energies. Our experience of fictional worlds, besides being a suspension of disbelief, can thus be interpreted as a sort of "liberated embodied simulation." When adopting an aesthetic attitude, our embodied simulation becomes liberated, that is, it is freed from the burden of modeling our actual presence in daily life (Gallese 2011, 2012; Wojciehowski and Gallese 2011; Gallese and Guerra 2015). Through an immersive state in which our attention is focused on the fictional world, we can fully deploy our simulative resources, letting our defensive guard against daily reality down for a while.

Finally, I posit that when engaged with fictional worlds, the contextual bodily framing—our being still—additionally boosts our embodied simulation. Our being still simultaneously enables us to fully deploy our simulative resources at the service of the immersive relationship with the fictional world, thus generating an even greater feeling of body. Being forced to inaction, we are more open to feelings and emotions. The specific and particularly moving experience generated when immersed in fictional worlds is thus likely also driven by this sense of safe intimacy with a world we not only imagine, but also literally embody.

When we relate to fictional worlds, our attitude toward their content can be characterized as a sort of "neotenic look," similar to the way we looked at the world during that early period of our development, in which, because of our poor motor autonomy, our interactions with the world were mainly mediated by the embodied simulation of events, actions, and emotions animating our social landscape. Probably we learn to calibrate gestures and expressions and to match them with experiences of pleasure or displeasure observing them in others thanks to embodied simulation and its plasticity.

When we relate to fictional worlds, as when contemplating art, our relative immobility is no longer the consequence of the immaturity of our sensorimotor development, but the outcome of our deliberate decision. However, immobility, that is, a greater degree of motor inhibition, probably allows us to allocate more neural resources, intensifying the activation of bodily formatted representations, and in so doing, making us focus more intensely on what we are simulating. Perhaps it is no coincidence that some of the most vivid fictional experiences we entertain, as those occurring during dreaming activity, are paralleled by massive inhibition of the muscle tone in our body.

During the aesthetic experience of fictional worlds, our experience is almost exclusively mediated by a simulative perception of the events, actions and emotions representing the content of fiction. For example, when watching a movie or reading a novel, we not only focus our attention on them, but our immobility enables us to fully deploy our embodied simulation resources and put them at the service of our immersive relationship with the narrative. This hypothesis can plausibly contribute to explaining the difference between our "aesthetic attitude" toward fictional worlds and our ordinary consciousness of prosaic reality.

Conclusion

I addressed and discussed the heuristic value of cognitive neuroscience for a novel understanding of human social cognition, by critically showing the limits and potentialities of the empirical investigation of the brain-body. I proposed that a comparative approach greatly reduces the risk of turning neuroscientific investigation into the search for a one-one mapping between concepts and distinct locations in the brain, because this epistemic strategy is most likely doomed to failure. I introduced the notion of neoteny and proposed that it can shed new light on the apparent discrepancy between the remarkable similarity between humans and chimpanzees' genetic endowment and their great social and cognitive dissimilarity. I proposed that the notion of embodied simulation within the frame of neoteny can account for the specific features of human social cognition and the building of human culture, showing that a new understanding of intersubjectivity can benefit from a bottom-up study and characterization of the nonpropositional and nonmetarepresentational aspects of social cognition (see also Gallese 2003, 2007).

One key feature of the new approach to intersubjectivity proposed here is the investigation of the neural bases of our capacity to be attuned to the intentional relations of others. At a basic level, our interpersonal interactions do not make explicit use of propositional attitudes. This basic level consists of embodied simulation enabling the constitution of a shared meaningful interpersonal space. The shared intersubjective space in which we live from birth is a basic ingredient of the specific human attitude for meaning making. Self and other relate to each other because they are opposite extensions of the same correlative and reversible "we-centric" space (Gallese 2003). Observer and observed are part of a dynamic system governed by reversible rules. By means of intentional attunement, "the other" is much more than a different representational system; it becomes a bodily self, like us.

The specific use of cognitive neuroscience proposed here leads to a new take on social cognition. This new take demonstrates on neurophysiological grounds the constitutive role played in foundational aspects of social cognition by the human body, when conceived of in terms of its motor potentialities. Of course, this covers only part of social cognition. However, embodied simulation also provides an epistemological approach, potentially useful for the empirical investigation of the more cognitively sophisticated aspects of human social cognition, such as the creation of fictional worlds. This new epistemological approach to social cognition can generate predictions about the intrinsic functional nature of our social cognitive operations, cutting across, and not being subordinated to, a specific mind ontology, such as that purported by the approach of classic cognitive science.

Acknowledgments

This work was supported by the EU Grant TESIS and by a grant by Chiesi Foundation to VG.

References

Ammaniti, M., and V. Gallese. 2014. *The Birth of Intersubjectivity: Psychodynamics, Neurobiology, and the Self*. W. W. Norton.

Angelini, M., M. Calbi, A. Ferrari, B. Sbriscia-Fioretti, M. Franca, V. Gallese, and M. A. Umiltà. 2015. Motor inhibition during overt and covert actions: An electrical neuroimaging study. *PLoS One* 10 (5): e0126800.

Aristotle. 2008. *De Anima*. Penguin Classics.

Aristotle. 2012. *The Art of Rhetoric*. Harper Press.

Bolk, L. 1926a. On the problem of anthropogenesis. *Proceedings of the Section of Sciences, Koninklijke Akademie van Wetenschappen te Amsterdam* 29:465–475.

Bolk, L. 1926b. *Das Problem der Menschwerdung*. Gustav Fischer.

Bolk, L. 1929. Origin of racial characteristics in man. *American Journal of Physical Anthropology* 13:1–28.

Cermolacce, M., J. Naudin, and J. Parnas. 2007. The "minimal self" in psychopathology: Re-examining the self-disorders in the schizophrenia spectrum. *Consciousness and Cognition* 16:703–714.

Chittka, L., and J. Niven. 2009. Are bigger brains better? *Current Biology* 19 (21): R995–R1008.

d'Errico, F., C. Henshilwood, G. Lawson, M. Vanhaeren, A.-M. Tillier, M. Soressi, F. Bresson, et al. 2003. Archaeological evidence for the origins of language, symbolism and music: An alternative multidisciplinary perspective. *Journal of World Prehistory* 17:1–70.

De Vos, J., and E. Pluth, eds. 2016. *Neuroscience and Critique*. Routledge.

Dissanayake, E. 1992. *Homo Aestheticus: Where Art Comes from and Why*. University of Washington Press.

Ferri, F., F. Frassinetti, M. Ardizzi, M. Costantini, and V. Gallese. 2012. A sensorimotor network for the bodily self. *Journal of Cognitive Neuroscience* 24:1584–1595.

Ferri, F., F. Frassinetti, M. Costantini, and V. Gallese. 2011. Motor simulation and the bodily self. *PLoS One* 6:e17927.

Freedberg, D., and V. Gallese. 2007. Motion, emotion and empathy in esthetic experience. *Trends in Cognitive Sciences* 11:197–203.

Gallagher, S. 2015a. Invasion of the body snatchers: How embodied cognition is being disembodied. *Philosophers' Magazine*, April, 96–102.

Gallagher, S. 2015b. Reuse and body-formatted representations in simulation theory. *Cognitive Systems Research* 34 (June): 35–43.

Gallese, V. 2000. The inner sense of action: Agency and motor representations. *Journal of Consciousness Studies* 7:23–40.

Gallese, V. 2001. The "shared manifold" hypothesis: From mirror neurons to empathy. *Journal of Consciousness Studies* 8 (5–7): 33–50.

Gallese, V. 2003. A neuroscientific grasp of concepts: From control to representation. *Philosophical Transactions of the Royal Society of London, Series B: Biological Sciences* 358:1231–1240.

Gallese, V. 2006. Intentional attunement: A neurophysiological perspective on social cognition and its disruption in autism. *Brain Research* 1079:15–24.

Gallese, V. 2007. Before and below theory of mind: Embodied simulation and the neural correlates of social cognition. *Philosophical Transactions of the Royal Society of London, Series B: Biological Sciences* 362:659–669.

Gallese, V. 2008. Mirror neurons and the social nature of language: The neural exploitation hypothesis. *Social Neuroscience* 3:317–333.

Gallese, V. 2011. Embodied simulation theory: Imagination and narrative. *Neuro-psychoanalysis* 13 (2): 196–200.

Gallese, V. 2012. Aby Warburg and the dialogue among aesthetics, biology and physiology. *Ph* 2:48–62.

Gallese, V. 2014. Bodily selves in relation: Embodied simulation as second-person perspective on intersubjectivity. *Philosophical Transactions of the Royal Society of London, Series B: Biological Sciences* 369 (1644): 20130177.

Gallese, V., and V. Cuccio. 2015. The paradigmatic body: Embodied simulation, intersubjectivity and the bodily self. In *Open MIND*, ed. T. Metzinger and J. M. Windt, 1–23. MIND Group.

Gallese, V., and C. Di Dio. 2012. Neuroesthetics: The body in esthetic experience. In *The Encyclopedia of Human Behavior*, vol. 2, ed. V. S. Ramachandran, 687–693. Elsevier Academic Press.

Gallese, V., L. Fadiga, L. Fogassi, and G. Rizzolatti. 1996. Action recognition in the premotor cortex. *Brain* 119:593–609.

Gallese, V., and A. Gattara. 2015. Embodied simulation, aesthetics and architecture: An experimental aesthetic approach. In *Mind in Architecture: Neuroscience, Embodiment, and the Future of Design*, ed. S. Robinson and J. Pallasmaa, 161–179. MIT Press.

Gallese, V., and M. Guerra. 2015. *Lo schermo empatico: Cinema e neuroscienze*. Raffaello Cortina Editore.

Gallese, V., C. Keysers, and G. Rizzolatti. 2004. A unifying view of the basis of social cognition. *Trends in Cognitive Sciences* 8:396–403.

Gallese, V., and G. Lakoff. 2005. The brain's concepts: The role of the sensory-motor system in reason and language. *Cognitive Neuropsychology* 22:455–479.

Gallese, V., and C. Sinigaglia. 2010. The bodily self as power for action. *Neuropsychologia* 48:746–755.

Gallese, V., and C. Sinigaglia. 2011a. What is so special with embodied simulation? *Trends in Cognitive Sciences* 15 (11): 512–519.

Gallese, V., and C. Sinigaglia. 2011b. How the body in action shapes the self. *Journal of Consciousness Studies* 18 (7–8): 117–143.

Goldman, A. I. 2012. A moderate approach to embodied cognitive science. *Review of Philosophy and Psychology* 3:71–88.

Goldman, A., and F. de Vignemont. 2009. Is social cognition embodied? *Trends in Cognitive Sciences* 13:154–159.

Goldman, A., and V. Gallese. 2000. Reply to Schulkin. *Trends in Cognitive Sciences* 4:255–256.

Gould, S. J. 1977. *Ontogeny and Phylogeny*. Belknap Press of Harvard University Press.

Gould, S. J. 1979. On the importance of heterochrony for evolutionary biology. *Systematic Zoology* 28 (2): 224–226.

Gould, S. J., and R. C. Lewontin. 1979. The spandrels of San Marco and the Panglossian paradigm: A critique of the adoptionist programme. *Proceedings of the Royal Society of London* 205:281–288.

Gualandi, A. 2013. *L'occhio la mano e la voce: Una teoria comunicativa dell'esperienza umana*. Mimesis Edizioni.

Heimann, K., M. A. Umiltà, and V. Gallese. 2013. How the motor-cortex distinguishes among letters, unknown symbols, and scribbles: A high density EEG study. *Neuropsychologia* 51:2833–2840.

Henshilwood, C., F. d'Errico, R. Yates, Z. Jacobs, C. Tribolo, G. A. T. Duller, N. Mercier, et al. 2002. Emergence of modern human behavior: Middle Stone Age engravings from South Africa. *Science* 295:1278–1280.

Hurley, S. 1998. *Consciousness in Action*. Harvard University Press.

Hurley, S. 2008. Understanding simulation. *Philosophy and Phenomenological Research* 77:755–774.

Iacono, A. M. 2005. Gli universi di significato e i mondi intermedi. In *Mondi intermedi e complessità*, ed. A. G. Gargani and A. M. Iacono, 5–39. Edizioni ETS.

Iacono, A. M. 2010. *L'illusione e il sostituto: Riprodurre, imitare, rappresentare*. Bruno Mondadori Editore.

Ingold, T. 2000. *The Perception of the Environment: Essays on Livelihood, Dwelling and Skill*. Routledge.

King, M. C., and A. C. Wilson. 1975. Evolution at two levels in humans and chimpanzees. *Science* 188:107–116.

Mazzeo, M. 2003. *Tatto e linguaggio: Il corpo delle parole*. Editori Riuniti.

Mazzeo, M. 2015. When less is more: Neoteny and language. *Cahiers Ferdinand De Saussure* 67:115–130.

Murata, A., L. Fadiga, L. Fogassi, V. Gallese, V. Raos, and G. Rizzolatti. 1997. Object representation in the ventral premotor cortex (area F5) of the monkey. *Journal of Neurophysiology* 78:2226–2230.

Pinker, S. 1994. *The Language Instinct*. HarperCollins.

Pinker, S. 1997. *How the Mind Works*. W. W. Norton.

Prochiantz, A. 1989. *La construction du cerveau*. Hachette.

Prochiantz, A. 2012. *Qu' est-ce que le vivant?* Seuil.

Raos, V., M. A. Umiltà, A. Murata, L. Fogassi, and V. Gallese. 2006. Functional properties of grasping-related neurons in the ventral premotor area F5 of the macaque monkey. *Journal of Neurophysiology* 95:709–729.

Reynert, P. 2016. Neuroscientific dystopia: Does naturalism commit a category mistake? In *Neuroscience and Critique*, ed. J. de Vos and E. Pluth, 62–78. Routledge.

Rizzolatti, G., L. Fadiga, V. Gallese, and L. Fogassi. 1996. Premotor cortex and the recognition of motor actions. *Brain Research: Cognitive Brain Research* 3:131–141.

Rizzolatti, G., L. Fogassi, and V. Gallese. 2000. Cortical mechanisms subserving object grasping and action recognition: A new view on the cortical motor functions. In *The New Cognitive Neurosciences*, 2nd ed., ed. M. S. Gazzaniga, 539–552. MIT Press.

Rizzolatti, G., L. Fogassi, and V. Gallese. 2001. Neurophysiological mechanisms underlying the understanding and imitation of action. *Nature Reviews: Neuroscience* 2:661–670.

Rizzolatti, G., and C. Sinigaglia. 2010. The functional role of the parieto-frontal mirror circuit: Interpretations and misinterpretations. *Nature Reviews: Neuroscience* 11:264–274.

Rochat, P. 2004. The emergence of self-awareness as co-awareness in early child development. In *The Structure and Development of Self-Consciousness*, ed. D. Zahavi, T. Grunbaum, and J. Parnas. John Benjamins.

Sakai, T., A. Mikami, M. Tomonaga, M. Matsui, J. Suzuki, Y. Hamada, M. Tanaka, et al. 2011. Differential prefrontal white matter development in chimpanzees and humans. *Current Biology* 21:1397–1402.

Slors, M. 2010. Neural resonance: Between implicit simulation and social perception. *Phenomenology and the Cognitive Sciences* 9:437–458.

Somel, M., H. Franz, Y. Zheng, A. Lorenc, G. Song, T. Giger, J. Kelso et al. 2009. Transcriptional neoteny in the human brain. *Proceedings of the National Academy of Sciences* 106 (14): 5743–5748.

Straus, E. 1960. *Psychologie der menschlichen Welt*. Springer.

Umiltà, M. A., T. Brochier, R. L. Spinks, and R. N. Lemon. 2007. Simultaneous recording of macaque premotor and primary motor cortex neuronal populations reveals different functional contributions to visuomotor grasp. *Journal of Neurophysiology* 98:488–501.

Umiltà, M. A., L. Escola, I. Intskirveli, F. Grammont, M. Rochat, F. Caruana, A. Jezzini, et al. 2008. How pliers become fingers in the monkey motor system. *Proceedings of the National Academy of Sciences of the United States of America* 105:2209–2213.

Wojciehowski, H. C., and V. Gallese. 2011. How stories make us feel: Toward an embodied narratology. *California Italian Studies* 2 (1). http://escholarship.ucop.edu/uc/item/3jg726c2.

Zeki, S. 1993. *A Vision of the Brain*. Blackwell Scientific.

17   Collective Body Memories

Thomas Fuchs

Human bodies are similar all over the world, but their habits, postures, and comportment are to a large extent shaped by culture. Cultures preordain and suggest certain ways of sitting, standing, walking, gazing, eating, praying, hugging, washing, and so on. In so doing, they induce certain dispositions and frames of mind associated with these bodily states and behaviors: for example, attitudes of dominance and submission, approximation and distance, appreciation and devaluation, benevolence or resentment, and the like. Cultural practices, rituals, roles, and rules shape the individual's *techniques of the body*, as Mauss (1935) termed them, and the resulting way the body moves and comports itself is one of the main carriers of cultural tradition. As Bourdieu notes, cultures are thus "treating the body as memory; they entrust to it in abbreviated and practical, i.e., mnemonic, form the fundamental principles of culture. The principles embodied in this way are placed beyond the grasp of consciousness" (Bourdieu 1977, 94). The main period for the transmission of these influences is of course early childhood and upbringing, which consists to a large extent of an "implicit pedagogy, capable of instilling a whole cosmology, an ethic, a metaphysic, a political philosophy, through injunctions as insignificant as 'stand up straight' or 'don't hold your knife in your left hand'" (ibid.).

This intimate connection between culture and embodiment is bound to a specific kind of memory, which usually escapes our conscious recollection or deliberate actualization—a system of embodied habits and skills acquired by the individual, which may also be termed *body memory* (Fuchs 2011a, 2012). This memory is of a kind quite different from the episodic memory by which we recollect and represent the past as such. Through repeated and typical interactions with others an individual habitus is formed, and with it the norms and rules of culture are inscribed into the body, yet in such a way that the resulting memory corresponds to an embodied and implicit knowing *how*, not to a knowing or remembering *that*.

The social interactions that shape the individual body memory usually follow certain patterns, styles, and rhythms (e.g., turn-taking), and they are often directed toward shared goals. Following Di Paolo and De Jaegher (this volume), we might also speak of "participation genres," such as joint play, shared meals, salutations, queuing, bedtime rituals, and the like.

Since such habitual or ritualized forms of embodied interaction are possible only in dyads or groups, the question arises whether we can also posit a superindividual level of memory formation, resulting in what may be termed *collective body memory*. This would be a crucial complement to the notion of "collective memory," which has been introduced by Halbwachs (1939) and further investigated by cultural anthropologists (e.g., Pennebaker, Paez, and Rime 1997; Assmann and Livingstone 2006), but which is mainly related to verbal tradition or explicit shared commemoration of the past.

The interbodily basis of collective memory is confirmed by the multifarious forms of ritualized and synchronized movements and performances, which contribute to building human culture. In his seminal work *Keeping Together in Time: Dance and Drill in Human History*, McNeill (1995) has collected compelling evidence that coordinated rhythmic movement—and the shared feelings it evokes—has played a profound role in creating and sustaining human communities. Synchronized action and chant facilitated group labor in rowing, tilling the soil, moving megaliths, and so on. From festival village dances or the chanting rituals of churches to the close-order drill of early modern armies, various forms of joint bodily movement have supported groups in their capacity for cooperation. This is based, above all, on shared bodily sensations and feelings, or what may be called *interbodily resonance* (Froese and Fuchs 2012), with the effect of weakening the psychological boundaries between the self and the group, and enhancing the sense of community and identity. More recently, these dynamics of social coordination and synchronized movement have also been explored from an enactive and dynamic systems perspective, emphasizing the coupling of interacting systems and the emerging autonomy of the interaction processes as such (De Jaegher and Di Paolo 2007; Fuchs and De Jaegher 2009; Schmidt and Richardson 2008; Wiltermuth and Heath 2009; Oullier and Kelso 2009; Valdesolo, Ouyang, and DeSteno 2010).

In what follows, I will investigate the idea of a collective form of body memory, which develops in dyads or social groups through repeated interactions and preordains a coordinated behavior of the members. This idea is closely related to the question whether there is an interbodily "we-experience" or even a kind of collective body, which could become the carrier of such memory formation. My main focus will be on a phenomenological approach, with frequent side-glances to enactive and dynamical systems aspects, which I regard as complementary. I will start with a definition and short explanation of individual body memory. As one of its subtypes, I will consider the phenomenon of intercorporeal memory, in order to then extend it in the direction of a dyadic and collective body memory. In the second section, I will take a closer look at several phenomena that may be attributed to collective body memory. These are, on the one hand, particular forms of interaction such as play and ritual, and, on the other hand, patterns of interaction and behavior in families, social classes, or cultural communities as a whole—often subsumed under the concept of *habitus*.

1   Individual, Dyadic, and Collective Body Memory

(a)   Body Memory

The capacity of conscious or explicit recollection, which is usually termed *episodic memory* (Tulving 1993), by no means exhausts the phenomenon of memory. Most of what we have experienced and learned is not made accessible to us in retrospect, but is reenacted through the practices of everyday life. We can define the entirety of established dispositions and skills as *body memory* that become current through the medium of the lived body without the need to remember earlier situations (Casey 2000; Fuchs 2000, 2011a, 2012; Summa 2011, 2012). It thus comprises all those habits, manners, skills, and practices that are performed prereflectively or "as a matter of course." It is a memory for patterns of movement such as walking or dancing, for the skillful handling of instruments such as a bicycle or a keyboard, for familiar *gestalts* of perception, for complex spatial situations (for example, finding one's bearings in a dwelling or a town), and last but not least for the habitual bodily interactions with others. This bodily memory, which was first considered by Maine de Biran ([1799] 1953), Félix Ravaisson ([1838] 1999), and Henri Bergson ([1896] 2007), does not "presentify" the past through explicit recollection, but rather reenacts it implicitly, as a grown and presently effective capacity.[1] While the term *implicit memory* as used in cognitive psychology (Schacter 1987, 1996; Rovee-Collier, Hayne, and Colombo 2000) covers some of its phenomena such as procedural or skills learning, the term *body memory* is more comprehensive and emphasizes its basis in the lived or subjective body.

Body memory is thus the ensemble of all habits and capacities at our disposal. It conveys the founding experience of "I can" (Husserl 1952, 253), an embodied knowledge or knowing how, or, in Merleau-Ponty's terms, the *operative intentionality* of the body (Merleau-Ponty 1962, 372, 382): I can dance a waltz because my lived body attunes of its own accord to the rhythm of the music and performs the movements. I can type with ten fingers, yet without being able to describe the position of the letters on the keyboard. I have long since forgotten the clear assignment of fingers and letters that I learned when I first learned to type. Now, the knowledge is "in my fingers," and they type of their own accord.[2] Bodily familiarity with

1. Bergson already emphasized this peculiar temporality of what he called *mémoire habitude* (as opposed to *souvenir-image* or episodic memory): "This consciousness of past efforts stored in the present is certainly a memory as well, but a memory fundamentally different from the first, always directed toward action, based in the present and looking only to the future. ... Indeed it does not represent our past, but enacts it" (*"cette conscience de tout un passé d'efforts emmagasiné dans le présent est bien encore une mémoire, mais une mémoire profondément différente de la première, toujours tendue vers l'action, assise dans le présent et ne regardant que l'avenir. ... À vrai dire, elle ne nous représente plus notre passé, elle le joue"* [1896] 2007, 87, my translation).

2. Though certainly not an embodiment theorist, Descartes already described this kind of memory: "Thus, for example, lute players have part of their memory in their hands, because the facility to move

things or performances means biographical forgetting, the descent of conscious deeds and experiences into a substrate from which consciousness has withdrawn, and yet which carries our everyday being-in-the-world.[3] I am familiar with an instrument, a face, or a situation as a whole because my capacity for perception and action comprises my earlier experiences in the form of types or patterns, without explicit remembering. Hence, one could say that body memory means my *lived past*. Moreover, in a sense this memory implies *a collective past*, for it is obviously shaped to a large extent by cultural practices, rituals, roles, and artifacts that the lived body adopts, or assimilates to, from birth on.

What is the locus of this embodied knowledge or body memory? Is this only a metaphorical term, and do we in fact have to locate it in the brain? According to the computational view of mind and brain, the process of learning writes bits of information into memory banks where they are stored and can be recalled at will. However, this representational and internalist view of memory does not fit the dynamic interaction with the environment that takes place when bodily skills or habits are reenacted. To be sure, this memory is based on specific patterns of neural activation derived from earlier experience; and, in contrast to correlates of episodic memory, these are mainly subcortically organized, that is, in the basal ganglia, cerebellum, and limbic system (Graybiel 1998; Ennen 2003). However, this does not imply any representational memory: instead of inner maps or models of external reality, the brain provides only the *open loops* of potential interactions. These loops are only closed to full functional cycles by suitable counterparts in the environment that the body currently connects with, leaving no role for separate representations (Fuchs 2011b).[4]

Granted, in these interactions, implicit protentions or anticipations play a crucial role: the hammer that I grasp will have a certain weight, the stairs I walk up will lead me to my apartment, and so on. These bodily protentions might also be related to concepts of "predictive processing" in the brain (see Kirmayer, this volume). Yet this is all part and parcel of the operative intentionality of the body whose connection with each environment opens a *procedural field* of possibilities, affordances, and probabilities. This field is not "represented"

---

and bend their fingers in various ways which they have acquired by habit, helps them to remember passages that require them to move their fingers in that way in order to play them" (see *Lettre à Meyssonnier*, 29.01.1640; Descartes 1996, AT III, 18–21, my translation).

3. William James made the fitting observation: "It is a general principle in psychology that consciousness deserts all processes where it can no longer be of use" (James [1890] 1950, 496).

4. The term *representation* suggests that the brain activities could, at least in principle, be separated from the cycle, as if they were reconstructing or modeling inside what is outside. But in a current sensorimotor coupling with the environment, there is no separate "inside" that could map, reconstruct, or represent the "outside." In such an ongoing circular process, no segment can "represent" or "stand for" another. Instead, the achievement in question is realized by the brain-body-environment system as a whole.

somewhere inside but extended before us, as our bodily being-toward-the-world (Merleau-Ponty 1962).[5]

Thus, if "memory" means not some kind of static inner depository, but *the capacity of a living being to actualize its dispositions acquired in earlier learning processes*, then this capacity is bound to the ongoing dynamic coupling between body and environment. An illustrative example is the attempt to find the keys for typing a certain word on an empty keyboard (where the letters have been removed from the keys) just by looking at it. Even for an experienced typewriter, this will be impossible—as mentioned above, one usually has no representational knowledge of the position of the letters. However, at the very moment of having one's fingers set on the keys, they project their capacity onto the keyboard, and one can write the word immediately, without thinking. Here the knowledge is clearly an embodied know-how *without knowing that*, and the memory may well be said to reside in the "hands-on-the-keyboard," or to put it more precisely, the memory is an emergent dispositional property of *the whole system of organism and keyboard connected to each other*. Thus, since the locus of this memory is not only the brain, "body memory" may not be regarded as a merely metaphorical term. Rather, it precisely describes the body in connection with the environment as the carrier of habit or skill memory. It is not static or reproductive but a dynamic memory, both in its formation through the body's interaction with the environment as well as in its flexible reactualization through similar interactions later on.

(b) Intercorporeal Memory

We can distinguish several types of individual body memory, for example, procedural, spatial, situational, traumatic body memory, and others (Casey 2000; Fuchs 2011a, 2012). I will focus here on one type, which may be termed *intercorporeal memory*, following Merleau-Ponty's notion of *intercorporéité* (1960) as a sphere of prereflective mutual bodily attunement (see also Moran, this volume). As we will see, this memory also enables the formation and tradition of collective patterns of interaction.

It is widely acknowledged that early childhood development is characterized by the incorporation of shared practices, which define the infant's sociocultural world. Infant research has shown that motor, affective, and social skills do not develop on separate tracks, but are

---

5. From this it follows that there is no such thing as "pure perception" that would give us mere objects without any possibilities or affordances. A hammer is "graspable," a staircase is "walkable," and a house is implicitly seen as having a backside (Husserl's "appresentation")—otherwise there would be no perception of the hammer, the staircase, or the house at all. Hence, seeing a hammer means perceiving *both* something present *and* "what-might-be." These possibilities are opened up by body memory: if one has learned to juggle, one sees a formerly neutral object as affording juggling (Gibson 1966). In other words, perception is always rich in possibilities, and these are not separable from it—as long as we do not imagine them in an *as-if* mode. The latter starts with what Kirsh (2009) has termed "projection," namely an imagination by which we deliberately overlay the perceptual field. Only an "as-if" mode of intentionality may rightly be called a representation.

integrated through the formation of affective-interactive schemas. Even the earliest experiences of how infants are held, comforted, guided, and reacted to by their caregivers are imprinted in their implicit or body memory, hence also displayed in their later actions and interactions. Repeated patterns of interaction soon become familiar and result in a prereflective, practical knowledge of how to get along with others—how to share pleasure, elicit attention, avoid rejection, reestablish contact, and so on. It may also be termed *implicit relational knowledge* (Lyons-Ruth et al. 1998) or *intercorporeal memory* (Fuchs and De Jaegher 2009; Fuchs 2012): a temporally organized, "musical" memory for the rhythm, dynamics, and vitality affects shaping interactions with others (Stern 1985; Amini et al. 1996).[6]

This primary intercorporeality has far-reaching effects: early interactions turn into implicit relational styles that form one's personality. As a result of learning processes, which are in principle comparable to acquiring motor skills, we later shape and enact our relationships according to the patterns extracted from our primary experiences. These implicit relational styles are also expressed in the habitual posture of the body. Thus, the submissive attitude toward an authority figure implies components of posture and motion (bowed upper body, raised shoulders, inhibited motion), components of interaction (respectful distance, low voice, inclination to consent), and of emotion (respect, embarrassment, humility). All our interactions are based on such integrated bodily, emotional, and behavioral dispositions, which have become second nature, like walking or writing. They are now part of one's habitus or "embodied personality structure" (Fuchs 2006). The shy, submissive attitude that we find in dependent persons—their soft voice, childlike facial expression, indulgence, and anxiousness—belongs to an overall pattern of comportment and attitudes that is part of their personality and even their identity. Thus, our basic attitudes, typical reactions, and relational patterns are crucially based on body memory.[7]

To summarize, early childhood development is characterized by the incorporation of shared practices, which shape the infant's habits of interaction and comportment. These embodied skills define the space of possible relations in which children grow up and later on live their adult life. In the dispositions and habits of the lived body, others are always implied: a person's typical patterns of posture, movement, and expression are comprehensible only as referring to actually present or imaginary others. Embodied personality structures can be regarded as *procedural fields of possibility* that are activated in contact with others and suggest certain types of behavior. They are therefore best accessible in the actual intercorporeal encounter: the lived body can be understood only given other embodied subjects.

6. Of course, the term *intercorporeal memory* could already be understood as denoting a superordinate memory of dyadic processes. However, in earlier papers (Fuchs 2011a, 2012) I have used it as an individual type of relational body memory or "implicit relational knowledge," and I will therefore maintain this usage. The superordinate memory will then be termed "dyadic body memory" (see below).

7. On the fundamental role of body memory for the diachronic identity of the self, see Fuchs 2016.

(c)   Dyadic Body Memory

As we have seen, each body forms an extract of its past history of experiences with others that is sedimented in intercorporeal memory. Can we speak of a superordinate dyadic or collective body memory as well? To answer this question in the affirmative, we just have to shift our focus somewhat, namely from a view on the individual to a view on the interactive history of a dyad or a group. For just as the intercorporeal experiences of an individual are transformed into body memory, the interactions *between* two persons also develop their own history. It manifests itself in shared patterns of interaction, which are actualized every time the two persons meet again. One may, for example, develop a specific style of interacting with a close friend, a particular way of talking, a special style of humor and so on, which are possible only with this person and turn up again even after years. In this case, the respective intercorporeal memories of the partners unite to form a *joint procedural field* that suggests and preordains certain typical postures, interactions, and interaffective experiences. Both body schemas are attuned to each other through resonant kinesthetic patterns and thus *interenact* the shared history: rituals of welcoming, joint repertoires of gestures, postures, movements, voice pitch, and even dialects, which one "falls into" in the presence of the other, as a kind of unintentional entrainment (Schmidt and Richardson 2008). We can call this a *dyadic body memory*.

Let us take another example, namely of a well-attuned pair of dancers whose hands and bodies find each other without guidance of the gaze, resonating with the rhythm of the music, and incorporating the dynamic flow of each other's bodies into how they modulate their own sway and movements. Both partners bring in their own procedural and intercorporeal capacities, and yet they behave and experience in a way that is possible only in the interaction. Together they create the spatiotemporal *gestalt* of the dance, which in turn draws them into its dynamics. Thus, they no longer experience themselves as clearly separate bodies, but rather *mutually incorporate* each other (Fuchs and De Jaegher 2009). Their kinesthetic body schemes literally extend and connect to form an overarching dynamic process (Froese and Fuchs 2012; Koch and Fischman 2011). Related examples include jointly sawing wood with a two-man saw (Christian and Haas 1949) or double sculls rowing (Lund, Ravn, and Christensen 2012), both forms of cooperation that lead over time to a harmonic, sinusoidal coordination of movements, with each partner tuning into a complementary activity that unconsciously compensates for the irregularities of the partner. Modifying Merleau-Ponty's notion, we might speak of an *operative we-intentionality*, since for skilled agents, the goal of the joint action is achieved through such habitual and largely prereflective bodily attunement.

Generally speaking, when two individuals interact in such ways, the coordination of their body movements, gestures, gazes, and so on can gain such momentum that it may even override the individuals' intentions. This is based on the general capacity of the body to connect with instruments as well as with other bodies in skillful interaction and thus to dynamically incorporate them. In mutual incorporation and resonance, both agents form an "extended body," as it were, which may even develop its own history. Rhythm and melody particularly support this incorporation by providing dynamic constraints for the movements of both partners. This process has been described at the systems level as the interaction gaining an autonomy of its own, or as the emergence of *participatory sense-making* (De Jaegher and Di Paolo 2007).

Where shall we localize this memory of joint dancing and other skillful or habitual inter-actions? On the one hand, the superordinate system or "extended body" of course has no natural substrate for forming a memory—it emerges only from the present connection of two bodies in which, based on each brain's neuroplasticity, the respective dispositions have formed. Each social memory must finally be based on the biological memory substrates of the individuals involved in order to become effective for their behavior. On the other hand, the "open loops" of these dispositions are especially preattuned to the corresponding loops of specific others. Only together are the individuals in a position to actualize and internact their reciprocally related memories, which justifies attributing the memory as an emergent dispositional property to the dyadic system or the dyad itself.[8]

(d) Collective Body Memory

Thinking about memory and identity in collective terms, we still tend to focus on verbal, representational, and other symbolic traditions. However, a great part of our collective mem-ory has been passed from one generation to the next through performative practices and specifically socialized bodies. "Every group ... will entrust to bodily automatisms the values and categories that they are most anxious to conserve. They will know well that the past can be kept in mind by habitual memory sedimented in the body" (Connerton 1989, 102).

Since Mauss's influential work on the techniques of the body (Mauss 1935), researchers have increasingly acknowledged and investigated the role of corporeality for carrying cul-tural memory (see Narvaez 2006 for an overview). On the one hand, cultural anthropology has long since criticized the idea of a natural or precultural body as "biological essentialism," emphasizing the history of the body and claiming embodiment as a cultural phenomenon. On the other hand, phenomenological approaches, in particular advanced by Csordas (1990, 1994, 1999), have moved away from representational theories that see bodies as mere sym-bols of cultural ideologies, inscribed textures, or enacted metaphors. Instead they argue for a primary level of meaning where experiential and expressive qualities of the body stand for themselves, thus bringing the prereflective layers of experience into focus. Cultural phenom-enology examines the unique ways in which the lived body unfolds in experiences of and cultural practices surrounding sickness, ritual, dance and sports, healing, and music. Thus,

---

8. A related question concerns the locus of cooperative agency. Stapleton and Froese (2015) have argued for a restricted notion of "collective agency": while cooperating, group members may well develop a shared lived perspective. However, this should not be conflated with a collective first-person perspective or "we-subjectivity," since subjectivity is necessarily bound to a living body.

embodiment is increasingly regarded as "existential ground of culture and self" (Csordas 1994, 6).

Against this background, body memory may serve as the mediator between embodiment and the history of culture, in particular when it is regarded from an interbodily point of view. For this, we have only to transfer the results of the last section to the embodied interactions of several persons or social groups. A *collective body memory* may then be defined as an ensemble of behavioral and interactive dispositions characterizing the members of a social group, which have developed in the course of earlier shared experiences and now prefigure similar interactions of the group. Here too, a procedural field of dynamic behavioral patterns emerges that induces the members to perform coordinated interactions and at the same time constitutes the meaning of their interactions. Similar to dyadic body memory, collective body memory is based, on the one hand, on the acquired dispositions of the individuals; on the other hand, it is actualized only through the interactions of the group as a whole. As we will see, it is also particularly suited to carry the identity of the group and make it tangible for its members.

The notion of collective body memory should be distinguished from other concepts of sociology and cultural studies. As already mentioned in the introduction, Halbwachs (1939) coined the term *collective memory* as the shared traditional knowledge of a group. This concept, however, does not refer to performative reenactments of former collective actions. The related notion of *cultural memory* (Pennebaker, Paez, and Rime 1997; Assmann 2011) includes bodily and performative components such as rituals, but also oral history, written documents, monuments, and other objective carriers of cultural tradition, which are not related to body memory and reenactments. What comes closer to collective body memory is the concept of *habitus* as introduced into anthropology by Mauss, and into sociology by Bourdieu. However, though describing what members of a group or culture have in common, it is still bound to the individual and seems less suited to illustrate the phenomena of group enactments such as games or rituals. I will therefore take the habitus as one, even though important, type of collective body memory (see below).

2 Some Phenomena of Collective Body Memory

Having outlined a general concept of collective body memory, I will now take a closer look at some of its appearances, namely games, family memory, rituals, and habitus.

(a) Games
Team sports such as soccer may serve as a first example of collective body memory. First, the game consists in a form, which is determined by rules, goals, and means. It is bound to individual capacities of dealing with the ball, but also to the embodiment of the playing field whose dimensions the player has to incorporate in order to handle the ball without much deliberation or calculation. The bodily dispositions of the player and the ever-changing

spatial configuration of the field are mutually implicated elements of an indivisible whole (Hughson and Inglis 2002). It is a form of consciousness that is not reflective, but rather a "field-consciousness" so well described by Merleau-Ponty:

For the player in action the football field is not an "object," that is, the ideal term which can give rise to an indefinite multiplicity of perspectival views and remain equivalent under its apparent transforma-tions. It is pervaded with lines of force (the "yard lines"; those which demarcate the "penalty area") and articulated in sectors (for example, the "openings" between the adversaries) which call for a certain mode of action and which initiate and guide the action as if the player were unaware of it. The field itself is not given to him, but present as the immanent term of his practical intentions; the player becomes one with it and feels the direction of the "goal." ... At this moment consciousness is nothing other than the dialectic of milieu and action. (Merleau-Ponty 1963, 168–169)

Moreover, in a well-attuned team, the players also have a sense of the joint positional play, for the routes the others will usually take, and for well-practiced combinations. The move-ments the single player performs on the field gain their meaning only against the back-ground of the movements of the team as a whole. Thus, the playing field and the soccer team together form a procedural field that induces the movements, directions, and dynamics of the players—of course, always in conflict with the opposing team. Having incorporated this procedural field, the player also becomes part of an extended or "collective body" with its peculiar flow and dynamics. The practical sense for the game that the experienced player has acquired is now at the same time a sense for the potentialities of the team's play, its immedi-ate future: it includes a *shared bodily protentionality* (see section 1a above), or in the words of Bourdieu:

A player who is involved and caught up in the game, adjusts not to what he sees but to what he fore-sees, sees in advance in the directly perceived present; he passes the ball not to the spot where his team-mate is but to the spot he will reach ... a moment later, anticipating the anticipations of the others. ... The "feel" for the game is the sense for the imminent future of the game, the sense of the direction of the history of the game that gives the game its sense. (Bourdieu 1990, 81–82)

The emergent procedural field implies not only an embodied sense of the future, but also a form of normativity: a player does not only perceive his teammate's intention of kicking the ball, but can perceive his kick as unsuitable for shooting at the goal. With repeated joint training, the mutual attunement of the players will increase, stabilizing their interactions according to the normativity of the field, and thus indicating the development of a collective body memory as an emergent property that preordains the individual actions. The more space and less resistance offered by the opposing team, the easier this shared memory will be reenacted in the game. Then the patterns of movements and combinations practiced in training will reemerge automatically, without deliberate purpose, reflection, or planning. If this does not succeed, however, it is the task of the trainer to change the tactics and possibly to actualize other dynamic patterns from the collective repertoire of the team. In this case, explicit cognitive strategies and implicit dispositions of the team work together.

(b) Family Memory

Even without codified rules, social groups over time develop their peculiar patterns and dynamics of interaction. A typical example is the collective memory of a family, which attributes specific roles, positions, and behavioral styles to its members. Thus, children are drawn into a topology of intercorporeal and affective resonance structures from birth on. Most families also develop specific rituals of shared meals, weekends, excursions, birthdays, and so on. This results in what may be called an *embodied family memory*: the behavior patterns and relations between the family members constitute a prereflective and invisible procedural field that is enacted each time the family gets together.

A way to make this field visible has been developed in the therapeutic method of the so-called family constellations (Steifel, Harris, and Zollmann 2002; Cohen 2006). They typically proceed as follows: One of the clients asks people from the therapeutic group to serve as representatives of his family. Then the client arranges them in a spatial constellation according to his intuitions about his family—standing close or distant, being turned toward or away from each other, and so on. The representatives have little or no factual knowledge about those they represent. Nevertheless, on the mere basis of their position within the constellation, they usually experience feelings, bodily sensations, or movement tendencies, which come very close to the experiences within the real family. Their statements then inform the further process of exploring the position, role, and feelings of the family members as well as the family dynamics as a whole.[9]

The mechanism behind this vicarious experience is not fully understood (see Lynch and Tucker 2005). However, we may well recognize the role of collective body memory in the process: the chosen configuration represents not a particular biographical situation but a constellation of felt relations that have sedimented in the client's lived body as an extract of uncountable experiences with his family. It refers to common orders of belonging or distancing, authority or inferiority, coalition and exclusion, and the like. Thus, body memory contains an invisible network of relations to the relevant persons of one's biography—persons that stand at our side or behind us, closer or more distant—and this should be understood quite literally in an embodied and spatial way. Hence, the client intuitively feels his sister closely at his right side, his father reassuringly behind his back, and so on. This invisible bodily and spatial network is made visible in the family constellation. It displays the collective body memory of the family as experienced by the representatives.

The concept of *lived space* (Fuchs 2007) may also be useful to explain the spatial dynamics of the constellation. Derived from Lewin's "topological" or "field psychology" (Lewin 1936), lived space may be regarded as the totality of the space that a person prereflectively "lives" and experiences, with its situations, relations, movements, and horizon of possibilities. This

---

9. Of course, the representatives' reports are subjective and contain some aspect of personal projection. However, the blending of projections with field resonance usually does not contaminate the process as a whole.

space is not homogeneous, but centered around persons and their bodies, characterized by qualities such as vicinity or distance, wideness or narrowness, connection or separation, attainability or unattainability, and structured by physical or symbolic boundaries. This results in more or less distinct domains such as one's own territory, home, sphere of influence, zones of prohibition or taboo, and so on.

Moreover, the lived space is permeated by "field forces" such as attraction and repulsion, elasticity and resistance, and the like. Competing attractive or aversive forces lead to typical conflicts, which may be regarded as opposing directions of possibility that the person faces. In analogy to physical fields, there are effects of "gravitation" and "radiation," caused for example by the influence of a significant other or by a dominant social group. Hence, lived space implies a unity of bodily, sensorimotor, affective, and intersubjective conditions and impacts, which are experienced and enacted in a prereflective, nonsymbolic spatial mode. Collective body memories are played out and enacted as lived spaces of individuals and groups; hence, the spatial structure of the family constellation with its peculiar effects may be regarded as a visualization of the topological field structure underlying the interrelations of the members.

(c) Rituals

Rituals are culturally prefigured social acts with high symbolic meaning that are governed by formalized rules and performed in ceremonial ways. They extend from simple types such as shared meals or salutations to so-called *rites de passage* (births, initiations, weddings, or funerals) (van Gennep 1909), rites of feasting or commemoration, and finally religious cults (Turner 1969; Bell 1997; Rappaport 1999). Their function lies, on the one hand, in regulating and smoothening social interactions in everyday life, and, on the other hand, in helping the group to cope with critical or precarious situations by embedding them in overarching social and mythical contexts. Group rituals may be conceived as arrangements of human bodies, often based on synchronized and rhythmic movements (dancing, singing, drumming, etc.) and turn-taking patterns. Their effect is mediated through the performance itself, which means that the concrete bodily enactment *evokes or creates* the jointly intended reality. At the same time, there is also a *mimetic relation* between former, present, and future ritual acts, which is anchored in the similarity and recurrence of bodily sensory actions and ceremonies. The more the different senses participate in the ritual (kinesthesis, touch, vision, hearing, smell, and taste), the more lasting its sedimentation in the individual body memory, and the more intense the shared experience of its reactualization. Rituals are thus essential parts of the collective body memory of a group, be it a clan, a tribe, or a larger community. They both express and enhance their members' sense of identity and togetherness.

Many rituals also serve as an explicit commemoration: they hark back to a primordial event or a mythical idol, a god, hero, or ancestor whose actions (e.g., an act of creation or a fight against a hostile power) are imitated and reenacted in the ritual. "Every religious festival, any liturgical time, represents the reactualization of a sacred event that took place in a mythical past, 'in the beginning,'" writes Eliade (1959, 69). The ritual performance enables the descendants or the believers to take part in the hero's life through a "mystical participation" (Lévy-Bruhl 1910) and to secure the continuous renewal of the primordial beginning through cyclical repetition. By bodily imitating what the mythical ancestors have done, reproducing their actions and gestures, one communes with them and shares their essences. The metaphysical basis of this participation is the mystical community of the essence of things, their identity over time and space. As a result, things may represent both themselves and another thing, be both past and present, be both here and there at the same time. However, this participation is based experientially on the capacity of the lived body to incorporate and reenact its former experiences as if they were immediately present. Body memory enables the *lived presence of the past*, thus establishing, as it were, an immediate communication between different times of one's life.[10] Collective body memory extends this communion to the ancient times of the group as a whole.

This timeless present realized by body memory is the foundation of religious rituals. For Christians, for example, it is created in the Holy Mass, which encompasses not only the ever renewed enactment of the Last Supper but also the intercorporeal presence of Christ himself. Both Jesus, the finite and historical person of a remembered past, and Christ, who transcends history, are present in the Mass at the same time. Moreover, through the sacrament the communicants become one body with Christ, con-substantial with him, for the bread and wine, which they taste and ingest, become identical with his body and blood. Indeed the community and the church as a whole literally participate in Christ's body: *corpus Christi mysticum*, the mystical body of Christ, was the term used since the twelfth century for the community of the Christian church, derived from St. Paul's comparison of Christ with the "head" and the Christians with the "limbs" of one body.[11] Thus, the collective body memory of the community mediates the ever renewed participation in Christ, and the past is resurrected through the shared intercorporeal present of the mass.

(d) Habitus

As a final form of collective body memory I will consider the concept of the *habitus*. It may be understood as a set of socially learned dispositions, skills, styles, tastes, and comportment that are often taken for granted or go unnoticed—one may also speak of a prereflective

10. Thus Proust, in his *In Search of Lost Time*, famously describes the Madeleine experience, the taste of a tea-soaked cookie reminding him of his childhood, as an overwhelming experience of timeless bliss in which, as in an experience of déjà-vu, the distant past and the present coincide into one unitary time (Proust 2003).

11. "For even as the body is one and yet has many members ..., so also is Christ. ... Now you are Christ's body, and individually members of it" (1 Cor. 12, 12–13); the church is "his body" (Eph. 1, 22–23). "Christ is the head of the church. He is the savior of his body" (Eph. 5, 23); he is the "head, from which all the body by joints and bands having nourishment ministered" (Col. 2, 19).

"social sense." The habitus is formed by the continuous sedimentation of shared experiences into the body memory and embodied personality structure of the individual. Thus, the skills, habits, and practices acquired throughout one's life find their proper intelligibility in the context of the respective life world and its social relations. Though the individual is the carrier of the habitus, it has been acquired in shared interactions and hence always remains implicitly related to actually present or imaginary others. Moreover, the combination of habitus-guided behaviors in interaction, for example of the roles of teacher and student, may be regarded as a superindividual embodied memory of interaction sequences. Considering this reciprocal relatedness of the individual habitus forms, it seems justified to include the notion in the concept of collective body memory.

Aristotle already used the concept of the habitus in his notion of *hexis*. In contemporary usage it was introduced into sociology by Marcel Mauss, who defined the habitus as those aspects of culture that are anchored in the body and in daily practices of individuals, groups, and societies (Mauss 1935). Bourdieu later re-elaborated the concept, defining it as the entire social appearance of a person, including his or her posture, manners, taste, clothing, attitudes, and general way of life (Bourdieu 1990). As a "system of internalized patterns," the habitus produces culture- or class-specific styles of thought, perception, and action that individuals take to be their own, but which they actually have in common with the members of their class. For the experienced observer, the habitus may even allow us to infer the rank and status of a person in society.

Clearly, we are dealing with a mnemonic and historical concept, but one of a special kind: "The habitus—embodied history, internalized as a second nature and so forgotten as history—is the active presence of the whole past of which it is the product" (Bourdieu 1990, 56). Precisely because it is not explicitly remembered or reflected upon, the habitus induces us to certain mindsets, outlooks on the world, as well as to particular styles of interacting with others as a matter of course. These enacted practices thus belong to a "cultural unconscious" that naturalizes certain behaviors, while making others seem "out of place" or even unthinkable. It suggests patterns of distance or nearness, pride or modesty, benevolence or competition, and the like—mediated, for example, through customary interpersonal distance (think of different norms for personal space in Northern or Southern Europe, or Europe and the United States), through body postures or movements, clothing (think of headscarves or burkas, to take a current example), and also through environmental affordances and artifacts such as architecture, furniture, means of transportation, and so on.

As mentioned above, the habitus is acquired through practical immersion in the life world, that is, through repeated interactive experiences, mimetic learning (e.g., watching elders or peers), and implicit routines in typical situations. It does not require purposeful instructions, deliberate imitation, or other kinds of explicit learning. Rather, it resembles learning one's mother tongue without having any explicit idea of grammar. By incorporating "schemes of being-with-others" (Stern 1985), infants already take over their attitudes and roles, thus chiming in with a social context that they cannot yet realize explicitly. No one can remember, for example, having consciously adopted a certain role in his or her family. The habitus becomes second nature, which effectively guides one's behavior, all the more as it is not conscious *as* a habitus.

As engendered by the immersion in the life world, the habitus is also important for our concepts of *social understanding*. The homogeneity of the habitus as the shared body memory of a community or culture entails that the common practices are immediately evident or foreseeable against the background of a given situation. This provides a primary, noninferential understanding of others without conscious transposition or perspective-taking (Reddy and Morris 2004; Gallagher 2008; Fuchs and De Jaegher 2009; Fuchs 2017). Growing up and being immersed in a shared practical context results in an implicit understanding of the "rules of the game" and of typical interactive sequences. Like the soccer players mentioned above, the members of a culture normally understand each other intuitively and know how to react appropriately without deliberation, anticipating the next moves without a need to resort to theory of mind or mentalizing procedures. *Common sense* is primarily a practical sense of embodied social habits and interactions that constitutes the prereflective background of social life (Fuchs 2001).[12]

Conclusion

Based on a phenomenological concept of body memory, I have introduced the concept of collective body memories. These develop in dyads or social groups through recurrent shared experiences and lead to spatial and temporal patterns of joint group behavior. As we have seen, this type of memory may be described from a phenomenological point of view:

• as a history of shared intercorporeality that is experienced by the participants as a sense of "chiming in" with a joint performance, game or ritual, and as a feeling of being "in the flow" of the cooperative process; and
• as the "social sense" or habitus of the members of a group that is based on countless intercorporeal experiences and provides for a smooth interaction and attunement in typical social situations.

On the other hand, collective body memories may also be described by an enactive or dynamical systems approach, considering the interacting agents as an integrated system that displays novel properties not reducible to the properties of its individual agents. We may call this *procedural emergence*. The agents then display patterns of interaction

12. Nevertheless, one may also regard *common sense* as the superordinate or "structuring structure," whereas the habitus in Bourdieu rather refers to the individual agent: "[T]he habitus refers principally to the structured nature of specific, individual agency, while common sense is, of course, a communal rather than an individual property" (Holton 2000, 88).

• that are reenacted in the current situation through nondeliberate coordination and synchronization of their actions;
• in which the overarching process gains a degree of autonomy over the individual dispositions; and
• which are either repeated in similar ways or show a dynamic development over time.

Thus, the phenomenon of collective body memories may be seen as a paradigmatic case of a combined phenomenological and enactive approach.

As examples of such memories, I have presented dyadic interactions such as dancing or rowing, the formation of a soccer team and its fluent interplay, the interrelational field of a family as visualized by the so-called family constellations, and the bodily enactments of social rituals. In such situations, the intercorporeal memories of the individuals unite to form overarching procedural fields. Moreover, the interactive processes develop an emergent dynamic involving the individuals in positions or behavior they would not participate in outside of the formation. Once the group joins again in a similar configuration and situation, their collective body memory is reactualized.

As a more general form of collective body memory I have presented the concept of the habitus or "social sense"—a set of dispositions, skills, styles, tastes, and behaviors that are acquired through the practical immersion in the life world. It is the result of the continuous sedimentation of shared experiences into the embodied personality structure of the individuals, thus manifesting the enculturation that is mediated through the body and its implicit memory. In this way, collective styles of intercorporeality and interaction are passed on from one generation to the next without becoming explicit—in an unconscious, collective history. Thus, Merleau-Ponty's notion of intercorporeality gains an additional, historical aspect: it means not only the primary familiarity of our bodies with each other, or their prereflective communication, but also the entanglement of human bodies in a shared history that is preserved in their collective memory.

These conceptual considerations are not meant to give an exhaustive account of types of collective body memory. One might think, for example, of culture-specific forms of childhood or adolescence, sexuality or aging, collective trauma or taboo, as various kinds of procedural fields, which are incorporated by individuals and dispose them to certain forms of shared behavior, interaction, or also restriction. A further question might be to what extent the external memory systems that are established in the course of cultural development (pictograms, artifacts, printing, computer technology, and the like) have a retroactive impact on individual and collective body memories, modifying and changing their habitual structure in important ways.

Finally, one might also pose the question of how the different forms of collective body memory described in this chapter are related to each other. Rituals and habitus, for example, are certainly closely connected, both in their development and their enactment. Frequently, rituals function as explicit and prominent ways for a culture or society to form shared habitualities and to endow them with normative significance. Over time, these rituals gradually sediment in the individual body memories and are then performed as "second nature" without explicit reenactment. On the other hand, novel forms of embodied interaction may spontaneously emerge in dyads or smaller groups, and, once stabilized, spread to larger communities where they are established as rites and norms. As we can see, body memory opens up a range of stimulating questions for further investigation in the field of embodiment, enaction, and culture.

Acknowledgments

I am grateful for helpful comments on earlier versions of this chapter by Christoph Durt, Sabine Koch, Michela Summa, Christian Tewes, and Samuel Thoma.

References

Amini, F., T. Lewis, R. Lannon, A. Louie, G. Baumbacher, T. McGuinness, and E. Z. Schiff. 1996. Affect, attachment, memory: Contributions toward psychobiologic integration. *Psychiatry* 59:213–239.

Assmann, J. 2011. *Cultural Memory and Early Civilization: Writing, Remembrance, and Political Imagination.* Cambridge University Press.

Assmann, J., and R. Livingstone. 2006. *Religion and Cultural Memory: Ten Studies.* Stanford University Press.

Bell, C. 1997. *Ritual: Perspectives and Dimensions.* Oxford University Press.

Bergson, H. [1896] 2007. *Matière et mémoire: Essai sur la relation du corps à l'esprit.* PUF.

Bourdieu, P. 1977. *Outline of a Theory of Practice.* Cambridge University Press.

Bourdieu, P. 1990. *The Logic of Practice.* Stanford University Press.

Casey, E. 2000. *Remembering: A Phenomenological Study.* Indiana University Press.

Christian, P., and R. Haas. 1949. *Wesen und Formen der Bipersonalität: Grundlagen für eine medizinische Soziologie.* Enke.

Cohen, D. B. 2006. "Family constellations": An innovative systemic phenomenological group process from Germany. *Family Journal* 14:226–233.

Connerton, P. 1989. *How Societies Remember.* Cambridge University Press.

Csordas, T. J. 1990. Embodiment as a paradigm for anthropology. *Ethos* 18:5–47.

Csordas, T. J. 1994. Introduction: The body as representation and being-in-the-world. In *Embodiment and Experience: The Existential Ground of Culture and Self,* ed. T. J. Csordas, 1–24. Cambridge University Press.

Csordas, T. J. 1999. Embodiment and cultural phenomenology. In *Perspectives on Embodiment: The Intersections of Nature and Culture*, ed. G. Weiss and H. F. Haber, 143–162. Routledge.

De Biran, M. [1799] 1953. *Influence de l'habitude sur la faculté de penser*. PUF.

De Jaegher, H., and E. Di Paolo. 2007. Participatory sense-making: An enactive approach to social cognition. *Phenomenology and the Cognitive Sciences* 6:485–507.

Descartes, R. 1996. *Œuvres de Descartes*. Ed. C. Adam and P. Tannery. Vrin.

Eliade, M. 1959. *The Sacred and the Profane*. Harcourt.

Ennen, E. 2003. Phenomenological coping skills and the striatal memory system. *Phenomenology and the Cognitive Sciences* 2:299–325.

Froese, T., and T. Fuchs. 2012. The extended body: A case study in the neurophenomenology of social interaction. *Phenomenology and the Cognitive Sciences* 11:205–236.

Fuchs, T. 2000. Das Gedächtnis des Leibes. *Phänomenologische Forschungen* 5:71–89.

Fuchs, T. 2001. The tacit dimension. *Philosophy, Psychiatry, and Psychology* 8:323–326.

Fuchs, T. 2006. Gibt es eine leibliche Persönlichkeitsstruktur? Ein phänomenologisch-psychodynamischer Ansatz. *Psychodynamische Psychotherapie* 5:109–117.

Fuchs, T. 2007. Psychotherapy of the lived space: A phenomenological and ecological concept. *American Journal of Psychotherapy* 61:432–439.

Fuchs, T. 2011a. Body memory and the unconscious. In *Founding Psychoanalysis: Phenomenological Theory of Subjectivity and the Psychoanalytical Experience*, ed. D. Lohmar and J. Brudzinska, 69–82. Kluwer.

Fuchs, T. 2011b. The brain—a mediating organ. *Journal of Consciousness Studies* 18:196–221.

Fuchs, T. 2012. The phenomenology of body memory. In *Body Memory, Metaphor, and Movement*, ed. S. Koch, T. Fuchs, M. Summa, and C. Müller, 9–22. John Benjamins.

Fuchs, T. 2016. Self across time: The diachronic unity of bodily existence. *Phenomenology and the Cognitive Sciences* 15.

Fuchs, T. 2017. Intercorporeality and interaffectivity. In *Intercorporeality: Emerging Socialities in Interaction*, ed. C. Meyer, J. Streeck, and S. Jordan. Oxford University Press.

Fuchs, T., and H. De Jaegher. 2009. Enactive intersubjectivity: Participatory sense-making and mutual incorporation. *Phenomenology and the Cognitive Sciences* 8:465–486.

Gallagher, S. 2008. Direct perception in the intersubjective context. *Consciousness and Cognition* 17:535–543.

Gibson, J. J. 1966. *The Senses Considered as Perceptual Systems*. Houghton Mifflin.

Graybiel, A. M. 1998. The basal ganglia and chunking of action repertoires. *Neurobiology of Learning and Memory* 70:119–136.

Halbwachs, M. 1939. *La mémoire collective.* PUF.

Holton, R. 2000. Bourdieu and common sense. In *Pierre Bourdieu: Fieldwork in Culture,* ed. N. Brown and I. Szeman, 87–99. Rowman & Littlefield.

Hughson, J., and D. Inglis. 2002. Inside the beautiful game: Towards a Merleau-Pontian phenomenology of soccer play. *Journal of the Philosophy of Sport* 29:1–15.

Husserl, E. 1952. *Ideen zu einer reinen Phänomenologie und phänomenologischen Philosophie. II. Phänomenologische Untersuchungen zur Konstitution. Husserliana,* vol. 4. Nijhoff.

James, W. [1890] 1950. *The Principles of Psychology,* vol. 2. Dover.

Kirsh, D. 2009. Projection, problem space and anchoring. In *Proceedings of the 31st Annual Conference of the Cognitive Science Society,* ed. N. A. Taatgen and H. van Rijn, 2310–2315. Cognitive Science Society.

Koch, S. C., and D. Fischman. 2011. Embodied enactive dance therapy. *American Journal of Dance Therapy* 33:57–72.

Lévy-Bruhl, L. 1910. *Les fonctions mentales dans les sociétés inférieures.* Presses Universitaires de France.

Lewin, K. 1936. *Principles of Topological Psychology.* Trans. F. Heider and G. Heider. McGraw-Hill.

Lund, O., S. Ravn, and M. K. Christensen. 2012. Learning by joining the rhythm: Apprenticeship learning in elite double sculls rowing. *Scandinavian Sport Studies Forum* 3:167–188.

Lynch, J. E., and S. Tucker, eds. 2005. *Messengers of Healing: The Family Constellations of Bert Hellinger through the Eyes of a New Generation of Practitioners.* Zeig, Tucker, and Theisen.

Lyons-Ruth, K., N. Bruschweiler-Stern, A. M. Harrison, A. C. Morgan, J. P. Nahum, L. Sander, D. N. Stern, and E. Z. Tronick. 1998. Implicit relational knowing: Its role in development and psychoanalytic treatment. *Infant Mental Health Journal* 19:282–289.

Mauss, M. 1935. Les techniques du corps. *Journal für Psychologie* 32:271–293.

McNeill, W. H. 1995. *Keeping Together in Time: Dance and Drill in Human History.* Harvard University Press.

Merleau-Ponty, M. 1960. Le philosophe et son ombre. In *Signes,* 201–228. Gallimard.

Merleau-Ponty, M. 1962. *Phenomenology of Perception.* Trans. C. Smith. Routledge & Kegan Paul.

Merleau-Ponty, M. 1963. *The Structure of Behaviour.* Trans. A. L. Fisher. Beacon Press.

Narvaez, R. F. 2006. Embodiment, collective memory, and time. *Body and Society* 12:51–73.

Oullier, O., and J. A. Kelso. 2009. Social coordination, from the perspective of coordination dynamics. In *Encyclopedia of Complexity and Systems Science,* ed. R. A. Meyers, 8198–8213. Springer.

Pennebaker, J. W., D. Paez, and B. Rime. 1997. *Collective Memory of Political Events: Social Psychological Perspectives.* Erlbaum.

Proust, M. 2003. *In Search of Lost Time: The Way by Swann's*, vol. 1. Trans. L. Davis and C. Prendergast. Penguin.

Rappaport, R. A. 1999. *Ritual and Religion in the Making of Humanity*. Cambridge University Press.

Ravaisson, F. [1838] 1999. *De l'habitude*. PUF.

Reddy, V., and P. Morris. 2004. Participants don't need theories. *Theory and Psychology* 14:647–665.

Rovee-Collier, C., H. Hayne, and M. Colombo. 2000. *The Development of Implicit and Explicit Memory*. John Benjamins.

Schacter, D. 1987. Implicit memory: History and current status. *Journal of Experimental Psychology* 13:501–518.

Schacter, D. 1996. *Searching for Memory: The Brain, the Mind, and the Past*. Basic Books.

Schmidt, R. C., and M. J. Richardson. 2008. Dynamics of interpersonal coordination. In *Coordination: Neural, Behavioral, and Social Dynamics*, ed. A. Fuchs and V. K. Jirsa, 281–308. Springer.

Stapleton, M., and T. Froese. 2015. Is collective agency a coherent idea? Considerations from the enactive theory of agency. In *Collective Agency and Cooperation in Natural and Artificial Systems*, ed. C. Misselhorn, 219–236. Springer International.

Steifel, I., P. Harris, and A. W. F. Zollmann. 2002. Family constellation: A therapy beyond words. *Australian and New Zealand Journal of Family Therapy* 23:38–44.

Stern, D. N. 1985. *The Interpersonal World of the Infant: A View from Psychoanalysis and Developmental Psychology*. Basic Books.

Summa, M. 2011. Das Leibgedächtnis: Ein Beitrag aus der Phänomenologie Husserls. *Husserl Studies* 27:173–196.

Summa, M. 2012. Body memory and the genesis of meaning. In *Body Memory, Metaphor, and Movement*, ed. S. Koch, T. Fuchs, M. Summa, and C. Müller, 23–42. John Benjamins.

Tulving, E. 1993. What is episodic memory? *Current Directions in Psychological Science* 2:67–70.

Turner, V. W. 1969. *The Ritual Process: Structure and Anti-Structure*. Aldine.

Valdesolo, P., J. Ouyang, and D. DeSteno. 2010. The rhythm of joint action: Synchrony promotes cooperative ability. *Journal of Experimental Social Psychology* 46:693–695.

van Gennep, A. 1909. *Les rites de passage*. Emile Nourry.

Wiltermuth, S. S., and C. Heath. 2009. Synchrony and cooperation. *Psychological Science* 20:1–5.

18 Movies and the Mind: On Our Filmic Body

Joerg Fingerhut and Katrin Heimann

Recent approaches in embodied, embedded, enactive, extended, and affective (4EA) cognitive science[1] argue that mental activity is best understood as relational: the mind is constituted by ongoing interactions between the organism and its environment and understanding the nature of those relations is therefore the main task of a science of the mind. The mind also is "integrative" in the sense that the tools we use and the environmental scaffoldings we enjoy co-constitute those relations (Menary 2007). One central idea of 4EA cognition is that the mental states that supervene on these relations are dependent upon context and are malleable to a certain extent. Theories focusing on cultural influences on the mind, on the other hand, have stressed cultural variations of our mental states due the influence of society and social groups (Prinz 2012) and have emphasized the enculturation of cognition and skills that can, for example, be shown by the way "patterns of practice" of a society or group correlate with certain perceptual discrimination abilities and neural response patterns (Roepstorff, Niewöhner, and Beck 2010; Hutchins 2011). The cultural impact on decision processes, mental states, the organization of our brains, and patterns of neural activity also encompasses the influence of material artifacts (as opposed to, e.g., social artifacts) of a society, including the products of popular culture and artworks with which our lifeworld is so replete. In this chapter, our focus will be on a very recent and rather pervasive kind of such an artifact: film, or more precisely, edited moving images,[2] and we will assess in what way film could figure in an embodied framework of the mind.

1. These approaches are sometimes thought to be largely complementary, and sometimes as yielding incompatible background theories or even incompatible assumptions with respect to the nature of the mind (Hutto and Myin 2013). It is not our aim here to assess the conceptual relation between them (but see Ward and Stapleton 2012; Fingerhut 2014 and the discussion in section 2 of the present chapter) or even to remap the field. We rather want to apply some of their claims to an understanding of how cultural artifacts and especially edited film bears on our mental states.

2. For convenience we will use "film" and "moving images" interchangeably, although the latter constitutes a generic term for more specific media references as "film," "video," "TV," etc. (Carroll 1996, xii). Some peculiarities of the medium, as well as the environmental context of viewing (e.g., movie theater vs. tablet), promote different psychological effects, yet the phenomena we address are by and large applicable to moving images more generally.

Given that the average American citizen now spends one-fifth of her lifetime engaging with real and fictional worlds via moving images (U.S. Bureau of Labor Statistics 2014), we need a deeper understanding of how this medium influences our habits of perceiving, thinking, and feeling.[3] 4EA cognitive science has already made ample reference to interactions between organisms and technologies (such as virtual realities or sensory substitution devices); yet film has largely been neglected. Here we will argue that an embodied approach to film can deepen our understanding of this medium, while at the same time providing the necessary means to understanding how film has already altered our embodied habits of perceiving and experiencing.

Film is often thought to be a media experience that is closer to real-world interaction than, for example, texts or static images (Bazin 1967; Kracauer 1960). In cognitive film theory researchers have argued that film's ability to affect us has its roots in a trend, established early on, of hiding its medial quality, which is grounded in the filmmakers' knowledge of our bodily habits in everyday perception. Adapting the presentational characteristics of film to such perceptual and bodily habits can be useful when it comes to improving film's technical means of involving the moviegoer, and film thus progresses in concealing the differences between film perception and the perceptual routines we apply in the extrafilmic world (Bordwell, Staiger, and Thompson 1985). It has therefore been suggested that filmic narrative devices, such as montage and specific camera techniques, have been developed "to match our cognitive and perceptual proclivities" by, for example, making use of the natural dynamics of attention and other structural features of human perception (Cutting and Candan 2013, 27; see also Cutting 2005; Carroll and Seeley 2013).

We will argue that it might be an undue simplification to see cinema as simply and smoothly approximating our given biological apparatus in the aforementioned sense. The converse might also be true: we have adapted to the technological and filmic means of the medium that—despite its ability to engage and immerse the viewer—still differs from everyday perception in substantial ways. Film therefore also expands our perceptual capacities. In film, we can, for example, instantly change perspectives (think of shot/reverse shots between two actors in a dialogue) or gain access to different scenes at the same time (think of the use of split-screens in *The Thomas Crown Affair* [1968], displaying different storylines that evolve simultaneously). By habituating ourselves to the medium of edited film, we may also have transformed some of our more general ways of experiencing the world—in ways that are scarcely understood.

In the first section of this chapter we will review different approaches to the ways that we cognitively engage with film (e.g. by highlighting certain neural responses to different filmic

---

3. Results of the American Time Use Survey are available online at http://www.bls.gov/news.release/atus.nr0.htm. Film is omnipresent in industrial cultures. The pervasive and far-reaching role of TV as a mass medium (also moving toward unification across cultures) is, for example, addressed by Wexler (2006, 237–241).

means; means that include the use of cuts, and of different camera and lens movements to portray scenes) and provide a basic embodied interpretation of recent research in this area. In the second section we will address philosophical claims regarding our embodied engagement with film stemming from phenomenological film theory and will provide an initial taxonomy of the roles visual cultural artifacts play in 4EA approaches to the mind. In the third section, we will focus on a more positive claim that constitutes the core of the present chapter, namely that familiarization with the filmic medium might change our experience of film as well as our extra filmic perceptual routines over time, a process that has led to the emergence of a filmic body. Cognitive film theory, as portrayed in the first section of this chapter, considers certain filmic techniques to be closer to our preexisting bodily habits than others, and it is because of this vicinity to our natural perceptual routines that such techniques succeed in creating seemingly more realistic situations (e.g., by engaging certain motor components of the brain). Or so they argue. We will entertain a thought that stands in opposition to this, namely that we entertain a filmic body in the movie context that adheres to its own rules. When we therefore engage with the medium in bodily terms (in order to have an illusory experience and immerse ourselves in its narrative), this engagement is not simply premised on what could be called our natural body (i.e., a fixed set-up of perceptual mechanism), but on novel skills and habits of perceiving that we have developed through our exposure to the conditions and syntax of film.

1   Cognitive Film Theory

1.1   Filmic Means in the Cognitive Science of Film

One of the most striking features of film (especially Hollywood film) is its capacity to engage the viewer in the world on screen. This is remarkable, because the world that we experience in film deviates from what we experience "in real life" in quite palpable ways. Just consider the effects of camera and montage, as two traditional devices of narrative film. When they are skillfully applied, we accept that we can see the world through what are normally nonaccessible perspectives, follow two (or more) different story lines via something called "crosscutting" between them,[4] and understand spatial and emotional relations in scenes where several people are shot from different perspectives. What is even more astonishing is that most of these narrative devices go unnoticed by the spectators, who are deeply immersed in the plot of the movie. In a study that tested our ability to consciously detect cuts (indicated

---

4. One famous example of this is that of the three story lines running in parallel in Fritz Lang's *M—Eine Stadt sucht einen Mörder* (1931). Such story lines can be very remote in time and space: the original *Star Wars* (1977) movies contain up to thirty edits per minute, often switching between spaceships, planets, or even galaxies without eliciting confusion in viewers.

by a key-press) an average of 15.8 percent of edits remained unnoticed—a phenomenon that the authors labeled "edit blindness" (Smith and Henderson 2008).

This immersion, however, is not easily achieved by any film, but is highly dependent on the refined skills of moviemakers in the creation of the images and their montage. Cognitive film theory examines the perceptual processes that underlie these effects and has argued in particular that the undisturbed engagement of the viewer with the plot is achieved by exploiting our naturally embodied habits of perceiving the world (Bordwell, Staiger, and Thompson 1985; Small 1992; Smith, Levin, and Cutting 2012). Film seems to have adapted the image stream to some of the real-life particularities of human vision and cognition in order to hide the other violations it makes with respect to the regularities of this very visual system.

One example of this is the exploitation of eye movements that, if evoked at the right time, can efficiently hide a cut. Saccades and blinks during normal vision create periods of retinal blindness interrupting the information flow in about one-third of our total viewing time (Henderson 2003). It has been known for some time that such eye movements can be externally evoked as startle or orienting responses using fast movements or contrast changes on the screen, as well as sudden tones. In addition, it has been found that viewers involuntarily time their blinks around moments of attentional rest, such as event breaks, and mirror actors' blinks (Oh, Jeong, and Jeong 2012; Nakano and Kitazawa 2010). It is likely that cuts made at these times remain largely unnoticed (Smith and Henderson 2008).

Yet this crucially depends on mapping and remapping processes in the spectators' brains, which are based on automatic and implicit predictions about the existence, location, and visual properties of objects not currently at the center of our attention. Only if the shot following a cut fits the predictions previously made about it will a masking be successful.[5] To match the spectators' natural predictions on these different levels, filmmakers use a set of rules that also have been described as montage guidelines for continuity editing (cf. Bordwell, Staiger, and Thompson 1985).

One of the most prominent examples of such a rule is the "180° rule." According to this rule, the initial shot of a scene draws an imaginary line, called the axis, through the center of the depicted action (which might be two people looking at, talking to, or fighting each other) that divides the action space in two: the first half is where the camera is located (placed within a circle orthogonally focusing on the 180° division line), while the second half is on the other side. According to this rule, the camera position can be varied between shots, as long as this line is not crossed. Indeed, previous research has shown that cuts complying with the 180° rule, even when they are not masked by a blink or saccade, very often go unnoticed, while noncontinuity jumps violating this rule are more easily detected (Smith and

5. In fact, it has been shown that the fulfillment of such expectations might even make another intervention (like provoking blinks or saccades) unnecessary for hiding the cut (see Smith and Henderson 2008).

Henderson 2008; Zacks, Speer, and Reynolds 2009). In line with the descriptions given above, it has been suggested that this difference depends on the fact that cuts not breaking this rule are still, perceptually speaking, *close enough* to normal vision to elicit the spatial updating processes also used for one's own eye movements. A breaking of the 180° rule (called a "cut across the line" in film language), on the other hand, violates our perceptual habits in a too fundamental way and consequently elicits different neural processes that make conscious detection more likely.

1.2   The Motor Component in Film Perception

Recent neuroscientific studies have expanded and refined understandings of how we differentially process continuity edits and noncontinuity edits of the same scene. An fMRI and an EEG study indicated that while all edits are registered in some way as a syntactic violation of the action happening in the movie, only continuity edits are followed by distinct remapping processes involved in overcoming the perspective change and thereby probably allowing the edit to go unnoticed (see Magliano and Zacks 2011, for the fMRI Study; Heimann et al. 2016, for the EEG study). On the other hand, Heimann et al. (2016) have also shown that "cuts across the line" are followed by later responses (a distinct late positivity in the event-related potential [ERP] signal) in the same region—responses that have also been observed in change blindness experiments. Koivisto and Revonsuo (2003) report that this component occurred only during trials in which the participants consciously noticed the applied change, suggesting that it marks a special awareness of their own perceptual processes.

While Magliano and Zacks (2011) propose that the different processing routes indicated by these findings might be invoked by attention-guided top-down processes (i.e., an immediately higher level of attention for noncontinuity edits changes the way they are further processed), Heimann et al. (2016) do not support this hypothesis (finding no differences in the common marker of visual attention, viz., the occipital alpha rhythm). However, they have observed additional differences in the activity of the motor cortex in response to different kinds of edits—an interesting finding given what previous research has suggested about such activations. There it has been assumed that the human cortical motor system is involved not only in processes preparing for action execution, but also in action observation. A certain type of neuron in motor regions, called "mirror neuron," is reported to be activated when an individual performs a goal-related action as well as when she sees that same action executed by another agent (Rizzolatti et al. 1996). Researchers have proposed that these action-perception links might also play a crucial role in our bodily response to and engagement with artworks, that is, in the perception of scenes and bodily movements or even of the traces of bodily actions depicted in static images and artworks, such as brushstrokes and cuts inserted into the canvas (Freedberg and Gallese 2007). Meanwhile, researchers have also experimentally shown that the respective motor activations occur in observations of traces of actions (letters, brushstrokes, etc.), and that they are enhanced by the individual's motor familiarity with the action observed (Calvo-Merino et al. 2006; Babiloni et al. 2009; Umiltà et al. 2012; Heimann, Umiltà, and Gallese 2013).

Transferring these insights to the realm of film, it could be claimed that our engagement with filmic editing techniques is also based on the action possibilities of our physical body in the real world. Heimann et al. (2016) also establish that while continuity edits are followed by a stronger activation in the left hemisphere of the brain—as expected during the observation of a right-handed action as it is executed by an actor in a watched scene—"cuts across the line" displaying the same scene from a slightly different angle are distinctively followed by a bilateral activation. This might indicate a sensorimotor disturbance due to the reversing of left and right of the people depicted in the scene (as happens in the "cut across the line"), which may in turn be responsible for the conscious awareness of these kinds of cuts.

These results, taken together, seem to support the claim that filmic effects on the spectator profoundly depend on expectations and principles anchored in our habits of everyday bodily interaction with the world. More precisely, the results connect such expectations to the activation of the motor cortex during action-observation. Gallese and Guerra (2012) have also posited this connection, and go on to claim that this neural link between an observed and an executed action might constitute the fundament of a spectator's vivid engagement with movies. Focusing on camera movement in film, Heimann et al. (2014) have shown that a dynamic camera approach enhances motor cortex activations during action observation, but only if a steadicam (not a zoom or a dolly cam) is used. This condition was also judged as giving the most realistic representation of the respective scene in post-trial questionnaires, as well as being most successful in making the spectators feel as if they themselves were approaching the scene (no measures regarding the level of interest in the film clips, or the aesthetic appeal of the different camera movements were taken). The authors therefore suggest that the motor cortex activation might be modulated by familiarity with the kind of visual stimuli provided by the moving image due to the camera technique applied. The human motor system reacts more strongly to pictures that, because of the use of a steadicam in their production, best simulate a prior motor experience. Films that aim at such engagement on the part of the spectator may therefore dwell on techniques that exploit some of the bodily habits we manifest in our perceptual interaction with the real world.

2   Film, Pictures, and the Embodied Mind

2.1   The Body in Philosophical Film Theory

Addressing our motor engagements with different filmic means can be seen as a direct contribution to an embodied approach to film. In this sense the advances that have become possible in the cognitive sciences could be seen as a refinement and further development of ideas in the philosophy of film, where in recent decades the spectator's body has become the focus of much theoretical attention. Sometimes this focus on bodily engagement is thought to stem from the advent of the post-classical or New Hollywood cinema of the late 1970s, when movies became "increasingly plot-driven, increasingly visceral, kinetic, and fast-paced, increasingly reliant on special effects" (Schatz 1993, 23). However, it might also be regarded as part of a move toward a more phenomenologically oriented philosophical film theory that aims to find links between theories of embodiment and film (see e.g., Sobchack 2004; Barker 2009). Indeed, this focus on the body accompanied and replaced some film theoreticians' focus on either the formal elements of film or wider questions concerning, for example, the ideology and political dimension of cinema. Much of this new interest is grounded in the basic insight that cinema affects us and creates its illusory effects by engaging our body and by arousing strong corporeal feelings of disgust, shame, desire, fear, and so on (see, e.g., Shaviro 1993; Gallese and Guerra 2012).

To capture the special kind of illusion in which film engages the viewer, Christiane Voss has introduced the idea of the *surrogate body* in cinema. With this concept, Voss wants to cover the specific ways in which the body of the spectator is a constitutive part of the architecture of film (Voss 2011, 2013). She therein follows earlier work that aimed to overcome the focus on semantic-reflexive engagement with movies and the passive role of the perceiving subject. Such work had already highlighted our sensory, visceral, synesthetic, and empathic bodily engagement with moving images (see, e.g., the concept of the "cinaesthetic body" in Sobchack 2004, 67). The body in Voss's theory becomes something that is shared between film and perceiver, something that emerges in the interaction between the two. Voss describes the role of the spectator as follows:

The spectator is neither object nor viewing subject of a technique of illusion that could be described independently of him or her. Rather, the film spectator constitutes, as a resonating body in need of further determination, the illusion-forming medium of cinema. (Voss 2011, 139)

This resonating body becomes necessary to overcome the abstractions of visual forms presented on a 2D screen, and does so by providing the screen with a "somatic space of meaning" (Voss 2011, 145). Yet the beauty of her concept lies not so much in an isolated acknowledgment of the contribution of the spectator's body (as a simplified reading of the above quote might suggest), but rather in the way it inverts the view of the relation between perceiver and film and thereby highlights how film itself engages and requires the viewer's body, as well as how both together constitute a reality and body of their own.[6] Spectators and the cinematic medium are no longer seen in isolation from one another; instead, their strong structural coupling—especially in the context of the cinematic

6. This brings such a concept close to that of *imagines agentes* that has been proposed for pictures more generally and that constitutes the core of a picture act theory, which "claims that the recipient of shaped forms is both the subject and object of their active and activating force" (Bredekamp 2014, 30–31).

experience in the movie theater—constitutes a new aesthetic space with its own rules and temporalities, no longer fully controlled by the perceptual processes and self-initiated bodily engagements of the spectator.

What is relevant for the purpose of our chapter is the idea that—in a sense that we will explore in more detail below—we share a body with film. And more than that: this shared body has to be understood in terms of engagements that are bound by rules (perceptual, cognitive, emotional) that pertain specifically to this medium. Let us illustrate this with an example that might also help us to avoid an overly simplistic reading of our central claim. The most obvious cases of sharing a body with film might be when one identifies with a character or when a "point of view" shot allows one to move *with* a character through a scene (think of the scene in Jim Jarmusch's *Dead Man* [1995] in which we ride with the dying character William Blake, played by Johnny Depp, down a steep hill in a vertigo-inducing way). Such experiences can be extremely strong, but this is only partially, or so we claim, because we identify with the respective characters and their perspective. Our bodily engagement is certainly influenced by such character identification and our embodied understanding of persons depicted in the film, but above and beyond that it is in a fundamental way mediated by the differential use of basic filmic means. As the studies on camera movements mentioned above indicate, we engage differently with depicted scenes based on the camera techniques used to produce the optic flow through which the scene presents itself. Some of these effects might be derived from the "simulative" potential of the camera techniques, imitating modes of perception and movement that we know from life "off screen." Yet these camera movements and cuts are not just movements and perspectival changes that we could normally undergo—they are always artificial to some extent. Crucially, their continuity with normal bodily engagement (i.e., the closeness to everyday interactions used to explain the effects in these studies) is not all there is to explain. We are already familiar with novel filmic means that go far beyond our standard perceptual repertoire (think of slow motion or time-lapse photography as clear examples of such extended perceptual means). Our claim is therefore that we become a different body in perceiving film, a body that we, for the sake of simplicity, will call the "filmic body." By treating our body—its perceived boundaries and its structures of perceptual and emotional engagement—as malleable, we can grasp the impact of edited film on our experience in a way that goes beyond what previous theories have imagined. We will explore this idea of a filmic body in section 3. In the following subsection we will lay some further ground for such an approach by introducing some claims from an embodied view of the mind regarding our interactions with cultural artifacts more generally.

2.2   4EA Picture Perception and Its Relation to Film
Within current 4EA (embodied, embedded, enactive, extended, affective) cognitive science and philosophy theorists have emphasized, with slightly different foci, that the mind should be understood as relational, that is, in terms of organism-environment interactions,

and as integrative, that is, as including tools and scaffoldings in the environment that co-constitute those relations. This implies, under some enactive readings of embodied cognition, that what we currently perceive and experience is determined by a history of a coupling of the living and striving organism with its environment (see Varela, Thompson, and Rosch 1991; Thompson 2007; Chemero 2009). Thompson has argued that we can distinguish three modes of ongoing bodily activity to capture what determines our mental life: (a) the body as involved in self-regulation; (b) the body as engaged in sensorimotor coupling with the world; and (c) the body as engaged in intersubjective interactions with other bodies (Thompson and Varela 2001). A full-fledged theory of the cognitive role of film (and fiction more generally) would therefore would have to include film's impact on our needs and strivings (body a), as well as its function as a quasi-social medium (body c). Yet, in the remainder of this chapter, we will focus especially on our sensorimotor engagement and the idea that our present experiences and perceptions are constituted by the implicit knowledge and implementation of sensorimotor rules,[7] rules that might be systematically altered or extended by the different media through which we gain access to the world. Therefore, elements that were previously seen as rather peripheral in the study of the mind now become focal: we access the world via our body (its sensory and motor systems); and the structure of this engagement—and its contribution to the constitution of mental states—becomes a central subject of analysis.

Important for our focus on cultural artifacts is that such bodily engagement is both: sometimes systematically limited and sometimes augmented and extended by the resources and possibilities for interaction provided by technology, artifacts or, more generally, the built environment in which we live. In some circumstances artifacts play such an active role in our epistemic and cognitive endeavors that it seems warranted to treat them as literal parts of the cognitive mechanisms that constitute mental states, that is, when they fulfill certain coupling conditions and are for example portable, reliable, trustworthy, and durable resources (defended as the "extended mind" thesis in Clark and Chalmers 1998). Although edited film, projectors, and screens (or the TV) do not immediately qualify given these original criteria (they might under some very specific conditions), we nonetheless want to highlight what could be considered the anthropological bedrock that underlies cases of extended cognition in the realm of human animals, namely that we are "profoundly

---

7. O'Regan and Noë's (2001) sensorimotor enactivism focuses directly on these sensorimotor contingencies. Thompson also emphasizes the importance of our sensorimotor way of being since it "comprises locomotion and perception, emotion and feeling, and a sense of agency and self" (Thompson 2007, 221). In a biologically oriented enactivism, this mode of bodily activity is always intertwined with the other two (see Fingerhut 2012 for discussion of this claim). In film theory, there have been some attempts to address basic evolutionary biological functions that are more directly related to survival and self-regulation, for example, in approaches that relate filmic means and storytelling to basic affective orientations and self-emotions (Grodal 2009a, 2009b).

embodied" creatures who are actively integrative in the sense that we are "forever testing and exploring the possibilities for incorporating new resources and structures deep into [our] embodied acting and problem-solving regimes" (Clark 2008, 73).[8]

Edited film has generally not played much of a role in these debates—either in talk of cognitive problem solving or in discussion about sensorimotor engagement. Yet at least our interaction with static images and our use of pictorial strategies (i.e., our ability to draw images and consult them) has been addressed and it might be helpful to turn to them in order to situate moving images in the discussion.

We might say that there are three ways in which pictures—and *mutatis mutandis* other material cultural artifacts—can be theoretically addressed within an embodied philosophy of mind and cognitive science: (1) in their function of providing an extended solution-space for cognitive problems; (2) in their capacity of externally embodying certain ideas, emotions, or meanings *as* objects; and (3) in their role in modulating our cognitive and perceptual access to the world. These aspects are not mutually exclusive, and cultural artifacts may exhibit an important influence along all of these aspects. The first two aspects require pictures to some extent to stand out *as* external objects—that is, to be available and perceivable as props that are perceivable or manipulable. Pictures, considered in this way, do not become fully integrated into our perceptual routines in the sense that they do not hide their mediality or their object-like character. On the contrary: in their availability to cognitive and perceptual scrutiny, they make novel elements externally assessable for the cognitive system (be it by providing new ways to address a cognitive problem, or by presenting novel contents and making explicit certain aspects of the way in which such contents present themselves); thus what makes them particularly valuable is their ongoing exteriority to the more confined cognitive system (see Fingerhut 2014 for discussion of this point).

We can elucidate the first aspect—pictures being something like an extended epistemic toolset—by considering how different kinds of depictions (representational paintings, photographs, scientific macro- or microscopic images) can offer a novel form of access to elements that were not previously within the reach of our senses. Take a famous example that illustrates this point: the debate at the end of the nineteenth century over whether a horse in full gallop ever has all four legs in the air (see e.g., Gericault's famous oil painting *The 1821 Derby at Epscom* [1892], which depicts all four legs as spread out and off the ground: two toward the front and two toward the back of the horse). This question was indeed satisfactorily settled only after the innovation of motion photography by the photographic pioneer

---

8. See also Menary 2007 for an extensive treatment of this idea. The integration of artifacts into cognitive routines comes in different degrees and complexities, and the literally extended cognitive mechanisms (e.g., one that includes technology that is permanent and fully integrated) might be the limiting case of this dimension (see Wilson and Clark 2008).

Eadweard Muybridge. Muybridge's photographic series of horses in motion made visible the specifics of quadruped motion and were used as evidence in scientific treatments of the topic (see Stillman 1882).[9]

The second aspect also focuses on the way that visual artifacts stand out as objects. But instead of focusing on cognitive solutions, it addresses the way in which images take a predominant role in the expanded realm of culture, so to speak. It highlights how ideas or emotions, once they are realized in artifacts, can at times gain a quasi-subjective force (think of certain pictures in the history of art or in religious contexts). This somehow seems to exceed the confined problem-spaces of standard cognitive science, yet such relations to artifacts are nonetheless an important aspect of how we experience the world and play a decisive role in how we conceive of ourselves as human agents. This alone should make them a valuable object of study also for this field. Questions regarding our emotional and aesthetic engagement with artifacts have not been the immediate concern of our chapter (although it also touches on them). Yet the third aspect introduced above, namely that of an altered *perceptual* access, is of particular interest for our examination of the role of film: how could visual artifacts have an impact on the very way we perceive?

Alva Noë has explored our engagement with pictures and what could be called a "media-specific form of perception"; he claims that "seeing pictures is a *distinct modality* of seeing" (Noë 2012, 110, italics added). One characteristic, according to Noë, pertains specifically to pictures: what they represent is at the same time *manifestly* absent. We therefore do not perceptually interact with the depicted in the same way that we do with other objects. This idea is couched in a *present-as-absent* mode of perceptual consciousness and has to be understood against the backdrop of a more general theory of experience that sees experience as a function of different modes of access to the world around us. Within this theory, we also have a visual sense of the presence of the back of a tomato because we know how to make contact with it based on our implicit knowledge of possible sensorimotor engagements (i.e., of the ways to move in order to achieve more direct access) with it (Noë 2009). Altering our bodily position with respect to a picture does not elicit this kind of visual pattern of sensorimotor contingencies with respect to the depicted object, but rather to the picture surface and therefore the medium as an object (we cannot look behind a depicted

9. This example is already close to moving images. Yet also the specifically filmic and therefore temporally integrated way of displaying content might facilitate cognitive access and therefore function as a tool for thinking. Think of image-guided surgery as used in medical research, or slow motion or time-lapse cameras as more mundane examples of how moving images contribute to new insights. Some conceptual knowledge regarding the means of production of such images (as well as static ones) is required to guarantee epistemic success. See also the discussion on whether we can immediately *see* through a microscope or rather acquire this capacity after some inferences that cancel out distortions due to the medium in Hacking 1985.

object, yet we can look at the back of the physical object that is used to portray it)[10]—and this, in a sense, constitutes the visually *manifest* aspect of the absence of the depicted object or scene.

Such differences regarding picture and object perception can be found at several levels of interaction. The one just described constitutes a rather fundamental difference (and led Noë to claim a specific "modality" of picture seeing), but looking more closely at additional, more fine-grained, bottom-up perceptual habituation effects might reveal further structuring features that picture perception exerts on us. Through their specific arrangement or via specific stylistic execution of scene-presenting features in pictures, we might, for example, become habituated to novel patterns of saccadic eye movements in searching a scene and more generally learn novel forms of embodied engagement, which natural scenes might not have afforded. Based on what we discussed with respect to the attention-guiding role of cinema earlier in this chapter, the media-specific rules of moving images could have an even stronger impact on our perceptual system compared to static images (although those have been part of our culture for a longer period of time); and it is this pervasive quality of film to which we will now turn.

3   Toward a Theory of Our Filmic Body

3.1   The Very Idea of a Filmic Body

Arrangements in pictures can be compared to the *mise-en-scène* in film, though the ways in which film differs should be apparent. Film is not fixed in time and spatial arrangement in the same way: it contains changing images, depicts moving objects, and produces an optic flow by itself. Although the general structure of "presence in manifest absence" to some extent pertains to moving images, and although film is also not spatially and temporally continuous in the same sense as the world is, one could argue that the experience of absence is sometimes overwritten by the illusionary character of film: the absence is not perceptually manifest in the same way as in static images. It could be argued that this is because when we watch a film, our overall self-initiated bodily movement, as well as our engagement with the surface of the filmic medium, is less pronounced than it is when we view still pictures.[11] This, along with other differences regarding our bodily engagement

10. Some have argued that surface properties and depicted objects also target two basic visual subsystems differentially (Matthen 2005; see also Nanay 2011 for an interpretation of such ideas under the heading of "twofoldness" in the experience of pictures). Noë (2012) is rather critical of a too-rigid separation of the subsystems as presented by Matthen (2005); and he also explores not so much the low-level features of our sensorimotor engagement with pictures but rather the role of understanding and of conceptual elements in constituting our "access space" and thereby in mediating our experience of presence.

11. We have argued (section 1.2) that Hollywood-style cinema is especially prone to avoiding the permanent engagement of the viewer with media-specific means (e.g., cuts), but this also holds for other

with film, has to be directly addressed if we are to understand the habituation of our filmic body. Based on this we will see how far such a concept (of a bodily pattern as an emergent property of organism and artifact) can go and in what ways it could provide the basis for future research.

As we have seen, both kinds of images, static and moving, differ from everyday scene-perception in substantial ways, but also from one another: whereas static pictures (after an initial gist perception) invite exploration, moving images to some extent execute the perceptual exploration of a scene for us. We therefore want to claim that the succession of moving images interlocks with our bodily (neuronal, muscular, etc.) responses in a way that constitutes the engagement of a filmic body, a body that has evolved from the interaction of organismic body and filmic medium. As noted, our biological body is largely at rest in the movie theater and does not—most of the time—self-initiate movements to generate differential feedback from the world or the screen. In addition, with respect to the bodily movements of the viewer that are still in place (such as saccades, blinks, and head movements), one could argue that they are to a large extent guided by the skillful knowledge of the filmmaker regarding motor-engagement and attention. Despite its many forms of artificiality, film exploits our bodily, perceptual habits in just such a way that it is picked up as a perceptual engagement with a scene. Yet, the fact that we defer to a succession of images as an approach to a scene in a quasi-perceptual, sensorimotor way should not disguise the fact that we have become habituated to altered bodily patterns and visual routines that go beyond any natural way of perceiving. It is these emergent patterns that form the basis of what we think of as our filmic body.

3.2   A Different Body Schema for Film?

We started this chapter by discussing cutting and camera movement as specific technical elements of film. What we now want to bring to the fore is that these techniques already indicate a form of extended vision. Through our engagement with film, we have generated a perceptual system of a filmic body (or even of filmic bodies), which involves a different set of sensorimotor contingencies depending on the peculiarities of film (or even different genres or styles). This system is separate from the one used off the screen precisely because of the different recurring patterns of sensorimotor contingencies evoked by its style. The differences from everyday scene perception are sensori*motor* in the sense that our motor engagement is systematically and differentially guided by this succession of images. To understand this impact of (cultural) resources on our cognitive system, it is helpful to look at studies that address the plasticity of our body representations. Standard examples in this respect are

---

properties of the screen. A case in point is the development of different aspect ratios in the history of film in order to offer a fully immersive cinematic experience by hiding the confining, external boundaries of the screen.

studies done on basic tool use but we will also consider in which ways insights from this field transfer to the cases under discussion here.

It has been repeatedly shown in nonhuman primates that after short periods of successful usage of simple handheld rakes and other tools those tools became integrated into the body schema of the user.[12] It has also been demonstrated that human tool use extends what could be considered "near space" out toward the exteriority (Longo and Lourenco 2006), and it has been assumed for a long time that the human body schema is malleable in this way (Head and Holmes 1911). This has led researchers to posit a shared cognitive machinery between apes and humans that facilitates body-schema plasticity and other aspects of our interaction with tools and artifacts (Vaesen 2012). Some cultural artifacts—especially those that are by and large seen as representational artifacts, such as pictures, texts, movies—initially seem to be very different from such basic tools such as rakes or sticks. Yet we want to argue that they indeed share some important commonalities: first of all, they also mediate our grasp of the world in a sense that goes beyond our basic biological setup. And, second, we might also (have to) adapt to them in a certain way, namely by integrating them into our (bodily) access to the world.

Interestingly, some researchers have recently called into question the strong similarity between humans and other animals regarding body-schema plasticity (as it is defended, e.g., by Vaesen 2012), and have emphasized rather fundamental differences: plasticity in humans occurs across different time courses, and not just after intense learning. In addition, and more strikingly, humans seem to be able to entertain multiple body representations that mediate different experiences related to different kinds of tools. These can be selectively activated depending on whether or not a tool is present (Serino et al. 2007; Longo and Serino 2012; Canzoneri, Magosso, and Serino 2012). Taking such results seriously, we can now argue that the same might hold for our interactions with media: we might be able to switch between different bodies (cinematic, augmented, virtual reality, etc.), and therefore build a repertoire of skilled bodies whose utilization depends on certain triggers and engagements provided by the respective mediums.

Our claim regarding filmic extended perception should not, therefore, be misunderstood as thinking of the camera as a literal part (i.e., a material extension) of our body. The camera's activity obviously lies in the past and we do not literally gain access to the depicted scene, even when we perceive it under certain conditions as belonging to our peripersonal space (i.e., as the space around our body that is within our reach). We should, on the other hand, nonetheless avoid a categorical separation between the perceptual apparatus and the

12. This has been shown, e.g., through recordings of single cells whose receptive fields extended after the use of a rake to grasp food that was not reachable with bare hands. For the concept of a "body schema" in those cases and a review of the literature, see Maravita and Iriki 2004.

technical means used in the production and display of edited film.[13] We therefore consider it a tenable position that we are not simple and plainly passive viewers (in the sensorimotor sense) in the cinematic context; rather, there is an active, bodily element in place, yet one that is partly carried out by the medium itself and by way of the transition of different images based on camera movements and situational changes through cuts. It is in this sense that we share a filmic body in the cinematic context, namely in that specific perspectives (as well as the transitions between them through cuts) might actually be readily incorporated into our body schemas as filmic patterns of bodily engagement with a scene and become reactivated and updated when we watch a movie.[14] Moving images only seemingly enable perceptions that are *as if* in real life. They immerse us, yet even the filmic techniques that remain unnoticed by the viewer render the optic flow distinctively different from normal perception. We live through scenes in a movie as experienced by our own eyes, yet those eyes (i.e., the visual apparatus) are, or so we want to argue, extended in the filmic context. They are extended in the sense that they include certain regularities (of the technical devices used to produce the images, of the devices used to display them, and with respect to the different kinds of montages) that pertain specifically to film.

The findings regarding motor activation in the filmic context fit nicely into this framework. Remember that one interesting finding in the studies discussed above was that the observation of pictures recorded using a steadicam correlated with a significantly stronger activation of the motor cortex. Heimann et al. (forthcoming) suggest two possible ways of explaining this. The first runs along the lines we discussed in section 1 of this chapter: the steadicam produces pictures of higher perceptual familiarity, and as such evokes a stronger reaction in the mirror system (an activation of the hand motor region induced by the observation of hand movements in the scenes presented, for example) and perhaps, in this way, a stronger engagement on the part of the spectator. The other possibility is that the stronger activation is caused by an additional activation of the motor regions of other body parts, for example, the muscles that are activated during walking. Indeed, as Heimann et al.

13. This is a fruitful outcome of the discussion in 4EA cognitive science, stemming from its emphasis on what were previously considered to be peripheral elements (i.e., the body and artifacts). It might also be instrumental in overcoming what is in our view an ill-advised dualism that has underwritten psychological theories of film from the very beginning: "What we need ... [is] an insight into the means by which the moving pictures impress us and appeal to us. Not the physical means and technical devices are in question, but the mental means. What psychological factors are involved when we watch the happenings on the screen?" (Münsterberg 1916, 39).

14. A "body schema" in this sense is a system of motor programs and habits of movement. We therefore use it in a looser way than, e.g., Maravita and Iriki (2004), building on the connotations employed in the embodied cognition literature, where it is described as contributing to the prenoetic structuring of our perception and experience and as a "system of sensory-motor capacities that function without awareness or the necessity of perceptual monitoring" (Gallagher 2005, 24).

(forthcoming) show, the motor cortex also reacts to movies simply displaying an empty room filmed with a moving camera (where no hand movements or people are visible). This can be read as supporting the second interpretation, indicating that camera movements themselves elicit neural activation patterns in the spectator related to movements of our own that would mirror the movement of the camera.[15] Building on this second interpretation, one could entertain the thought we introduced above that such patterns of engagement can be altered by exposure and habituation to filmic approaches to certain scenes. This can happen even over short timespans (e.g., in the way a feature film establishes a specific optic flow that allows us to generate certain sensorimotor expectations for the duration of the film) or over the ontogenetic development of a culturally embodied subject (via exposure to edited film more generally).

Our proposal goes beyond the idea that moving images similar to the optic flow we know from real life engage the motor system in the usual way.[16] That successions of images that deviate more strongly from our perceptual habits—such as "cuts across the line," "jump cuts," and artificial camera movements—initiate differential responses of the motor system (therefore being more likely to be detected as deviations) now has to be integrated into a larger explanatory scheme. From the above-mentioned perspective (of a filmic body schema habituated to the filmic medium that has evolved through learning), it could be that we are even capable of incorporating movements and montages that differ to a greater degree from everyday perceptions, thereby appropriating the activity of the camera even in difficult cases. Our filmic bodily responses have therefore to be seen as a mixture of both our natural habits and learned regularities of the filmic means. These cases might include "cuts across the line"—but also depictions of slow motion or time-lapse (think of Darren Aronofsky's depiction of drug-induced experiences in *Requiem for a Dream* [2000]).

3.3 Film Comprehension as a Learned Skill

Experimental findings support the claim that we do not immediately understand filmic means and techniques, but that they first have to be integrated into our skill set for watching film. Only after they have become integrated and after one has become an experienced viewer, they stop being immediately visible when they occur (e.g., as described in the edit

15. Questions like "Where am I?" or rather "Do I move *with* the camera?" are also interesting for a theory of film perception. Even if we are fully immersed in a camera's perspective, we nonetheless do not simply switch positions according to every shot in the film. Our perspective on a scene, e.g., is determined by what has been called the *establishing shot*, i.e., the shot at the beginning of a scene that provides us with a vantage point on what is to come. We do not, therefore, simply switch to the viewpoint of the people participating in a dialogue filmed in a shot/reverse shot pattern, for example.

16. Even if the optic flow mimics everyday perception our motor engagement might be stronger in the filmic condition than in the real world. This might be because we do not produce the relevant efference copies we normally do in self-movement, which we could then subtract from the overall optic flow.

blindness experiments of Smith and Henderson, 2008). To begin with, developmental research comparing the reactions of children to those of adults has collected evidence that media competence is not naturally given, but is acquired over time. While it has been shown that even young children are able to follow the plot of a movie in general (Comuntzis 1987), and thus have a basic media competence, several papers report major differences in the precise perceptual behavior of adult and child movie spectators (Ohler 1994; Munk et al. 2012; Acker and Tiemens 1981). Acker and Tiemens (1981), for example, have shown that toddlers have no accurate understanding of the application of a zoom or "close up," interpreting them as a growth of the object filmed rather than a result of camera manipulation. As we reported above, adults who are asked to press a button every time they see a cut in a movie take significantly longer to do so for continuity edits than for montages that violate continuity rules (Smith and Henderson 2008). However, Ohler (1994) did not find consistent results in young children (of four, six, and eight years of age). Reaction times similar to those in adults were found only in the group of eight-year-olds. Younger children indeed showed the opposite behavior: the detection of noncontinuity edits took them more time. It has been argued that this could be a result of the higher workload that these stimuli represent for smaller children—an explanation still in line with the thesis that such stimuli are always perceived as rather unfamiliar. It might also be possible, however, that the actual reason for the difference is that what is perceived as familiar or unfamiliar depends on training in the medium rather than similarity with real-world experiences alone. This view is additionally supported by findings indicating that children apparently need more guidance in understanding the medium, not commanding the full media expertise from the beginning. Munk et al. (2012) have shown that children's eye movements during TV watching, in comparison with adults' eye movements, are more frequent (with four-year-old children displaying the highest frequency), indicating that the attention of inexperienced spectators is in general less clearly directed by the medium. Wass and Smith (2015) have furthermore shown that TV programs developed especially for children show a much higher number of strong attention-guiding techniques than programs made for adults, indicating an awareness of this requirement on the part of the program makers. It seems likely that such techniques are intentionally used to facilitate rule or habit acquisition. And it is through such a habituation that our filmic body develops—carrying traces of interaction with the real world as well as from TV and cinematic experiences.

Referring to findings in children, of course, is problematic. The differences from adult perceivers might be also explained by their still-developing perceptual capacities or other cognitive functions. To further explore how an initial and intuitive media response might look, a focus on media-inexperienced adult audiences might be better suited. Previous research in this area in fact supports our claims. In a series of experiments, researchers presented movie clips to a remote group of people from Turkey who had no previous exposure to edited film (Schwan and Ildirar 2010; Ildirar and Schwan 2015). They showed severe difficulty in interpreting stories that were depicted using a number of continuity-editing techniques that are very familiar to more experienced moviegoers, such as the establishing shot or the shot/reverse shot pattern. It thus seems likely that these syntactic elements, and therefore the 180° rule too, represent acquired habits that depend on previous film exposure. In the same vein, Worth and Adair (1972) have shown that movies made by media-inexperienced adult audiences do not follow a film language that we would normally expect given our background in continuity-editing film styles, but rather follow their own rules.

Taken together, these studies support the idea that the structured, technological, medium-specific ways in which edited film operates might also be based on learned skills rather than a natural predisposition due to the closeness of the medium (and the kinds of optic flow it produces) to human perception. They therefore provide additional motivation to see our cognitive apparatus as altered by the ongoing evolution of the medium of film and our skills as depending also on some contingent properties of cultural artifacts that we have been exposed to. This in turn leaves space for the thought that also the cuts and camera movements that now still grab our attention might in the future become integrated into our perceptual understanding rendering therefore other properties of the medium more visible.

3.4 Future Directions and Experiments

We think that by building on the neuroscientific studies reported above, as well as other paradigms recently developed in cognitive neuroscience, we can considerably expand on the reported results and explore the ways in which spectators gain modes of perception that are sustained by a filmic body.

First, the hypothesis that cognition adapts to film has to be related to certain ontogenetic attunement processes. One could look into whether the above-mentioned EEG responses to different camera movements and montages change over time, differentiating between late and early responses, and also indicating processes that are unconscious and might not show a behavioral effect. Given the above-mentioned difficulties with studies on children, ideally such studies would involve comparisons not only between children and adult viewers, but also between media-inexperienced and experienced populations. Furthermore, it should be explored whether prolonged media exposure, even after initial habituation, leaves experienced viewers liable to alterations in their responses. This might be possible by comparing normal adults and film experts, or by comparing an audience before and after exposure (over different time periods) to media specifically designed for this purpose (i.e., by deploying excessively "artificial" coded stylistic means).[17] If media expertise can be understood as an

17. We have omitted considerations of the aesthetic evaluation of moving images from the present chapter. However, they could be related to establishing and violating certain expectations regarding filmic means (based on our filmic body) that are either more generally in place or become instantiated over the course of a movie. That is, if the way in which a film exposes or hides its mediality constitutes one axis on which such an evaluation might be based (as we think it should), such habituation effects have to be included in explanations of the aesthetic appreciation of film.

attunement process, we should expect to see that (a) some responses occur only with media-experience and (b) the same or other responses might still be alterable in adulthood owing to further exposure to the media and evolution in style.[18]

Second, additional paradigms that focus on an exploration of the filmic body could investigate how and to what degree space and time perception and experience are modulated by such a bodily state. An initial idea would be to test whether certain involvements on the part of the motor system could modulate the way the close space on screen becomes the near space of the perceiver and to what extent this depends on the stylistic means employed (think of the "point of view" shots mentioned above, as for example in the famous scene in Hitchcock's *Notorious* [1946], where the camera approaches what is for the character a significant object). Possible paradigms could be derived from previous research on the borders of peripersonal space, measuring the bodily response of approaching sounds (Canzoneri, Magosso, and Serino 2012) or sounds close to the body (Bassolino et al. 2010) that show whether what has been called our "peripersonal space" might extend to what is perceived on the screen.

Last, there are already some indications that the technical elements of moving images permeate our mental states more generally. Murzyn (2008) and Hoss (2010) have reported that exposure to edited film in childhood can have an effect on our dreams or dream reports. In contrast to younger people, who grew up in the age of color television and only seldom report dreaming in black-and-white, people who had a different kind of exposure report differently: from a population of adults older than fifty-five, who lived through the era of black-and-white cinema and TV, a quarter report dreaming in black-and-white still today. Indeed it is not hard to imagine that a stretching and shortening of time, zooms, or cuts can be and are built into our dreams or even our imaginative and problem-solving processes,[19] which should be the subject of future experiments. Such experiments should include the detailed exploration of reported experiences as well as a search for certain markers of filmic means in the brain activity of subjects that might occur outside filmic contexts.

18. The behavioral studies discussed in the previous paragraph indicate that young children and media-inexperienced audiences do not yet experience noncontinuity edits as strong violations of what could be called a perceptual habit—possibly because they do not yet command in a strong sense a media-related habit. This might be reflected in differences in early ERP responses referring to the registration process. On the other hand, extreme media-expertise might not change the registration of a violation, but the response to it (that is, it might diminish a strong startle-effect, for example), possibly leading to differences in the later stages of stimuli-processing between normal adults and film experts, also measurable by ERP.

19. Among these might be improved attention strategies, or better task-switching abilities (learned through short edits and "cross cuts"), and the like. Thanks to Jesse Prinz for pointing these out.

Conclusion

We have argued in this chapter that we learn to see film by integrating filmic means into our bodily routines of perceptual interactions and that we develop what we called a filmic body in this process. Interdisciplinary projects and experimental designs based on insights from a 4EA framework of the mind will help us further explore the nature of filmic engagement. This might help us identify some previously neglected roles that filmic styles and narrative devices for moving images can play—opening up a wide variety of theoretical engagements with aesthetic encounters with film that we did not even touch on in this chapter. More generally, we suggested that film might be a paradigm case that can crucially contribute to understanding the cultural nature of perception and of the mind. Although what we suggest in no way replaces the need for historical film studies and behavioral developmental research, it also might require the identification of neural markers for certain filmic means, markers whose correlations with spectators' experiences and whose role in broader embodied engagement can then be further explored.

In concluding and instead of providing a summary of the main points of this chapter, we want to highlight in what way this chapter's first section (on cognitive film theory) and the third (on the idea that we evolve a filmic body) relate. We have emphasized the medium-specific embodied engagement that is based on a set of perceptual regularities. This in no way makes the research into the behavioral responses and neural realizations superfluous; it simply augments the assumption of cognitive film theory that certain filmic techniques engage us in ways of perceiving that are closer to preexisting bodily habits than others (creating seemingly more realistic situations by fostering a specific embodied engagement on the part of the spectator). What our position crucially adds to this is the following thought: this preexisting body should not be misunderstood as a natural body whose unaltered perceptual machinery processes visual information in a culture-independent manner; rather it is a body that is already encultured and can immediately latch onto certain medium-specific ways of perceiving and experiencing. In this sense we have argued against a prevalent claim of cognitive film theory, namely that we "didn't evolve to watch movies, movies have evolved to match our cognitive and perceptual proclivities" (Cutting and Candan 2013, 27). We have argued instead that we have also evolved alongside the medium of film (and other cultural media for that matter). What follows is that on a broader level both the syntax of film language and film cognition have to be seen as plastic products of ongoing mutual influence between films and embodied agents, thereby moving the medium toward novel filmic means and us toward novel experiences.

The latter claim follows from our understanding of embodied cognition as mainly sustaining a relational view of the mind: the structure of those relations is malleable in the sense that we can integrate certain means of access to our environment (including those mediated by culture and by material artifacts) into our bodily routines. A theoretical engagement with filmic means and the changes they may inflict on our cognitive system will provide further corroboration of this point and enhance our understanding of the embodied and enculturated mind.

Acknowledgments

We are indebted to Jesse Prinz, Andreas Roepstorff, and Vittorio Gallese for support and helpful comments on a draft of this chapter, as well as the editors for their thorough feedback.

References

Acker, S., and R. Tiemens. 1981. Children's perceptions of changes in size of televised images. *Human Communication Research* 7:340–346.

Babiloni, C., C. Del Percio, P. M. Rossini, N. Marzano, M. Iacoboni, F. Infarinato, R. Lizio, et al. 2009. Judgment of actions in experts: A high-resolution EEG study in elite athletes. *NeuroImage* 45:512–521.

Barker, J. M. 2009. *The Tactile Eye: Touch and the Cinematic Experience*. University of California Press.

Bassolino, M., A. Serino, S. Ubaldi, and E. Làdavas. 2010. Everyday use of the computer mouse extends peripersonal space representation. *Neuropsychologia* 48 (3): 803–811.

Bazin, A. 1967. *What Is Cinema?* Trans. H. Gray. University of California Press.

Bordwell, D., J. Staiger, and K. Thompson. 1985. *The Classical Hollywood Cinema: Film Style and Mode of Production to 1960*. Routledge.

Bredekamp, H. 2014. The picture act: Tradition, horizon, philosophy. In *Bildakt at the Warburg Institute*, ed. S. Marienberg and J. Trabant, 1–32. De Gruyter.

Calvo-Merino, B., J. Grèzes, D. E. Glaser, R. E. Passingham, and P. Haggard. 2006. Seeing or doing: Influence of visual and motor familiarity in action observation. *Current Biology* 16:1905–1910.

Canzoneri, E., E. Magosso, and A. Serino. 2012. Dynamic sounds capture the boundaries of peripersonal space representation in humans. *PLoS One* 7:3–10.

Carroll, N. 1996. Defining the moving image. In *Theorizing the Moving Image*, 49–74. Cambridge University Press.

Carroll, N., and W. P. Seeley. 2013. Cognitivism, psychology, and neuroscience: Movies as attentional engines. In *Psychocinematics: Exploring Cognition at the Movies*, ed. A. P. Shimamura, 57–75. Oxford University Press.

Chemero, A. 2009. *Radical Embodied Cognitive Science*. MIT Press.

Clark, A. 2008. *Supersizing the Mind*. Oxford University Press.

Clark, A., and D. Chalmers. 1998. The extended mind. *Analysis* 58 (1): 7–19.

Comuntzis, G. M. 1987. Children's comprehension of changing viewpoints in visual presentations. Poster presentation, Visual Communication Conference.

Cutting, J. E. 2005. Perceiving scenes in film and in the world. In *Moving Image Theory: Ecological Considerations*, ed. J. D. Anderson and B. Fisher Anderson, 9–27. Southern Illinois University Press.

Cutting, J. E., and A. Candan. 2013. Movies, evolution, and mind: From fragmentation to continuity. *Evolutionary Review* 4 (3): 25–35.

Fingerhut, J. 2012. The body and the experience of presence. In *The Feeling of Being Alive*, ed. J. Fingerhut and S. Marienberg, 167–199. De Gruyter.

Fingerhut, J. 2014. Extended imagery, extended access, or something else? Pictures and the extended mind hypothesis. In *Bildakt at the Warburg Institute*, ed. S. Marienberg and J. Trabant, 33–50. De Gruyter.

Freedberg, D., and V. Gallese. 2007. Motion, emotion, and empathy in aesthetic experience. *Trends in Cognitive Sciences* 11:197–203.

Gallagher, S. 2005. *How the Body Shapes the Mind*. Clarendon Press.

Gallese, V., and M. Guerra. 2012. Embodying movies: Embodied simulation and film studies. *Cinema: Journal of Philosophy and the Moving Image* 3:183–210.

Grodal, T. K. 2009a. *Embodied Visions: Evolution, Emotion, Culture, and Film*. Oxford University Press.

Grodal, T. K. 2009b. Film aesthetics and the embodied brain. In *Neuroaesthetics*, ed. M. Skov and O. Vartanian, 249–260. Baywood Publishing.

Hacking, I. 1985. Do we see through a microscope? In *Images of Science*, ed. P. Churchland and C. Hooker, 132–152. University of Chicago Press.

Head, H., and G. Holmes. 1911. Sensory disturbances from cerebral lesions. *Brain* 34 (2–3): 102–254.

Heimann, K., M. A. Umiltà, and V. Gallese. 2013. How the motor-cortex distinguishes among letters, unknown symbols and scribbles: A high density EEG study. *Neuropsychologia* 51:2833–2840.

Heimann, K., M. A. Umiltà, M. Guerra, and V. Gallese. 2014. Moving mirrors: A high density EEG study investigating the effect of camera movements on motor cortex activation during action observation. *Journal of Cognitive Neuroscience* 26:2087–2101.

Heimann, K., S. Uithol, M. Calbi, M. A. Umiltà, M. Guerra, and V. Gallese. 2016. "Cuts in action": A high density EEG study investigating the neural correlates of different editing techniques in film. *Cognitive Science*.

Heimann, K., S. Uithol, M. Calbi, J. Fingerhut, M. A. Umiltà, M. Guerra, and V. Gallese. Forthcoming. Embodying the camera: A high density EEG study investigating the effect of camera movements on motor cortex activation during the observation of an empty room.

Henderson, J. M. 2003. Human gaze control during real-world scene perception. *Trends in Cognitive Sciences* 7 (11): 498–504.

Hoss, R. 2010. Content analysis on the potential significance of color in dreams: A preliminary investigation. *International Journal of Dream Research* 3 (1): 80–90.

Hutchins, E. 2011. Enculturating the supersized mind. *Philosophical Studies* 152 (3): 437–446.

Hutto, D. D., and E. Myin. 2013. *Radicalizing Enactivism*. MIT Press.

Ildirar, S. and S. Schwan. 2015. First-time viewers' comprehension of films: Bridging shot transitions. *British Journal of Psychology* 106 (1): 133–151.

Koivisto, M., and A. Revonsuo. 2003. An ERP study of change detection, change blindness, and visual awareness. *Psychophysiology* 40:423–429.

Kracauer, S. 1960. *Theory of Film: The Redemption of Physical Reality*. Oxford University Press.

Longo, M. R., and S. F. Lourenco. 2006. On the nature of near space: Effects of tool use and the transition to far space. *Neuropsychologia* 44 (6): 977–981.

Longo, M. R., and A. Serino. 2012. Tool use induces complex and flexible plasticity of human body representations. *Behavioral and Brain Sciences* 35 (4): 229–230.

Magliano, J. P. and J. M. Zacks. 2011. The impact of continuity editing in narrative film on event segmentation. *Cognitive Science: A Multidisciplinary Journal* 35:1–29.

Maravita, A., and A. Iriki. 2004. Tools for the body (schema). *Trends in Cognitive Sciences* 8 (2): 79–86.

Matthen, M. 2005. *Seeing, Doing, and Knowing: A Philosophical Theory of Sense Perception*. Oxford University Press.

Menary, R. 2007. *Cognitive Integration: Mind and Cognition Unbounded*. Palgrave Macmillan.

Munk, C., G. D. Rey, A. K. Diergarten, G. Nieding, W. Schneider, and P. Ohler. 2012. Cognitive processing of film cuts among 4- to 8-year-old children: An eye tracker experiment. *European Psychologist* 17 (4): 257–265.

Münsterberg, H. 1916. *The Photoplay: A Psychological Study*. Appleton.

Murzyn, E. 2008. Do we only dream in colour? A comparison of reported dream colour in younger and older adults with different experiences of black and white media. *Consciousness and Cognition* 17 (49): 1228–1237.

Nakano, T., and S. Kitazawa. 2010. Eyeblink entrainment at breakpoints of speech. *Experimental Brain Research* 205:577–581.

Nanay, B. 2011. Perceiving pictures. *Phenomenology and the Cognitive Sciences* 10:461–480.

Noë, A. 2009. Conscious reference. *Philosophical Quarterly* 59 (236): 470–482.

Noë, A. 2012. *Varieties of Presence*. Harvard University Press.

Oh, J., S. Y. Jeong, and J. Jeong. 2012. The timing and temporal patterns of eye blinking are dynamically modulated by attention. *Human Movement Science* 31:1353–1365.

Ohler, P. 1994. *Kognitive Filmpsychologie: Verarbeitung und mentale Repräsentation narrativer Filme.* MAkS Publikationen.

O'Regan, J. K., and A. Noë. 2001. A sensorimotor account of vision and visual consciousness. *Behavioral and Brain Sciences* 24 (5): 939–973.

Prinz, J. J. 2012. *Beyond Human Nature: How Culture and Experience Shape the Human Mind.* W. W. Norton.

Rizzolatti, G., L. Fadiga, L. Fogassi, and V. Gallese. 1996. Premotor cortex and the recognition of motor actions. *Brain Research: Cognitive Brain Research* 3:131–141.

Roepstorff, A., J. Niewöhner, and S. Beck. 2010. Enculturing brains through patterned practices. *Neural Networks* 23 (8): 1051–1059.

Schatz, T. 1993. The new Hollywood. In *Film Theory Goes to the Movies*, ed. J. Collins, H. Radner, and A. Preacher Collins, 8–36. Routledge.

Schwan, S., and S. Ildirar. 2010. Watching film for the first time: How adult viewers interpret perceptual discontinuities in film. *Psychological Science* 21 (7): 1–7.

Serino, A., M. Bassolino, A. Farnè, and E. Làdavas. 2007. Extended multisensory space in blind cane users. *Psychological Science* 18 (7): 642–648.

Shaviro, S. 1993. *The Cinematic Body.* University of Minnesota Press.

Small, E. S. 1992. Introduction: Cognitivism and film theory. *Journal of Dramatic Theory and Criticism* 6 (2): 165–172.

Smith, T. J., and J. M. Henderson. 2008. Edit blindness: The relationship between attention and global change blindness in dynamic scenes. *Journal of Eye Movement Research* 2 (2): 1–17.

Smith, T. J., D. Levin, and J. E. Cutting. 2012. A window on reality: Perceiving edited moving images. *Current Directions in Psychological Science* 21 (2): 107–113.

Sobchack, V. 2004. *Carnal Thoughts: Embodiment and Moving Image Culture.* University of California Press.

Stillman, J. D. B. 1882. *The Horse in Motion as Shown by Instantaneous Photography.* J. R. Osgood.

Thompson, E. 2007. *Mind in Life: Biology, Phenomenology, and the Sciences of Mind.* Harvard University Press.

Thompson, E., and F. J. Varela. 2001. Radical embodiment: Neural dynamics and consciousness. *Trends in Cognitive Sciences* 5 (10): 418–425.

Umiltà, M. A., C. Berchio, M. Sestito, D. Freedberg, and V. Gallese. 2012. Abstract art and cortical motor activation: An EEG study. *Frontiers in Human Neuroscience* 6:311.

U.S. Bureau of Labor Statistics. 2014. *American time use survey.* http://www.bls.gov/tus.

Vaesen, K. 2012. The cognitive bases of human tool use. *Behavioral and Brain Sciences* 35 (4): 203–218.

Varela, F. J., E. Thompson, and E. Rosch. 1991. *The Embodied Mind: Cognitive Science and Human Experience*. MIT Press.

Voss, C. 2011. Film experience and the formation of illusion: The spectator as "surrogate body" for the cinema. *Cinema Journal* 50 (4): 136–150.

Voss, C. 2013. *Der Leihkörper: Erkenntnis und Ästhetik der Illusion*. Fink.

Ward, D., and M. Stapleton. 2012. Es are good: Cognition as enacted, embodied, embedded, affective, and extended. In *Consciousness in Interaction*, ed. F. Paglieri, 89–104. John Benjamins.

Wass, S. V., and T. J. Smith. 2015. Visual motherese? Signal-to-noise ratios in toddler-directed television. *Developmental Science* 18 (1): 24–37.

Wexler, B. E. 2006. *Brain and Culture: Neurobiology, Ideology, and Social Change*. MIT Press.

Wilson, R. A., and A. Clark. 2008. How to situate cognition: Letting nature take its course. In *Cambridge Handbook of Situated Cognition*, ed. P. Robbins and M. Aydede, 55–77. Cambridge University Press.

Worth, S., and J. Adair. 1972. *Through Navajo Eyes: An Exploration in Film Communication and Anthropology*. Indiana University Press.

Zacks, J. M., N. K. Speer, and J. R. Reynolds. 2009. Segmentation in reading and film comprehension. *Journal of Experimental Psychology: General* 138:307–327.

19 Painful Bodies at Work: Stress and Culture?

Peter Henningsen and Heribert Sattel

The objective of this chapter is twofold: to present data on significant cultural influences on pain-related psychosocial workplace conditions, which belong to the core issues of psychosomatic medicine, and to discuss the conceptual consequences for a cultural neuroscience of pain. To achieve this aim, we start with an introduction of the basic characteristics of chronic pain at work, including the current view on risk factors for such pain, and then discuss the influence of pain-related psychosocial workplace conditions.

Chronic pain, as a common consequence of tissue damage, encompasses the experience of the pain sensation itself and a whole universe of related emotions, thoughts, pain behaviors, and suffering, which are at least partly visible to others in the environment. Remarkably, tissue damage is not a necessary precondition. The current biopsychosocial view on risk factors for chronic pain typically concentrates on intraindividual risk factors and includes genetic dispositions, injuries, and stressful life events, among others. However, an embodied approach emphasizing the "body being in the world," which integrates cultural perspectives, seems more appropriate.

In recent years, epidemiologic work has shown the relevance as well of group-level psychosocial risk factors for chronic pain, especially in terms of psychosocial workplace conditions. Lack of social support at work, injustice, high job strain, and effort-reward imbalance are especially important here.

Nevertheless, even this group-level perspective does not capture all relevant differences: studies in different societies or cultures have shown significant cultural influences on the relation between psychosocial workplace conditions and chronic pain, from both an "etic" and an "emic" perspective.

The current view on the link between culture and pain involves different aspects. Culturally shaped ways of world-making influence the interpretation and labeling of and the treatment strategies for distress; however, there is now, as a new component, sufficient knowledge on the relational biology of pain. This entails more than looking at the group-level social and affective neuroscience of pain processing; rather, on this level of investigation, there is also a cultural modulation of emotion and pain processing, which means that culture also determines the differences in neural processes underlying the same psychological phenomenon of emotion and pain experience.

We discuss the consequences of these epidemiologic and neurobiological findings in the present chapter.

1 The Experience of Pain: A Common Human Condition

Pain is an aversive and complex subjective perception, which—in its acute form—serves as a warning signal and usually elicits intentional states and behaviors, which in turn motivate or intend to bring about the termination of the assumed causes of pain. Withdrawing from pain-associated situations aims to protect against damage, helps to heal affected body parts, and intends to help avoid similar future experiences. The intensity of pain normally ranges from inconvenient to seemingly unbearable. The relevance of this in the context of working conditions and working persons is twofold: on the one hand, pain is often considered as a consequence of distinct distressing physical or psychosocial working conditions, and on the other hand, pain itself directly influences the individual's work ability and performance.

The International Association for the Study of Pain (IASP) defines pain as "an unpleasant sensory and emotional experience associated with actual or potential tissue damage, or described in terms of such damage" (IASP 2011). The last part of this definition expresses in a somehow cumbersome manner that tissue damage is not required for the experience of pain. Recently and consequently, a category called "functional pain" has emerged (Creed et al. 2010). However, the experience of pain is always a subjective/individual experience, even if its cause can be clearly objective and physical, such as a broken bone. As such, pain is influenced by a multitude of psychic factors. This underlines the outreaching associations and consequences of experiencing pain for the mental or psychological domain: pain involves cognitive sequelae as we anticipate and attribute these potential experiences. Moreover, it affects our social sphere through psychosocial processes, for example, when we try to share the experience with relevant others. Finally, cultural styles are likely to be shaped by specific attitudes and behaviors toward the expression of pain, which in turn influence the experience of pain sensations again (Bates, Edwards, and Anderson 1993).

The simplistic but common medical assumption of a straightforward path from physiologic events, through bodily sensations and subjective experiences, to individual symptom reports, must be left behind when embodiment and culture are introduced. In their particular cultural context, these experiences shape the concept of embodiment as "bodily being in the world" (Csordas 1994, 4). A wealth of studies has shown that the way physiologic processes are transferred into sensory experience and in their subsequent verbal report relies widely on the cultural context of meaning. Physiologic perturbations initiate a "nested series of cognitive schemas involving knowledge about sensations, symptoms, illnesses or other models of affliction, and broader socio-moral notions of self and personhood" (Kirmayer 2008, 319).

Because the experience and expression of pain differ enormously between cultures, one might conclude that these can be influenced easily and rapidly by psychological interventions or changing life conditions. However, the opposite is more accurate: "Pain confirms the deep presence of culture in the body" (Scarry 2007, 65). Whenever immigrants have to leave their homeland and start anew in an outland, confronting themselves with a progressive change and loss of their cultural habits, "the last to be lost is the cultural disposition to express or not to express physical pain in a certain way" (65).

2 Chronic Pain

According to the actual diagnostic criteria of the *International Classification of Disorder* (10th ed.; ICD-10; World Health Organization 1992) and the *Diagnostic and Statistical Manual of Mental Disorders* (5th ed.; DSM-5; American Psychiatric Association 2013), pain is considered as chronic when its symptoms occur for six months or longer. Expert associations define pain as chronic after a three-month duration (IASP 2011). Alternatively, chronic pain has been

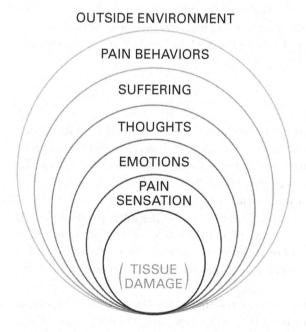

Figure 19.1
Chronic pain and its consequences on expanding experiential levels.

described as "pain which persists past the normal time of healing" (Turk and Okifuji 2000, 20). During this period of predominantly continuous presence of pain, its function as a warning signal fades into the background, and the medical focus shifts from the search for possible causes toward the treatment of the chronic pain itself. The treatment of chronic pain, often labeled as chronic pain syndrome, is considered as complex, costly, and often of limited effect (Barsky, Orav, and Bates 2005). A core feature of chronic pain is that it is particularly associated with physical and psychological stress. The constant discomfort potentially leads to anger and frustration, both for the affected individual and for those in his or her social environment (American Psychological Association 2015). The symptom or symptoms have to be "distressing and/or result in significant disruption of daily life" (American Psychiatric Association 2013) to be considered symptoms of chronic pain.

Altogether, the enduring experience of chronic pain sensations in particular encompasses a whole universe of related emotions, thoughts, pain behaviors, and suffering, which are at least partly visible and transferred to the sufferer's social environment; tissue damage is no longer a necessary precondition. Typically, individually distinctive vicious circles result, in which the experience of chronic pain is followed by anxiety, which are capable of impairing self-efficacy (Burke, Mathias, and Denson 2015) and evoking additional physical conditions unfavorable for recovery, such as sleep problems. The potential sources that charge this vicious circle stem from different domains: health (general health worries, frustrating contacts with health professionals, and medication worries), employment (general worries, work cover worries, and financial worries), and private life (lack of enjoyment, and family and relationship worries).

3 Prevalence of Chronic Pain

The particular societal impact of chronic pain emanates from a combination of its circumscribed consequences and its frequent occurrence. In 2003, the prevalence of chronic pain in the United States, Europe, and Australia was reported to range from 10.1 percent to 55.2 percent of the population (Harstall and Ospina 2003). The prevalence varied even more broadly when based on studies from Asia, in which chronic pain was reported in 7.1 percent (Malaysia) to 61 percent (Cambodia and Northern Iraq) of the adult population (Zaki and Hairi 2015); the majority of these studies followed the IASP criteria (daily or nearly daily pain for at least three months and impact on everyday life). The enormous spread of these figures across various countries points toward the relevance of genetic and cultural influences.

The prevalences of pain symptoms in various working populations are comparable. According to a large survey in the German working population, frequent occurrence of neck pain, low back pain, or headache during the last twelve months was reported by 48, 46, or 34 percent of about 20,000 respondents, respectively (Wittig, Nöllenheidet, and Brenscheid 2013). In a large study in France, 20 percent of employees reported multisite musculoskeletal symptoms lasting for at least thirty days within the last twelve-month period (Parot-Schinkel et al. 2012).

4 Pain at Work and the Global Burden of Disease

The "Global Burden of Disease" study and its various revisions intended to "quantify non-fatal health outcomes across an exhaustive set of disorders at the global and regional level" (Vos et al. 2012, 2164). One of the main outcomes of this study regarded years of life lived in less than ideal health (years lived with disability; YLDs). The leading specific causes of YLDs were, among others, low back pain, neck-pain tension-type headache, and migraine. Three of the ten most prevalent sequelae of diseases were pain related. In 2010, musculoskeletal diseases predominantly associated with pain accounted for 21.3 percent of all YLDs, with only mental and behavioral disorders causing a slightly higher global burden of disease (Becker and Kleinman 2013). For both conditions, the often chronic course of the symptoms can be considered as a major cause of the burden. Consequently, chronic pain consumes a large amount of health care resources around the globe (Harstall and Ospina 2003).

5 Individual Risk Factors for Chronic Pain

In the explanation of an occurrence of chronic pain, three main aspects can be distinguished: the vulnerability to experiencing pain, the trigger of the first experience associated with the chronic pain in question, and the maintaining factors for the continuing experience (Henningsen 2014). On the individual level, the potential factors for vulnerability include genetic influences (e.g., gender, genotype, and epigenetic implementation) and environmental factors, such as early stressful life events (e.g., trauma, abuse, and loss of significant others). These factors usually interact and create an individual pattern of psychological health dispositions and beliefs. The triggers for the first experience of the episode in question may be actual injuries and diseases (often of a smaller nature than expected to explain the severity of the resulting symptoms) or stressful life events, such as work-related or private conflicts, injustices, or loss of significant others. The individual maintaining factors potentially stem from physiologic qualities (e.g., acquired pain-processing structures and memory for pain), paralleled by psychological (e.g., catastrophization, avoidance behavior, and depressiveness) or psychosocial (e.g., secondary morbid gain and wish to retire) processes. In the context of work-related pain, physical factors (e.g., awkward postures, strenuous lifting, and carrying weights) may play a role (IJzelenberg and Burdorf 2005).

Recent advances in the study of genetic and epigenetic processes could broaden this viewpoint, adding a neurobiological perspective. This perspective extends the listed simple, observable factors to include potential mechanisms, allowing the explanation of these risk factors at a molecular, cellular, or nervous system network level (Denk, McMahon, and Tracey 2014). In combination with individual epigenetic mechanisms corresponding with chronic

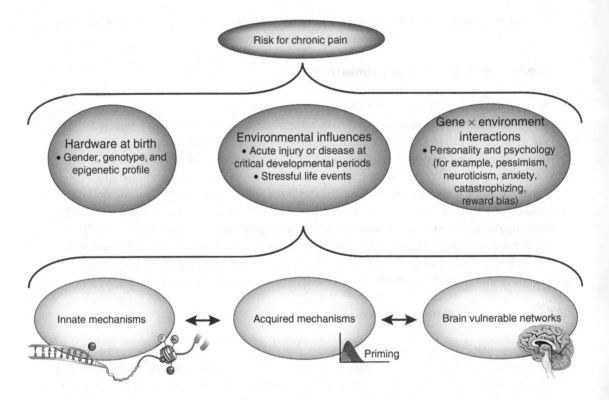

Figure 19.2

Risk factors and related mechanisms for chronic pain. Reprinted by permission from Macmillan Publishers Ltd: Nature Neuroscience; Denk, McMahon, and Tracey 2014, © 2014.

pain, a full-blown biopsychosocial model will potentially result from these findings, offering manifold leverage points for therapy on all observed levels (Descalzi et al. 2015).

6   Group-Level Risk Factors for Pain at Work

Other risk factors for pain at work can be better analyzed and interpreted at a group level, beyond individual genetics and the environment, because these are nested within organizations and/or shared by cohorts of working persons.

This applies to physical workload, for example, for industrial workers, in which manual material handling, awkward back postures, or strenuous limb positions, such as working with arms above the shoulder level, may come into play. This holds, too, for psychosocial aspects, which seem to be equally important risk factors for chronic pain at work. IJzelenberg and Burdorf (2005) showed high rates of work-related physical as well as work-related psychosocial loads. On the one hand, a high perceived physical load at baseline was correlated with a significantly higher risk for the presence of pain and health care use six months later (odds ratios: 1.67 and 1.85, respectively). On the other hand, with regard to psychosocial loads, high perceived job strain and low levels of social support from the supervisor had even stronger associations with all the assessed negative outcomes at follow-up. Odds ratios of approximately 2 or even higher indicated that the presence of these work-related psychosocial phenomena at baseline made the occurrence of pain, sick leave, and health care use twice as likely.

A small number of theoretical models from the realm of occupational psychology has attempted to explain those specific, group-level work-related psychosocial risk factors, with good empirical proof. The effort-reward imbalance model ("gratification crisis" model; Siegrist 1996) started from the assumption that the occupational status of an employee is constituted by recurrent incidents of contributing and performing, of being rewarded or esteemed, and of belonging to some significant group. Work characterized by high efforts and low resulting rewards could elicit unpreferred physiologic and psychological reactions in an employee (e.g., physiologic strain reactions, disappointment, and absenteeism). Individual differences, for example, in the level of commitment to the job, potentially modulate the effects of the imbalance. Simon et al. (2008) showed that effort-reward imbalance was the most dominant risk factor for working disability in nursing professionals (with extremely high odds ratios exceeding 5, after controlling for physical and other psychosocial risk factors). In contrast, physical exposure to lifting and bending showed only small associations (odds ratios below 1.6).

In the widely known (job) demand-control model (Karasek and Theorell 1990), job strain is assumed to be a result of the combination of low scope of job decision making (low autonomy) and high job demands (high workload, marked monotony, and other work stressors). The authors added that lack of social support at work apparently further increased health risks. Moreover, high-level role conflicts were related to job strain and presence of wrist/hand, shoulder, and back pains (Eatough, Way, and Chang 2012). Role conflicts potentially emerge for employees who receive incompatible role expectations from other members of the organization. As an antidote, the authors identified "safety-specific leadership," in which the leaders emphasize safe performance and injury prevention and reward safety-related compliance.

Evidence suggests that job strain explained by either effort-reward imbalance or the demand-control model is modulated by work-life imbalance or work-family conflicts (Sembajwe et al. 2012). This association was explicitly confirmed for pain (Sembajwe et al. 2013). Independently, Hämmig et al. (2011) examined the workforces of four large Swiss companies (insurance, transportation, banking, and health care). The presence of a work-life conflict was strongly associated with pain, as was physical strain at work, workload, and job autonomy. The association between a work-life conflict and pain dwindled somewhat after controlling for all the other risk factors assessed in the study but remained significant.

A more specific issue is the influence of the perception and attribution of injustice after the onset of a pain condition, for example, after an injury (Sullivan, Scott, and Trost 2012). In such case, in which an employee expresses a significant loss due to his or her symptoms, blames others, or persists in a sense of unfairness and/or irreparability of loss, the prognosis for recovery is poor.

Evidence for a prospective association between psychosocial stressors and the emergence of pain was also recently well established in a high-quality meta-analysis (Lang et al. 2012). This systematic review of baseline-adjusted prospective longitudinal studies aimed for an integrated estimation of the lagged effect of psychosocial risk factors on musculoskeletal pain problems, again in industrialized work settings. The majority of the integrated studies reported psychosocial stressors in a symptom-free population at baseline and musculoskeletal problems at least one year later. The authors analyzed the distinctive domains of high job strain (high demands or low control), low social support (as granted by supervisors or colleagues), and, finally, low job satisfaction, low job security, and monotonous work. For musculoskeletal pain-related problems, three categories were chosen: lower back symptoms; neck, shoulder, and/or upper back symptoms; and upper extremity symptoms. Analyses were made only when at least five studies reported data on the resulting psychosocial load–musculoskeletal problem combination. The authors concluded that a majority of the psychosocial stressors had a small but significant effect on the development of future musculoskeletal problems. Consequently, organizational interventions have to be developed to reduce psychosocial stressors, which are likely to precede negative health outcomes in employees.

7   Cultural Influences on the Relation between Psychosocial Workplace Conditions and Chronic Pain

After the introduction of an organizational nesting level, it is not implausible to seek for other, and even more overarching, factors related to working conditions. Culture is considered to be such a factor and has been studied by applying epidemiologic methods. Cultural membership may be defined in this context simply as having citizenship in a geographic country, being a member of a particular ethnicity, or sharing a language with a certain population. In this sense, Coggon et al. analyzed data from the "Cultural and Psychosocial Influences on Disability" (CUPID) study (Coggon, Ntani, Palmer, et al. 2013a; Coggon, Ntani, Vargas-Prada, et al. 2013b). Within this meticulously executed and controlled study, 12,426 participants from 47 occupational groups (mostly office workers or nurses) in 18 different countries reported whether or not they suffered from disabling musculoskeletal pain; the prevalence of those symptoms varied between 2.6 percent and 42.6 percent. This extensive study again confirmed well-known risk factors (such as occupational physical activities or psychosocial work-related factors) and established an association with adverse health beliefs. Even after controlling for all established risk factors, a difference of up to eightfold in prevalence remained for groups of employees carrying out quite similar occupational activities. These differences were not related to societal factors as systems of compensation for work-related illness or financial support for health-related incapacity for work.

The authors discussed whether a different understanding of the respective terms for the complaints in different languages and cultures might have been a reason for the observed variation. To avoid such bias, they considered aspects of translation thoroughly, and the study focused on pain with a significant impact on everyday activities. Nevertheless, even among occupational groups from the same country and language, respectively, the prevalence of pain still differed extremely between the examined groups of workers. The authors did not suggest an explanation for this particular and the aforementioned cultural variability.

Taken all together, this study depicts the differential contribution of the classical epidemiologic perspective (etic perspective) on neck, back, and wrist pain in globalized working environments. The authors do not give an explanation for the cultural variation of the presence of pain, which may reflect a general limitation of classical epidemiologic techniques.

8   Etic and Emic Perspectives: Where Culture Really Comes into Play

These particular problems are reflected in the discussion of the "etic" and "emic" perspectives. These two terms are derived from the linguistic terms "phonetic" and "phonemic." "Phonetic" describes objectively differentiable sounds of the human language and how they are produced, transmitted, and perceived. In contrast, "phonemic," more recently renamed as "phonologic," refers to how such sounds encode a specific intended meaning within or across languages. These different linguistic approaches have been transferred to and generalized for social and behavioral sciences. Here, "etic" indicates that issues are described with the claim of objectivity and studied by a researcher who places him- or herself outside or over the particular social or cultural group, claiming to take an independent stance. Alternatively, on the "emic" side, facts and circumstances can be studied by a researcher focusing on the perspective of a subject within the social or cultural group of interest, representing a more phenomenological stance.

9   Advantages and Handicaps of Both Approaches

An etic approach requires a descriptive system that is equally valid for all cultures of interest. This implies the potential representation of both similarities and differences between individual cultures (Helfrich 1999). Moreover, the definitions of the objects of investigation and the respective methods of comparisons applied must be equivalent. In our case, these would be health, pain, and physical and psychosocial work-related loads. The methods of comparisons would be based on counting and estimating those factors to represent their potentially thinkable individual manifestations. It seems unlikely that these claims can be fulfilled a priori. Moreover, an etic approach cannot explain the phenomenon of culture but instead studies culture as a factor that potentially explains different symptoms, attitudes, cognitions, and behaviors. Finally, such approach cannot deal well with the heterogeneity and dynamic character of cultural agglomerations (Patel 2001).

An emic approach considers "culture" not as an external, objective, and—within a particular culture—uniform factor whose effects on the individual can be examined. On the contrary, it places culture at the core of being an acting human in society (Tylor [1871] 1974). In this sense, human behavior is shaped by reasons stemming from the acting person. Accordingly, this should be understood through his or her perspective rather than studied by using objective techniques from a third-person perspective (Helfrich 1999). Culture as a label for presumably homogeneous populations is often meaningless, not least because of significant intracultural variation. Culture can be characterized by a fractal nature (Roepstorff 2013): cultural differences can be found at all levels of an investigation, when comparing Asians and Europeans, Dutch people and Germans, people from regions within a country, or different subcultures; as one takes a closer look, cultural differences always seem to replicate. As a conclusion, the emic approach "attempts to reconstruct the experiential world of the individual through his/her own reports and explanations" (Helfrich 1999, 136).

10   Katakori

As an example, the emic perspective is indispensable in understanding *katakori*, or congealed shoulders. This syndrome is well known in Japan, where it is described as the "most common, most banal and most everyday" complaint, but it is practically unknown everywhere else (Kuriyama 1997, 127). *Kata* is the Japanese term for "shoulder," whereas *kori* can be translated somewhat incompletely as "stiff" or "congealed"; it additionally conveys a "lack of flow" of vital energy. *Katakori* can be described as "subjectively, an extremely uncomfortable feeling, accompanied by dull, heavy pain in the muscles of the neck and shoulders," with fatigue, poor posture, and/or psychological stress as its chief causes. "One is tempted to say: katakori is mental tension manifest as muscular tension" (Kuriyama 1997, 128).

According to Shigehisa Kuriyama, interpersonal phenomena are better suited to explain the specific characteristics of *katakori*: *kori* is comparable to the stiffness and punctiliousness between old friends, who, despite reconciling after an intensive dispute and falling out, somehow no longer find that flow of deep friendship again. He identified *katakori*-causing situations, which are related to social stress in an impermeable society, and protracted, highly adverse activities. Viewed from the perspective of the Japanese society, *katakori* cannot simply be compared to "Western" forms of shoulder aches, although it has been introduced in medical international research recently as "neck and shoulder pain" (Nabeta and Kawakita 2002).

Katakori is a prevalent problem among employees in Japan. Iizuka et al. (2012) found *katakori* in 68.1 percent of 484 nursing staff. The occurrence of *katakori* was associated with, among other things, psychological stress. A study of 2,022 Japanese workers showed that lack of support from colleagues or supervisors at work was related to the presence of *katakori* (Fujii et al. 2013).

10   Repetitive Strain Injury

Another culturally bound work-related pain syndrome is repetitive strain injury (RSI), which was first reported in an Australian publication (Stone 1982). The author related increasingly extended and accelerated industrial working processes, for example, the growing use of keyboards, to a huge increase in RSI, which a few years later was referred to as "the epidemic of the 1990s" (Thompson and Phelps 1990). From the very beginning, the term RSI was used to denote a group of conditions that were already well known in clinical practice (e.g., epicondylitis or carpal tunnel syndrome) (Woodward 1987); however, the name became increasingly popular because of its suggestively causative label. RSI mainly consists of upper limb, hand, or wrist pain symptoms. Accordingly, it is used as an umbrella term; specific diagnoses (such as the above-mentioned examples) can be established for a proportion of the affected patients, but not in all. Therefore, the complaints of the latter also have to be labeled as nonspecific (van Tulder, Malmivaara, and Koes 2007). The cultural relevance of RSI is related to the genesis of the term, which at first was linked strictly to Australia and was unknown in other countries.

In this context, it should be stated that the aforementioned CUPID study (Coggon, Ntani, Palmer, et al. 2013a) intensively discussed how a symptom can be present in the "public spirit." The study assessed back or arm pain and whether the participants knew someone outside work with such pain. Having heard of RSI or equivalent terms was found to be related to a slight but statistically significant risk of perceiving one's own hand/wrist problems. The authors added that the responses of individuals with pain might be biased because the perception of their own symptoms could have potentially primed the perception of similar pain in other people.

11   Interim Conclusion

Studies of painful bodies at work can start from an individualistic perspective on risk factors. This establishes the basis for an investigation, as well as for individual patient-centered treatments. However, this perspective is insufficient, because relational factors are potentially even more influential in explaining pain at work. These factors are the psychosocial characteristics of the workplace at a group level; the remaining "unexplained rest" partially reflects cultural factors. All these can be integrated by applying an etic and/or an emic perspective, which complement each other (Helfrich 1999).

The current view on the link between culture and pain involves different aspects. Cultural epidemiology provides a conceptual framework of how expectations and the interpretation of distress are transferred to and distributed within a population. Culturally shaped ways of world-making influence the interpretation, the labeling of, and the treatment strategies for distress. Both patients and health professionals must be viewed in the context of their—caveat, joint or separate—cultures, considering all the involved culturally specific microcomponents: precisely how distress is labeled; how treatment strategies are developed and put into practice; which issues of stigma, legitimization, and social positioning must be considered; and so on.

12 Relational Biology of Pain

The term "relational biology of pain" reflects, on a first level, that the relational nature of the experience of pain is subserved by biological mechanisms because they are examined by looking at the group-level social and affective neuroscience of pain processing. Depending on the experimental paradigm, research question, and method, some influences of the relational context on pain experience are seen as extrinsic to the neurobiological processing of pain itself (e.g., the neurobiological correlates of cognitions and intentions influencing pain experience); others are intrinsic (e.g., the mirror neuron mechanisms of activation of the individual's own pain matrix through the observation of pain in others; see Henningsen and Kirmayer 2000; Denk, McMahon, and Tracey 2014).

On a second level, cultural factors come into play, and their neurobiological correlates are also both extrinsic and intrinsic to the neurobiological processing of pain itself (the same holds true also for the neurobiological processing of emotions). Recently, for instance, Immordino-Yang (2013) showed substantial differences in the processing of social emotions between participants of Chinese and American descent. The participants completed an extensive interview about true and compelling social stories intending to induce compassion and admiration. After each section, the participants were asked, "How does this person's story make you feel?" Then the participants were subjected to an fMRI scan while the essentials of each story were presented again, and the self-reported strength of their current emotional reaction and psychophysiologic data were recorded. Although all participants reported equally strong and frequent emotions during the experiment, both groups differed regarding emotion-related embodied sensations. These occurred frequently in and were mentioned spontaneously by the American participants but not at all in the Chinese participants. Comparing the presentations of admiration-inducing social stories with strong emotional response with those of less-extraordinary stories without a particular emotional response, the patterns of activation were found to be substantially similar for both groups. However, the American group only showed corresponding activation in the lateral frontal motor planning cortices and the lateral parietal somatosensory cortices.

After showing these culture-specific neuronal and behavioral differences, Immordino-Yang raised the question, "What do cultural differences in brain functioning mean for the ways individuals experience the world?" (2013, 42). At least partially, she answered this question one year later, in a report on brain activity related to psychical pain and disgust, visceral states, and culture (Immordino-Yang, Yang, and Damasio 2014). By using a comparable experimental approach applying compassion- and admiration-inducing stories, the researchers again studied a Chinese and an American group, complemented by a bicultural East Asian/American group. Starting from existing psychological evidence that socioemotional feelings within a specific cultural context are cognitively constructed, the researchers explored the anterior insula, in which visceral states, as well as the construction and processing of social feelings, are represented. The authors found that these processes were more independent than expected and that, in both cultural groups, the strength of feeling was related to different structures of the anterior insula. They concluded that neuroplasticity allows the brain to form its own network in a culture-specific manner as a consequence of the presence of corresponding culture-specific learning processes. Interestingly, intermediate findings resulted from the bicultural East Asian/American group.

The influential role of culture is supported by an examination of emotional pain, too (Cheon et al. 2013). The authors contrasted Western and East Asian cultures, with the former characterized by more pronounced individualism, and the latter by collectivistic social models, with greater emphasis on the connectedness of the self to others. The extent of relational interdependency in each participant was assessed independently by using a self-construal scale that comprises the "other-focusedness" index (Hardin, Leong, and Bhagwat 2004). The participants were studied by applying an fMRI paradigm that consisted of the passive viewing of others in painful and nonpainful situations. Higher extents of other-focusedness were related to heightened neural responses within the affective pain matrix, in particular in Korean compared with Caucasian-American participants. Although the specific cultural properties of relational interdependence, as well as the reference groups to which the claimed collectivistic model relates, might be discussed controversially, the observed specific differences between the members of the two cultural groups remain significant.

It might be too simple to base a cultural neurobiology solely on neuroimaging and comparable designs. Again, Immordino-Yang and colleagues (2014) studied the correlations between socioemotional feelings and neuronal activity and found a significant influence of culture. Surprisingly, in her experiment, similar levels and locations of neuronal activity across participants were related to culturally modulated differential psychological processes. Accordingly, she claimed that cultural studies of relational biology should always include psychological measures to detect these potential differences.

These considerations are relevant to our understanding of chronic pain at work in different cultures. These phenomena should not be described as mere direct correlates but should be considered as composite and complex dispositions toward pain in different cultures.

Context, interaction, and experience determine within such a model not only the mechanisms of actual pain experience but also the individual risks, for example, through epigenetics and dispositions. The extension from a social neuroscience toward cultural biology based on a "cultural brain" (Kirmayer 2006), understood in this sense, might constitute a profound dynamic multilevel explanatory framework. Starting from a very similar perspective addressing cultural neuroscience, Han et al. (2013) underlined the indispensable interdisciplinarity of such research and deducted its origin from cultural psychology, social cognitive neurosciences, and research on neuroplasticity and gene-environment interactions. The combination of these approaches allows the study of cultural group differences in neural activities, the relations between cultural values and brain activity, and the modulation of neural activities by culture-specific psychological priming. These approaches offer the chance to avoid a global reductionist replacement of psychosocial and cultural explanations because they allow valid causal associations on psychological (first-person meanings), social, and cultural levels as well. Further, they extend rather than replace the multilevel empirically based pluralism that already exists in high-level research on the etiology of medical illnesses (Kendler and Campbell 2009). A particular concern of future research should be to overcome simplifications of culture as an inflexible set of traits and characteristics (Mateo et al. 2012).

Conclusion

Chronic pain at work is determined by more than one constituent, and we can identify individual, organizational, and cultural risk factors for such pain. However, simple mechanistic or reductionist explanations for the interplay of these dimensions will not hold, for example, when considering that increasingly globalized working conditions do not necessarily "produce" globalized, uniform distress reactions. Recently, there has been a "discovery of (neglected) culture" toward explaining differences in pain (at work) and in its biological correlates, laying the groundwork for a cultural biology. By deepening our conceptual understanding of the interplay of individual and organizational agents based on biological processes, we may be able to develop a nonreductionist extension toward culture, improving our ability to explain the phenomenon of pain at work on multiple levels (Henningsen 2015).

References

American Psychiatric Association. 2013. *Diagnostic and Statistical Manual of Mental Disorders*. 5th ed. American Psychiatric Association.

American Psychological Association. 2015. *Coping with Chronic Pain*. http://www.apa.org/helpcenter/chronic-pain.aspx (accessed July 25, 2016).

Barsky, A. J., E. J. Orav, and D. W. Bates. 2005. Somatization increases medical utilization and costs independent of psychiatric and medical comorbidity. *Archives of General Psychiatry* 62:903–910.

Bates, M. S., W. T. Edwards, and K. O. Anderson. 1993. Ethnocultural influences on variation in chronic pain perception. *Pain* 52 (1): 101–112.

Becker, A. E., and A. Kleinman. 2013. Mental health and the global agenda. *New England Journal of Medicine* 369 (1): 66–73.

Burke, A. L., J. L. Mathias, and L. A. Denson. 2015. Psychological functioning of people living with chronic pain: A meta-analytic review. *British Journal of Clinical Psychology* 54 (3): 345–360. doi:10.1111/bjc.12078.

Cheon, B. K., D. M. Im, T. Harada, J. S. Kim, V. A. Mathur, J. M. Scimeca, T. B. Parrish, H. Park, and J. Y. Chiao. 2013. Cultural modulation of the neural correlates of emotional pain perception: The role of other-focusedness. *Neuropsychologia* 51 (7): 1177–1186.

Coggon, D., G. Ntani, K. T. Palmer, V. E. Felli, R. Harari, L. H. Barrero, S. A. Felknor, et al. 2013. Disabling musculoskeletal pain in working populations: Is it the job, the person, or the culture? *Pain* 154 (6): 856–863.

Coggon, D., G. Ntani, S. Vargas-Prada, J. M. Martinez, C. Serra, F. G. Benavides, K. T. Palmer, et al. 2013. International variation in absence from work attributed to musculoskeletal illness: Findings from the CUPID study. *Occupational and Environmental Medicine* 70 (8): 575–584.

Creed, F., E. Guthrie, P. Fink, P. Henningsen, W. Rief, M. Sharpe, and P. White. 2010. Is there a better term than "medically unexplained symptoms"? *Journal of Psychosomatic Research* 68 (1): 5–8. doi:10.1016/j.jpsychores.2009.09.004.

Csordas, T. J. 1994. *Embodiment and Experience: The Existential Ground of Culture and Self*, vol. 2. Cambridge University Press.

Denk, F., S. B. McMahon, and I. Tracey. 2014. Pain vulnerability: A neurobiological perspective. *Nature Neuroscience* 17 (2): 192–200.

Descalzi, G., D. Ikegami, T. Ushijima, E. J. Nestler, V. Zachariou, and M. Narita. 2015. Epigenetic mechanisms of chronic pain. *Trends in Neurosciences* 38 (4): 237–246.

Eatough, E. M., J. D. Way, and C. H. Chang. 2012. Understanding the link between psychosocial work stressors and work-related musculoskeletal complaints. *Applied Ergonomics* 43 (3): 554–563.

Fujii, T., K. Matsudaira, N. Yoshimura, M. Hirai, and S. Tanaka. 2013. Associations between neck and shoulder discomfort (katakori) and job demand, job control, and worksite support. *Modern Rheumatology* 23 (6): 1198–1204.

Hämmig, O., M. Knecht, T. Läubli, and G. F. Bauer. 2011. Work-life conflict and musculoskeletal disorders: A cross-sectional study of an unexplored association. *BMC Musculoskeletal Disorders* 12 (1): 60.

Han, S., G. Northoff, K. Vogeley, B. E. Wexler, S. Kitayama, and M. E. Varnum. 2013. A cultural neuroscience approach to the biosocial nature of the human brain. *Annual Review of Psychology* 64:335–359.

Hardin, E. E., F. T. Leong, and A. A. Bhagwat. 2004. Factor structure of the self-construal scale revisited implications for the multidimensionality of self-construal. *Journal of Cross-Cultural Psychology* 35 (3): 327–345.

Harstall, C., and M. Ospina. 2003. How prevalent is chronic pain? *Pain Clinical Updates* 11 (2): 1–4.

Helfrich, H. 1999. Beyond the dilemma of cross-cultural psychology: Resolving the tension between etic and emic approaches. *Culture and Psychology* 5 (2): 131–153.

Henningsen, P. 2014. Schmerzen und funktionelle Körperbeschwerden [Pain and functional bodily complaints]. In *Psychische und psychosomatische Gesundheit in der Arbeit: Wissenschaft, Erfahrungen und Lösungen aus Arbeitsmedizin, Arbeitspsychologie und Psychosomatischer Medizin*, ed. P. Angerer, J. Glaser, H. Gündel, P. Henningsen, C. Lahmann, S. Letzel, and D. Nowak, 156–168. Hüthig Jehle Rehm.

Henningsen, P. 2015. Still modern? Developing the biopsychosocial model for the 21st century. *Journal of Psychosomatic Research* 79 (5): 362–363.

Henningsen, P., and L. J. Kirmayer. 2000. Mind beyond the net: Implications of cognitive neuroscience for cultural psychiatry. *Transcultural Psychiatry* 37 (4): 467–494.

Iizuka, Y., T. Shinozaki, T. Kobayashi, S. Tsutsumi, T. Osawa, T. Ara, H. Iizuka, and K. Takagishi. 2012. Characteristics of neck and shoulder pain (called katakori in Japanese) among members of the nursing staff. *Journal of Orthopaedic Science* 17 (1): 46–50.

IJzelenberg, W., and A. Burdorf. 2005. Risk factors for musculoskeletal symptoms and ensuing health care use and sick leave. *Spine* 30 (13): 1550–1556.

Immordino-Yang, M. H. 2013. Studying the effects of culture by integrating neuroscientific with ethnographic approaches. *Psychological Inquiry* 24 (1): 42–46.

Immordino-Yang, M. H., X. F. Yang, and H. Damasio. 2014. Correlations between social-emotional feelings and anterior insula activity are independent from visceral states but influenced by culture. *Frontiers in Human Neuroscience* 8.

International Association for the Study of Pain (IASP). 2011. *Classification of Chronic Pain*. 2nd ed. http://www.iasp-pain.org/PublicationsNews/Content.aspx?ItemNumber=1673 (accessed July 25, 2016).

Karasek, R. A., and T. Theorell. 1990. *Healthy Work: Stress, Productivity, and the Reconstruction of Working Life*. Basic Books.

Kendler, K. S., and J. Campbell. 2009. Interventionist causal models in psychiatry: Repositioning the mind-body problem. *Psychological Medicine* 39 (6): 881–887.

Kirmayer, L. J. 2006. Beyond the "new cross-cultural psychiatry": Cultural biology, discursive psychology, and the ironies of globalization. *Transcultural Psychiatry* 43 (1): 126–144.

Kirmayer, L. J. 2008. Culture and the metaphoric mediation of pain. *Transcultural Psychiatry* 45 (2): 318–338.

Kuriyama, S. 1997. The historical origins of "katakori." *Nichibunken Japan Review* 9:127–149.

Lang, J., E. Ochsmann, T. Kraus, and J. W. Lang. 2012. Psychosocial work stressors as antecedents of musculoskeletal problems: A systematic review and meta-analysis of stability-adjusted longitudinal studies. *Social Science and Medicine* 75 (7): 1163–1174.

Mateo, M. M., M. Cabanis, N. C. de Echeverría Loebell, and S. Krach. 2012. Concerns about cultural neurosciences: A critical analysis. *Neuroscience and Biobehavioral Reviews* 36 (1): 152–161.

Nabeta, T., and K. Kawakita. 2002. Relief of chronic neck and shoulder pain by manual acupuncture to tender points: A sham-controlled randomized trial. *Complementary Therapies in Medicine* 10 (4): 217–222.

Parot-Schinkel, E., A. Descatha, C. Ha, A. Petit, A. Leclerc, and Y. Roquelaure. 2012. Prevalence of multisite musculoskeletal symptoms: A French cross-sectional working population-based study. *BMC Musculoskeletal Disorders* 13 (1): 122.

Patel, V. 2001. Cultural factors and international epidemiology depression and public health. *British Medical Bulletin* 57 (1): 33–45.

Roepstorff, A. 2013. Why am I not just lovin' cultural neuroscience? Toward a slow science of cultural difference. *Psychological Inquiry* 24 (1): 61–63.

Scarry, E. 2007. Pain and the embodiment of culture. In *Pain and Its Transformations: The Interface of Biology and Culture*, vol. 4, ed. S. Coakley and K. K. Shelemay, 64–66. Harvard University Press.

Sembajwe, G., T. H. Tveito, K. Hopcia, C. Kenwood, E. T. O'Day, A. M. Stoddard, J. T. Dennerlein, D. Hashimoto, and G. Sorensen. 2013. Psychosocial stress and multi-site musculoskeletal pain: A cross-sectional survey of patient care workers. *Workplace Health and Safety* 61 (3): 117–125.

Sembajwe, G., M. Wahrendorf, J. Siegrist, R. Sitta, M. Zins, M. Goldberg, and L. Berkman. 2012. Effects of job strain on fatigue: Cross-sectional and prospective views of the job content questionnaire and effort-reward imbalance in the GAZEL cohort. *Occupational and Environmental Medicine* 69 (6): 377–384.

Siegrist, J. 1996. Adverse health effects of high-effort/low-reward conditions. *Journal of Occupational Health Psychology* 1 (1): 27.

Simon, M., P. Tackenberg, A. Nienhaus, M. Estryn-Behar, P. M. Conway, and H. M. Hasselhorn. 2008. Back or neck-pain-related disability of nursing staff in hospitals, nursing homes and home care in seven countries: Results from the European NEXT-Study. *International Journal of Nursing Studies* 45 (1): 24–34.

Stone, W. E. 1982. Repetitive strain injuries. *Medical Journal of Australia* 2 (12): 616–618.

Sullivan, M. J., W. Scott, and Z. Trost. 2012. Perceived injustice: A risk factor for problematic pain outcomes. *Clinical Journal of Pain* 28 (6): 484–488.

Thompson, J. S., and T. H. Phelps. 1990. Repetitive strain injuries: How to deal with "the epidemic of the 1990s." *Postgraduate Medicine* 88 (8): 143–149.

Turk, D. C., and A. Okifuji. 2000. Pain terms and taxonomies of pain. In *Bonica's Management of Pain*, 3rd ed., ed. J. D. Loeser, S. H. Butler, and C. R. Chapman, 13–23. Lippincott, Williams & Wilkins.

Tylor, E. B. [1871] 1974. *Primitive Culture: Researches into the Development of Mythology, Philosophy, Religion, Art, and Custom.* Gordon Press.

van Tulder, M., A. Malmivaara, and B. Koes. 2007. Repetitive strain injury. *Lancet* 369 (9575): 1815–1822.

Vos, T., A. D. Flaxman, M. Naghavi, R. Lozano, C. Michaud, M. Ezzati, and J. Abraham. 2012. Years lived with disability (YLDs) for 1,160 sequelae of 289 diseases and injuries, 1990–2010: A systematic analysis for the Global Burden of Disease Study 2010. *Lancet* 380 (9859): 2163–2196.

Wittig, P., C. Nöllenheidet, and S. Brenscheid. 2013. *Grundauswertung der BIBB/BAuA-Erwerbstätigenbefragung 2012* [Basic Analyses of the BIBB/BAuA Working Population Survey 2012]. BIBB/BAuA. http://www.baua.de/de/Publikationen/Fachbeitraege/Gd73.html (accessed July 25, 2016).

Woodward, C. 1987. Repetitive strain injury: A diagnostic model and management guidelines. *Australian Journal of Physiotherapy* 33 (2): 96–99.

World Health Organization. 1992. *International Statistical Classification of Diseases, 10th Revision (ICD-10).* World Health Organization.

Zaki, L. R. M., and N. N. Hairi. 2015. A systematic review of the prevalence and measurement of chronic pain in Asian adults. *Pain Management Nursing.* 16 (3): 440–452.

20   Embodiment and Enactment in Cultural Psychiatry

Laurence J. Kirmayer and Maxwell J. D. Ramstead

Cultural psychiatry aims to understand the implications of cultural diversity for psychopathology, illness experience, and intervention. The emerging paradigms of embodiment and enactment in cognitive science provide ways to approach this diversity in terms of bodily and intersubjective experience and narrative practices. In turn, cultural psychiatry provides striking examples of how cultural variations in ways of life and social contexts shape embodied experience. While evolutionary history reaches all the way up from brain circuitry to cultural forms of life, culture reaches all the way down to neuroplastic circuitry and epigenetic regulation; hence, human biology is fundamentally cultural biology and human environments are social environments, constituted by relationships with others and with cooperatively constructed institutions and practices. The implication for psychiatry is that current efforts to explain the mechanisms of mental disorders in terms of brain circuitry must be complemented with models of the social interactions that shape both the content and process of psychopathology.

Recent work has begun to apply embodied and enactivist approaches to understanding mental disorders (Colombetti 2013; Fuchs 2009; Fuchs and Schlimme 2009; Zatti and Zarbo 2015). We believe that cultural psychiatry stands to gain a great deal from these new paradigms. This chapter will outline an approach to the cultural neurophenomenology of mental disorders that focuses on the interplay of culturally shaped developmental processes and modes of neural information processing that are reflected in embodied experience, narrative practices that are structured by ideologies of personhood, culturally shared ontologies or expectations, and situated modes of enactment that reflect social positioning and self-fashioning. Research on metaphor theory suggests ways to connect the approaches to embodiment and enactment in cognitive science with the rich literature on the cultural shaping of illness experience in current medical and psychological anthropology. The resulting view of cultural enactment has broad implications for psychiatric theory, research, and practice, which we will illustrate with examples from the study of the phenomenology of delusions.

1   The Perspectives of Cultural Psychiatry

Cultural psychiatry calls on the resources of cognitive and social science to provide a more comprehensive model of normal functioning and affliction that can include the wide variations in illness experience seen across cultures and communities. Accounting for this variation requires theory that spans individual neurobiology, psychology, and sociocultural processes (Kirmayer and Gold 2012; Kendler 2014). Older models have tended to approach culture and social context in terms of lists of individual traits, internalized representations, or beliefs. The emerging paradigms of embodiment and enactment provide new ways to think about the influence of context on behavior and experience. In particular, they offer approaches to organism-environment interaction that emphasize the co-emergence of mind and culture over evolutionary, developmental, and everyday timescales (Leung et al. 2011; Seligman, Choudhury, and Kirmayer 2016). A key element of these processes for psychiatry is the intersubjective grounding of experience through modes of embodied interpersonal interaction, cooperation, and collaboration (Fuchs and De Jaegher 2009).

The engagement with cognitive science can work in the other direction as well. Cultural psychiatry can contribute to the study of intersubjectivity by providing compelling examples of pathology that reveal some of the taken-for-granted, implicit assumptions of regnant theories, and expose the fault lines in the everyday construction of reality (Gold and Gold 2014). A specific symptom or syndrome can provide a focus for interdisciplinary analysis. Attention to cultural variation serves to foreground social context and encourages theorists to test models against the diversity of human experience. Of course, psychiatry not only provides privileged access to uniquely instructive kinds of experiences but, as a clinical discipline, also presents a set of tasks with their own epistemic challenges and social consequences. The clinical engagement with suffering brings with it the urgent imperative to acknowledge and respond to the other—their face, voice, presence, or perspective—in an ethics of intersubjective encounter (Kirmayer 2015a). The mandate to take helpful action requires tracing the effects of theoretical models through clinical and social practices to their impact on individuals and communities.

Ethnographic research and intercultural clinical work have found both commonalities and substantial variability in the ways that people experience and express mental disorders (Kirmayer 2006; Larøi et al. 2014). This has sometimes been framed in terms of a distinction between *pathogenesis*, involving putative underlying mechanisms of mental disorder that are assumed to be based on universal aspects of human neurobiology, and *pathoplasticity*, the particular symptoms or behavioral expressions of distress that vary with cultural knowledge, beliefs, and interpretations (Kleinman 1987). For example, auditory hallucinations are a common symptom of certain types of psychoses such as schizophrenia, but the specific content of hallucinations clearly reflects local culturally mediated experience: people in some contexts hear the voices of the gods and spirits of their tradition while others hear the whispers of government spies or aliens from outer space (Gold and Gold 2014; Larøi et al. 2014).

While the distinction between pathoplasticity and pathogenesis can be applied to any mental disorder, some types of problems (or some aspects of particular disorders) are viewed as more plastic or malleable than others. Dissociative disorders exhibit the most extreme form of this cultural plasticity, since both their form and content seem to be structured by cultural models, scripts, or templates (Seligman and Kirmayer 2008). There is a rich historical and ethnographic literature on dissociative phenomena (hysteria, trance, possession) that illustrates the cultural shaping of syndromes and their rise and fall with particular social and cultural circumstances (Kirmayer and Santhanam 2001).

Despite its intuitive appeal, there are several problems with the distinction between pathoplasticity and pathogenesis. First, it assumes we can clearly distinguish universal mechanisms of psychopathology from locally variable expressions. But cultural contexts may influence processes at many levels in the transition from neurobiological abnormality through cognitive processes to behavior and experience. Each of these levels counts as a link in a causal chain. Moreover, the apparent universality of some symptoms or disorders may reflect not biological universals but social or existential commonalities. Second, individuals interpret and respond to their own symptoms with culturally varied coping strategies that may change the course of the illness, amplifying or reducing symptoms and distress. These feedback or "looping" effects may be what makes a transient or mild symptom become a severe and persistent health problem (Ryder and Chentsova-Dutton 2015). Third, there are added levels of looping effects that extend beyond the individual to interpersonal relationships, networks, and larger social processes that exert powerful influences on the causes, course, and outcomes of distress (Hacking 1995; Kirmayer 2015b). Many potentially pathological symptoms—including not only anxiety and depression but also hallucinations and delusions—occur in mild, self-limited forms throughout the general population. What distinguishes pathology from such normal variation may be the sociocultural feedback amplification of underlying neurobiological defects and/or the failure of compensatory cognitive-social mechanisms to restore participation in consensual reality. Both cognitive-interpretive and social-interactional processes may play decisive roles in determining which symptoms become persistent, distressing, and disabling mental health problems. Hence, psychopathology cannot be understood completely in neurobiological or individual terms but requires a broader social and cultural perspective (Kirmayer and Gold 2012).

Cultural psychiatry has made use of concepts of culture developed in the social sciences, particularly anthropology (Kirmayer 2006). While work on culture in the 1960s emphasized the metaphor of the text, privileging hermeneutic strategies borrowed from the study of literature, current medical anthropology emphasizes that culture is embodied as well as discursive. In addition to its embodiment in individual experience and habitus, culture has material dimensions that reside in social institutions and are embodied and enacted in social practices. This focus on patterns of habitus, institutions, and practices emphasizes the conservative nature of culture through its historical continuity. But cultures are also always in flux. New cultural forms and practices are continuously produced through processes of intercultural encounter and hybridization at the boundaries of individuals and groups or institutions.

Ethnographic research aims to capture the interplay between discourse, social institutions, and practices by examining everyday interactions through participant observation and in-depth interviews that explore individual experience. Cultural psychiatry employs these methods to characterize the nature of psychopathology in context. This process of contextualization can be applied both to basic research questions as well as to clinical assessment and intervention. Clinical inquiry into the context of illness experience is guided by tacit theories of what is at stake for individuals, which draw from knowledge of biomedical, socioeconomic, moral, and political exigencies. The contextual view then raises a challenge for narrowly biological or psychological theories of the person that do not account for the deep embedding of all experience in local social worlds (Kirmayer and Gold 2012; Kirmayer and Crafa 2014).

2   Embodiment, Enactment, and Expectation

The embodiment paradigm in cognitive science examines the ways in which cognition and experience depend on the body, understood both as a physical object—what phenomenologists following Husserl (1990) call the *Körper*—and as it is consciously experienced (*Leib*, or lived body). Cognitive processes are described as embodied in that they are coupled to and depend on bodily processes both for their content and for aspects of their underlying structure (Varela, Thompson, and Rosch 1991; Thompson and Varela 2001; Gallagher 2005; Thompson 2007; Chemero 2009; Shapiro 2011).

Embodiment has received much attention and empirical study in cognitive and social psychology as a way to understand the emergence of more abstract modes of thought (e.g., Barsalou 2008; Glenberg 2010). Sensorimotor processes provide an underlying structure upon which more elaborate and abstract forms of cognition can be scaffolded through developmental processes of learning and skill acquisition (Williams, Huang, and Bargh 2009). This scaffolding can happen in several ways: *psychophysically*, through the ways that bodily structure and physiology structure perception and cognition; *cognitively*, through the ways that sequences, habits, or patterns of bodily action and experience provide analogical structures that organize more complex action, thought, and experience; and *socially*, through the ways that the body serves as the medium for interpersonal interaction and communication (Kövecses 2015).

Situated cognition is closely related to the embodiment paradigm (Rowlands 2010). Cognition is situated in that it always occurs within, depends on, and is partly constituted by the social and environmental contexts from which it receives background meaning and toward which it is oriented (Clark 1997; Clark and Chalmers 1998; Clark 2008; Robbins and Aydede

2009). There is much debate in the recent literature on the ontological entailments of situated cognition, with authors describing cognition as merely *embedded*, literally *extended*, or basically environment-involving and *extensive* (see Hutto and Myin 2013; F. Adams and Aizawa 2001; Rupert 2004; Sutton et al. 2010; Clark, 2008). These views share the common intuition that cognition draws significance from, and participates in, wider circles, which engage other cognitive agents and sociocultural institutions through ordinary cognition, language, and social practices.

Enaction (or enactment) is closely related to, although conceptually distinct from, the embodiment paradigm. Enactivist theorists emphasize the co-constitution of organism and environment as well as the dynamic constitution of meaning in experience through ongoing cycles of perception and action, with the explicit recognition that cognition necessarily involves active engagement with the environment, both implicit and explicit (conscious and nonconscious, agentic and automatic) (Noë 2005; Thompson 2007; Froese and Di Paolo 2011; Di Paolo and Thompson 2014). The enactive approach emphasizes the point of the view of the organism itself, as a cognitive agent that encounters the world in terms of *affordances*, that is, possibilities for action that are "ready to hand" (*zuhanden*) and that shape behavior (Varela 1999; Chemero 2009). Older cybernetic theories of hierarchical information processing (e.g., Miller, Galanter, and Pribram 1960) and more recent models of predictive processing (which we discuss below) posit that this loop between perception and action is the basic building block of cognition. The body is both the locus of sensation, perception, and feeling and also the vehicle of action on the world through ongoing sensorimotor loops, which result in dense coupling between individual and environment.

Radical embodied and enactive approaches take this conception of cognition as dynamic coupling a step further to argue that these processes are sufficient to account for much behavior and experience without the need for representational processes and resources (Thompson 2007; Chemero 2009; Hutto and Myin 2013). New work on neural computational modeling supports this view. As Piccinini (2015) explains, computations (digital, analog, neural) need not bear any semantic content. Instead, computation can be defined mechanistically as the manipulation of computational (rather than representational) vehicles, without reference to satisfaction conditions (e.g., "This pattern of neural spiking represents that object"). Although representational (semantic) information processing does entail computational implementation, computation per se does not entail that representations are being processed; there can be computation without representation. So there are good reasons to be minimalistic about representational processes; representation may not be as pervasive as earlier representational theories of mind would claim. However, crucially for an enactivist cultural psychiatry, nothing *precludes* neural computations from acquiring representational content, notably through histories of dense causal coupling, which ground neural computations (Hutto and Myin 2013; Hutto and Satne 2015). Indeed, intensively social and cultural forms of cognition might warrant or even mandate representations.

Recent work on predictive processing promises to offer a computationally and neurologi-cally plausible framework for "minimal" models or representations. Predictive processing models are a family of models in cognitive science based on a more general view of the brain as a prediction machine (Frith 2007; Bar 2011; Friston et al. 2014; Hohwy 2013; Clark 2016). In this view, the brain is modeled as a complex dynamical system the main function of which is to predict the ongoing stream of sensory information that impinges on its sensory receptors. When differences occur between predictions and actual sensory input ("prediction error"), the system acts to reduce the discrepancy. The actions taken to minimize predictive error are of two complementary kinds: behavior in the environment (that is, making the world more like the prediction), or changes in the statistical models that generate the predic-tions implemented in the brain's various neural networks. In this family of models, neural computations are modeled as bidirectional computational cascades involving cortical and subcortical networks. In this cascade, top-down predictions of future events by internal cog-nitive models, propagated through backward connections, are continuously updated and modified through Bayesian inference, to "explain away" the prediction error signals that are propagated through forward connections (or lead to predictively guided action sequences). Rather than simply mirror the causal structure of the environment, as in older representa-tionalist models, these generative neural models encode probabilistic information about dis-tal causes of sensory stimuli, which can be leveraged to guide intelligent, socially and culturally appropriate modes of behavior in dynamically changing and meaningful material environments. Indeed, through top-down influences, these networks can implement the goal-oriented control structures posited by Miller, Galanter, and Pribram (1960) in their cybernetic model of hierarchical information processing.

Predictive processing provides a neural-computational model of the mechanism required to account for the scaffolding of the encultured mind through enactment. Three elements of the predictive processing account are relevant to understanding the enaction of culture (Clark 2016). First, the changes in neural networks involved in predictive processing consti-tute *generative* models, which do not simply mirror the causal structure of the environment but rather provide a statistical model of potential input from the environment in terms of probability distributions (predictions or expectations). The use of such generative models moves us from a view of the brain as a passive, bottom-up feature detector, toward a model of the brain as an endogenously active, "phantastic organ" (Hohwy 2013; Friston et al. 2014). This cognitive architecture provides building blocks for imagination because it can generate nonactual sensory states, and thereby explore nonactual perceptual and cognitive spaces. Predictive processing thus points the way toward an enactivist view of imaginative processes (which include cultural symbols, patterned practices such as storytelling, mental time travel, and other forms of mental projection) together with perceptual modes of encounter with social and material environments.

Second, predictive processing provides a neural-computational implementation of social and cultural *affordances*, that is, expectancies, prescriptions, and possibilities for action in context (Rietveld and Kiverstein 2014; Bruineberg and Rietveld 2014; Ramstead, Veissière, and Kirmayer 2016). This aspect of the predictive processing approach fits with the phenomenology of everyday life, in that the world presents itself to us experientially as possibilities and potentialities for action (Dennett 2013). Predictions are not simply descriptive, but also *prescriptive*, because one of the basic ways to minimize prediction error is through *action*. As prediction error signals propagate throughout the cortical hierarchy, they are minimized both through changes in plastic neural connections *and* through action, depending on the context and the available behavioral repertoire. The computations performed by brain-bound predictive networks might then be said to acquire their representational contents through various forms of embodied and sociocultural enactment (Hutto and Myin 2013; Hutto and Satne 2015). Generative models thus extract and recapitulate the salient context-sensitive features of the material (sociocultural) environment and, having acquired semantic content through enactment, are leveraged to specify and engage available cultural affordances.

Third, predictive models present a framework with which to understand the interactions between brain and culture, revealing important causal pathways from culture to cognition and back. Human environments are largely socially and culturally constructed, designed (intentionally or through forms of coevolution) to produce and foster specific kinds of cognition in specific contexts. The fast (and cognitively "frugal") responses at the level of the brain and body, in selection, tuning, and action, are made possible—and shaped or sculpted—by a rich social scaffolding. This scaffolding can be understood as the slow, multigenerational construction of "cognitive niches," that is, social contexts or environments dense with materially encoded expectations (Odling-Smee, Laland, and Feldman 2003; Wilson and Clark 2009). Social contexts call for, make possible, and facilitate particular modes of cognition. These contexts sediment cultural knowledge and allow new generations to install sets of cultural expectations through patterned cultural practices, such as storytelling and other forms of narration. The social context also provides material circumstances that can be reconfigured through cooperative action. We thus have reentrant loops of social information processing, in which predictive brains pattern their environments to elicit further predictions in other brains.

Predictive processing fits well with embodied, situated, and enactivist accounts of cognition, providing a minimal, dynamical model that unifies perception, action, and imagination in rolling computational cycles. Predictive processing offers cultural psychiatry resources to begin to understand the forms of neural computation that enable the cultural scaffolding of cognition and experience. As we shall see, culturally shaped sets of expectations, embodied and enacted in material artifacts, places, and practices, can shape the mechanisms, course, and outcome of illness by altering the experience and interpretation of symptoms, syndromes, and behaviors.

3   Bridging Bodily and Discursive Levels of Experience through Metaphor

The interplay of physiological, cognitive, and social processes of embodiment and enactment can be approached through the study of metaphor, which mediates bodily and discursive levels of elaborating experience (Kirmayer 1992, 2008). A growing body of work
demonstrates how conceptual metaphors arise from basic bodily or sensorimotor experiences. Lakoff and Johnson (1980) argued that complex concepts are built on a scaffolding of
metaphors that can be traced back developmentally to early bodily experiences of similarity
across sensory modalities and to basic actions that accompany the infant's exploration of the
world and acquisition of language. Similarly, Barsalou (2008) has elaborated an approach to
cognition based on modality-specific perceptual symbol systems, which are grounded in specific sensory-perceptual experiences. Grounded cognition places concrete instances of bodily
experience at the base of more abstract thinking, which retains an associative logic based on
the sensory-perceptual features of prototypes.

A wealth of experimental research shows the ways in which particular metaphoric fram-
ings of experience result in predispositions to perceive the world and act in specific ways
(Gibbs 2006; Meier et al. 2012). There is two-way traffic between linguistic tropes and bodily
experience, with bodily experience giving rise to particular metaphors and metaphors evoking specific bodily experiences (Körner, Topolinski, and Strack 2015). This two-way traffic
may involve multiple mechanisms and is consistent with predictive processing models that
suggest that the same networks are used to generate and interpret action and perception
(Clark 2016), as well as "neural reuse" models, which argue that the brain dynamically rede-
ploys computationally differentiated circuitry in soft-assembled transient processing ensembles (Anderson 2014).

A parallel line of work in anthropological theory has approached embodiment in terms of
body practices or techniques (following Mauss [1935] 1979) and the use of the body as a
medium of social communication. Much work in psychological anthropology, inspired by
the phenomenological philosophy of Merleau-Ponty ([1945] 2012), has explored the ways in
which embodied experiences arise and are shaped by particular cultural histories and contexts (e.g., Csordas 1990, 1994; Desjarlais and Throop 2011). Culture is inscribed on the body
through body practices, which are socially situated and reflect both current contexts and collective history. Bodily enactments include expressive gestures, verbal narration, and cooperative social action, from everyday improvised interactions to the highly structured activities
of collective ritual.

Enactment grounds meaning and experience in action—both interaction with the physical environment and with the social world, which is composed of other actors and institutions (Thompson 2007; Froese and Di Paolo 2011). This is the context for situated practices
that are guided by norms, expectations, and affordances. The situated actions enabled by
social contexts include modes of metaphorically construing and narrating experience and
presenting one's self, identity, aspirations, commitments, and desires.

Narration is central to culturally mediated experience. Narratives may be elaborated by extending or enacting a metaphor and may in turn give rise to new metaphors (Kövecses 2015). Developmentally, narration is learned through social practices of communication and conversation (Hutto 2008), which become internalized in modes of self-monitoring and cognition that structure memory, identity, and action plans. Narratives follow conventional forms that are learned through the everyday practices of conveying information and experience, explaining perspectives, organizing memory, justifying actions, positioning oneself, and persuading others. The rhetorical functions of narrative involve interpersonal influence through giving compelling accounts that change how situations are perceived and evaluated and that enjoin others to take action.

4   The Metaphoric Mediation of Illness Narratives

Experience is always preceded by and embedded in cultural systems of meanings and practices, which influence modes of attention and interpretive frames or models. Cultural models may be organized in many ways, including collective symbols, images or representations, and forms of cooperative activity. Illness experience, no matter how strongly influenced by pathophysiology, is also given shape by culturally mediated processes of attention, interpretation, and enaction (Kirmayer 2008, 2015b).

In seeking to understand illness experience and behavior, medical anthropology has privileged narrative, eliciting stories that provide causal explanations or explanatory models. Kleinman (1978) advanced an influential approach to explanatory models in terms of the person's explicit notions of cause, course, outcome, and proper treatment, much the same way that physicians reason about disease. Explanatory models of illness may reference local theories of body (ethnophysiology), person (ethnopsychology), the social world, and other realms recognized by religious or spiritual traditions (Kirmayer and Bhugra 2009). However, it appears that people use other kinds of knowledge structures in reasoning about illness, including contiguity, analogy, and metonymy. Based on ethnographic fieldwork on lay illness models, Young (1981, 1982) argued that the explanatory model approach was based on a highly rationalistic account of everyday experience. In reality, illness narratives usually include reasoning based on temporal contiguity ("chain complexes") and prototypes (models, images, or representations used to reason analogically about current illness experience), as well as the causal attributions central to explanatory models (Kirmayer, Young, and Robbins 1994). Chain complexes may be structured by procedural learning and affordances that individuals are not able to describe explicitly but which are indicated or suggested in their account. While procedural knowledge may be difficult to recount, it can be observed in context or measured in experimental settings. Similarly, prototypes acquired through past experience or media exposure, or even generated in imagination, may provide a tacit structure to narrative based on analogy or metaphoric elaboration. Individuals may not be able to explain the links in their accounts causally but only by appeal to the salience of the prototype that welds together disparate elements of action and experience into a compelling whole. Based on Young's critique, the McGill Illness Narrative Interview (MINI) was designed to elicit narratives that might include these different modes of reasoning (Groleau, Young, and Kirmayer 2006). Research with the MINI has documented the complex and evolving nature of illness narratives and the role of salient prototypes and chain complexes (e.g., Groleau and Kirmayer 2004).

   Illness experience depends on mutually constitutive processes of embodiment and enactment. As suggested in figure 20.1, illness experience and self-fashioning occurs in loops or cycles in which the embodied ground of experience interacts with larger sociocultural discursive forms. Metaphor theory provides a way to bring together embodied and enactivist

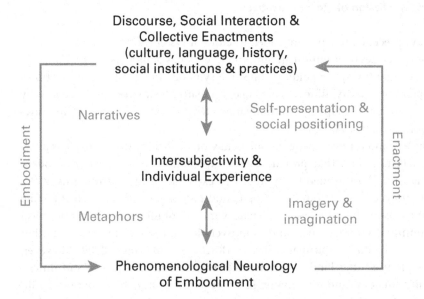

Figure 20.1

Embodiment and enactment in experience. Diagram of multilevel looping effects studied by cultural neurophenomenology (Kirmayer 2015b). Embodied experiences give rise to metaphors, which structure and shape individual experience, and to narratives, which are amplified, stabilized, and extended through collective enactments. In turn, discursive practices give rise to new metaphors and modes of embodiment. These cycles of enactment and embodiment occur on multiple timescales: (I) in the historical process of producing culture, novel metaphors become the taken-for-granted building blocks of common sense; (II) over the course of individual development, collective narratives and modes of cooperative enactment structure individual autobiographical narratives and identity; and (III) in the ongoing improvisations of everyday life, extended metaphors and narratives are tools for self-fashioning and positioning. The discursive level itself is structured metaphorically and can give rise to new tropes (Danesi 2013; Kövecses 2015).

views of behavior with representational processes of imagery, imagination, and more abstract model building. This has particular relevance for our understanding accounts of illness experience in which individuals may present not coherent narratives but arresting symptoms and fragmentary experiences, apprehended through the analogical constructions of metaphor.

Illness experience is an intersubjective, temporal, dynamic process shaped by culture (Kirmayer 2015b). Within cultures, participants share expectations, which may be thought of as tacit, prediction-driven interpretive frameworks, and practices that allow them to cooperate in co-constructing versions of experience that are mutually intelligible and that serve to position participants vis-à-vis each other and background values. More elaborate models of illness experience are structured by narratives of self and personhood which make explicit reference to local ontologies.

5   Embodiment, Enactment, and Intersubjectivity in Delusion

In his novel *The Echo Maker*, Richard Powers (2006) describes the experience of Mark Schuter, a young man who suffers a brain injury that leaves him with Capgras syndrome—a delusional disorder characterized by the experience that familiar people have been replaced by imposters. The novel chronicles Mark's transformations of consciousness and the parallels between the ways he reconstructs his identity and the ongoing efforts of others to fashion their own identities in the face of major life events and moral quandaries. One key narrative concerns the corrosive impact of Mark's delusion on his sister Karen, who comes to take care of him during his slow recovery. Through multiple narrative strands, the novel explores the tension between embodied experiences, driven by "instinctual" processes of the brain, which reflect our evolutionary history, and individuals' own efforts at self-fashioning as reflective, moral beings actively striving to construct, enact, and live by coherent narratives.

The novel begins with the image of sandhill cranes that return each year to the same site in Wisconsin as part of their annual migration. This migratory behavior is driven by instinct—based on inborn neural structures primed to learn the migration route that structures the lives of the sandhill cranes. "The birds are presented as symbols of an archaic prehistory that still lives on in our present-day. Their flights of migration are determined by old patterns that have engraved themselves in their brains and bodies" (Herman and Vervaeck 2009, 413). The sandhill cranes are called "echo makers" in a local Native American language because they echo each other in their calls. Powers elaborates this metaphor of echoing: the cranes echo their own past by returning each year to the same place; in so doing, they enact a narrative of migration and return. In parallel, humans echo their own histories and each other's stories and, in so doing, co-construct their own narratives of identity.

The birds enact a form of temporal continuity (through instinct, habit, learning) that contrasts with Mark's breakdown in self-identity. Mark has lost the affective core of his personal history because the archaic brain circuits that would signal familiarity have been disrupted. The developmental process of mirroring others carries us forward from "instinct" (preestablished patterns of action) to habit (acquired patterns that also run more or less automatically), to reflection (as expressed through narratives of self and other). This developmental progression culminates in the processes of self-narration that are, perhaps, what makes us most human; but the self is a fragile, unstable attainment that requires constant echoing from others to maintain. If we lose this echo, we may experience a radical destabilization, a vertiginous loss of the sense of self. This is what happens to Mark's sister Karen in response to his denial of her identity—but it also happens in more mundane ways in response to other kinds of misrecognition that occur in other relationships in the novel.

As a distinct symptom, Capgras delusion has been explained by dual-process neuropsychological theories that posit separate circuitry for facial recognition and feelings of emotional recognition/familiarity, on the one hand, and distinct, presumably higher-order mechanisms for resolving anomalies (i.e., incongruences between recognition and familiarity) in experience, on the other (Connors and Halligan 2015; Coltheart, Langdon, and McKay 2011; Ellis and Lewis 2001). When facial recognition occurs in the absence of feelings of familiarity, violating tacit acquired expectations, one has the odd or uncanny experience of seeing someone familiar but feeling as though he or she is a stranger. This uncanny experience can be interpreted in various ways depending on available knowledge and cultural expectations: robots, clones, or alien body-snatchers. Under ordinary circumstances, the experience might be interpreted by the individual undergoing the delusional ideation as a sign of mental dysfunction, but when other mechanisms of salience are disturbed, the uncanny feeling persists and leads to more bizarre explanations (Sass and Byrom 2015).

Recent efforts to explain delusions in terms of predictive processing point to the potential consequences of specific kinds of malfunctions in neural computational mechanisms (e.g., Fletcher and Frith 2009; Frith and Friston 2012; R. Adams et al. 2013; Clark 2016). Delusions have been modeled as resulting from unchecked, self-generated error signals. One suggestion is that misestimations of the precision or reliability of signals in a particular predictive system might lead to systematic prediction errors that can never quite get "canceled out," and thus retain high salience as unresolved discrepancies. This spurious but persistent error signal is then propagated throughout the hierarchy, eventually leading to delusional ideation functioning (albeit pathologically) to suppress the signal. Indeed, the unresolved error signal, labeled as highly salient, might lead to (sometimes bizarre) alterations in higher-order expectations (which correspond to tacit beliefs). The very same mechanisms that allow for neural plasticity and perceptual learning thus exhibit a corresponding fragility and susceptibility in experiences of delusion. Different pathways might produce similar effects, and research on neuromodulatory mechanisms is underway to disentangle the sundry causal strands. For instance, Pezzulo (2014), following Seth and colleagues' (2012; Seth 2013) interoceptive predictive model of conscious experience, suggests that Capgras syndrome might be the effect of faulty interoceptive prediction circuitry, leading to affective states mismatched to perceptual experience. Ordinary perceptual inference ("These are my friends ...") combined with faulty affectively laden interoceptive predictions ("... but they feel like strangers ...") might lead to faulty, even bizarre, higher-order inferences ("... so they must be imposters!"). Although such delusional beliefs would usually have low or null prior probability and, as such, would not be leveraged to guide perception, misestimations of reliability, faulty affective (interoceptive) prediction circuits, or some other mechanism, each leading to spurious but high-salience error signals, might mandate such inferences, making delusional ideation computationally inevitable. Moreover, because the spurious error signal is generated by faulty circuitry, the error remains epistemically encapsulated despite its propagating across the hierarchy. This is because no body of *external* information can be leveraged to annul the *self-generated* error signal, leading to the peculiar (and notorious) epistemic encapsulation and resistance to counterevidence of delusional ideation. The predictive processing model thus helps to explain how delusional ideas can be maintained with conviction despite external counterevidence or obvious conflict with other beliefs or commitments. If it is the basic circuitry underlying the predictive processes of action and perception that has gone awry, the person may experience the strange ideas as obvious, insistent, or compelling. Explicit theories will be generated by the afflicted person after the fact, employing different cognitive systems, in attempts to explain or justify his or her experience to others.

Capgras syndrome is usually theorized in terms of disturbances of the individual's cognitive function, but Powers's novel points to important social dimensions. The processes of face recognition and person recognition are interactive, recursive, and intersubjective: we see others as familiar and they see and feel us seeing them that way even as they have their own experience of familiarity. In everyday life, this interaction would be mutual, and the response of each person to the other would calibrate or reinforce the shared experience of recognition. When the recursive cycle of mutual recognition breaks down, the experience is difficult for both participants. This experience of being misrecognized, or told that one is a replacement, is troubling for the nonafflicted person. Indeed, when the relationship is important and close contact with the delusional person continues, this misrecognition can create a profound sense of self-estrangement. Capgras syndrome thus provides a compelling example both of the embodied and enacted nature of experience and of the intersubjective stabilization of modes of selfhood through modes of interaction and narration.

6 Cultural Ontologies and Constructions of Normativity

Enactive predictive processing architectures specify the computational scaffolding required for situated cognitive agents to acquire and install cultural representations and explicate the interface between neural computation and shared cultural content. This gradual developmental process of semantic enculturation, in turn, ramifies in conscious experience, notably by shaping taken-for-granted background knowledge and by modulating the individual's tacit assumptions and expectations. The specific content of delusions reflects the social structures to which predictive networks are coupled; over the developmental time course of enculturation (or, for migrants, acculturation), these networks encode the available cultural idioms and symbols present in the sociocultural environment of the agent (Bhavsar and Bhugra 2008; Gold and Gold 2014). The background makes ideas that appear odd to an outsider seem natural or inevitable to participants in that cultural world. Clinicians recognize this normativity, and when an idea is shared within a cultural community, it is usually not interpreted as a delusion even if it appears strange and unwarranted to an observer external to the culture (Rashed 2013; Adeponle, Groleau, and Kirmayer 2015).

Defining what counts as abnormal or pathological is a central issue for psychiatry. Psychiatric discourse on normalcy and disorder is always already set against the everyday mechanisms that construct the tacit norms that guide interpretation and behavior both at the levels of discourse and interpersonal interaction (Hacking 2004). Culture provides these background expectations or tacit dimensions to experience (Geertz 1983). Cultural norms and frameworks are internalized and enacted not simply as individual habits but also as forms of coordinated social interaction, including institutional routines and practices. In these interactions, there is a reliable expectation that others will respond to complement or complete one's own actions. On the basis of shared expectations, each participant can play his or her part in the flow of social interaction. Engaging these tacit structures leads to enculturation as the person acquires background assumptions and expectations that are salient and materially embodied in cultural artifacts, places, and practices (Searle 1995; Clark 2016).

More than this, others make epistemic commitments and take action on the basis of assumptions that make their interpretations of events seem obvious or apparent, real, and reliable. While epistemic claims serve rhetorical and pragmatic functions, they are also grounded in a shared ontology. The epistemic frame that allows for the enactment of culture includes both a folk theory of knowledge, expressed through participation in (and respect for) particular institutional forms of authority, a more or less explicit shared ontology that recognizes certain kinds of entities or actors in the world (persons, gods, spirits), and bodily dispositions, habitus, or sets of expectations that guide attention and perception (Lock 1993). Everyday action and experience are informed, shaped, and constrained by a cultural background that includes a folk epistemology and shared ontology.

Shared ontologies need not be elaborate or formulated as metaphysical theories; they can be acquired through the installation of specific sets of cultural expectations, embodied in cultural symbols and places, and enacted through everyday habits and rituals. Predictive processing models suggest that culturally specific ontologies might be implemented by (mainly implicit) sets of predictions or expectations encoded in the brain's neural networks. As we have seen, predictive processing regimes implement affordances, which exist on a continuum from lower-level, representation-poor (even contentless) sensorimotor potentialities to full-blown, representation-rich cultural and social prescriptions for modes of perception and action. The ontologies we acquire from enculturation—which are employed to appropriately parse and make sense of the world, and to act in situationally appropriate ways—are thus reflected in neural generative models implemented in the brain.

A cultural ontology, then, can be described as a shared style or mode of expecting the world to be a certain way, and to afford certain possibilities for action while foreclosing others. These ontologies, as expectations, are continuously installed across the individual's lifespan through participation in cultural practices. The participation of members of a social group or community in such cultural enactment dynamically constitutes the shared meaning of their local world, and further grounds, scaffolds, and ultimately makes possible individual enactments of meaning, by prescribing and normalizing certain modes of experience and action while proscribing (and perhaps even pathologizing) certain others. For example, people in many parts of the world take part in daily ritual observances in which they make offerings to gods by leaving food, spices, or other materials on statues that symbolize or stand for the god. These are places of presence, in which actions can be communicated to the gods in the hope of beneficial effects. Throughout India, for example, many people make regular offerings to Hindu gods and sometimes experience the god as present through various kinds of manifestation.

On September 21, 1995, a statue of Ganesha in New Delhi was observed to drink milk from an offering bowl (Singhal, Rogers, and Mahajan 1999; Vidal 1998). News of the phenomenon quickly spread both through the country and globally. By the evening of that day, several million people had tried to offer milk to Ganesha, and many reported the statue drank. There was a 30 percent increase in sales of milk throughout India (and in parts of the UK) that day and shortages in many places. News media presented scientific explanations for the phenomenon in terms of the capillary action of stone but these had little effect on the spread of the phenomenon. Similar episodes occurred in 2006 in Uttar Pradesh and in 2010 in Trinidad.

The story of Ganesha drinking milk illustrates the ways in which culturally consonant ideas and experiences can spread and be ratified by interactions involving individuals, families, communities, and global networks. This is related to cultural enactment in several ways: (1) through specific *contexts* or *places* or *settings* (e.g., the shrine and statue), in which the situation affords certain possibilities of action and experience, calling forth certain responses as a result of encultured expectations; (2) through prescribed *practices* (ritual actions or scripted patterns of behavior) or habitual sequences of thought and action that may be expected and routinized and hence be accompanied by varied degrees of self-awareness, background knowledge, or conviction; and (3) through the larger social processes mediated by networks of relationships to others and to social institutions that ratify particular modes of action and experience. One consequence of these structural determinants of enactment is that one can do things with or without conviction (just responding to the context or to other's expectations), but, once carried out, one's actions loop back and reinforce one's conviction. The enactment of a shared ontology by a community further reinforces its credibility. Social psychology has a long tradition of studying how beliefs and commitments can be stabilized and persist even in the face of contradiction or dramatic counterevidence; this may occur through both cognitive (dissonance reduction) and social (conformity or sanctions) mechanisms (Festinger, Riecken, and Schachter 1966).

The plausibility of statues of Ganesha drinking milk derives from the social grounding of normativity in shared settings, practices, and social processes, showing how individuals' actions-in-context become expectations and convictions through shared sets of embodied meanings and cultural background knowledge (that is, shared ontologies). In the case of the drinking statues, the backdrop to the phenomenon includes everyday practices of making offerings, a particular ontology that understands the statues not simply as representations of the gods but as sites of the gods' living presence, and familiar mythic narratives—performed in both religious and artistic settings—that portray the gods' agency. A fuller account would also have to consider the impact of the ongoing tensions between Hindu religious and secular/scientific worldviews, which adds an overtly political dimension to the individual's experience and epistemic claims (Vidal 1998). All of these cultural discourses shape agent expectancies, contributing to the plausibility of the statue drinking and the perceptual vividness of the anticipated event.

Of course, many perceptual experiences seem to come to us fully formed and stamped with their own authenticity, certainty, and conviction (Wittgenstein 1969). They do not seem to be the origins of norms but the expression of them. In part, this is because we cannot perceive certain things until we adopt the relevant norms. Culture can influence the shape of this process at several levels: (1) as a discursive process of constructing narratives or recruiting metaphors that fit within larger systems of understanding and experience and which therefore lend credence and conviction to experience through coherence; (2) as an experiential process of noticing certain bodily felt, emotional, or other phenomena that are inherently "solid," unequivocal, or even numinous precisely because of their embodied qualities; (3) as a bodily habitus by which one meets the world so that certain forms of perception are always already present; (4) as a mechanism of rewiring specific brain circuitry so that feelings of conviction are activated and doubt bypassed (or, in other instances, creating feelings of doubt and uncertainty about what should be perceived as natural, normal, and unquestioned in experience, as in the example of Capgras syndrome); and finally, (5) as an epistemic stance that the individual adopts more or less consciously, which may be influenced by all of the preceding mechanisms.

7   Toward a Cultural Neurophenomenology of Psychopathology

The story of the milk-drinking statues is a dramatic instance of the "epidemiology" of collective representations (Sperber 1996). But the same process goes on continuously to reshape background knowledge and to foreground prescribed forms of experience in every cultural community or network. There is nothing inherently pathological in such cases, though they may give rise to particular kinds of social problems (further exacerbated by the new forms of networking and coupling of minds made possible by the Internet; see Halpin 2013; Veissière 2016). The looping process that links experiential conviction to forms of cultural normativity, underwritten by shared ontologies, can help us understand other examples of the social spread of unusual experiences, including certain forms of psychopathology (Hacking 1998). Together with models of metaphoric embodiment and enactment, the systematic study of looping effects (Hacking 1999, 2002) provides a way to forge a cultural neurophenomenology (Kirmayer 2015b). We will illustrate this prospect with an example from ethnographic research on mass psychogenic illness.

In rural Nepal, in recent years, numerous episodes of spirit possession affecting groups of schoolchildren and adults have been reported in the local and national media (Sapkota et al. 2014). Incidents of negative spirit possession affecting clusters of individuals, sometimes referred to as "mass hysteria" or "mass psychogenic illness," often occur among groups of schoolchildren and women. These episodes are understood as a kind of affliction and are clearly distinguished from the deliberate, positive experiences of spirit possession sought by religious adepts and healers as part of culturally sanctioned and normative spiritual practices or healing rituals. The afflicted person may exhibit pseudo-seizures resembling tonic-clonic convulsions, writhe on the ground, and speak in an altered voice, while striking out at others. Within the local ontology, these alterations in speech and behavior are understood as evidence that the person is possessed by a spirit and will require spiritual intervention to be cured of the affliction. The episode of possession may occur in public settings (for children, in school; for adults, at a shrine or other public place), and others who are present and watch or interact with the afflicted person may also become affected by the condition themselves either immediately or sometime later.

Spirit possession is experienced as involuntary, and the afflicted individual usually disavows awareness and control after the episode. Psychologically, possession can be understood as involving dissociative mechanisms, in which cultural models and scripts that are cognitively compartmentalized and held out of awareness govern behavior (Kirmayer and Santhanam 2001; Seligman and Kirmayer 2008). The disavowal of causation and control may serve to protect afflicted persons from responsibility or blame for their behavior. At the same time, possession may serve as a coping strategy through its expressive and communicative functions, which mobilize social support and conflict resolution.

Spirit possession occurs in the context of particular cultural ontologies. Everyday life in rural Nepal is shaped by culturally prescribed expectations about spirits and their power to influence illness and well-being. In the Nepali local ontology, spirits are present in all living beings (humans, animals, trees, etc.) as well as nonliving things or places (e.g., mountains, streams, village sites, or landmarks). There is a shared conviction that spirits can migrate back and forth from people to other beings or sites. Thus, a spirit can come to occupy and trouble an individual, causing physical illness, madness, and other kinds of misfortune. Many people engage in regular religious, spiritual, or folk practices to promote personal, social, and spiritual well-being, also understood to reduce the risk of attacks by malign spirits.

The person in Nepal who undergoes negative possession has compelling experiences that arise from bodily processes that may be normal or disordered, related to quotidian concerns or to severe personal and social stressors. Cultural models provided by local ontologies—and encoded as sets of expectancies—are ready to hand to guide the form that this suffering takes (Kirmayer and Bhugra 2009). The enactment of possession then becomes a model and metaphor for both self and others, and, if conditions are ripe, it can spread in ways that increase its likelihood of occurrence.

Evidence suggests that those affected by negative spirit possession have endured forms of social adversity and psychological distress that may have contributed to their vulnerability (Sapkota et al. 2014). Some of the stressors that might contribute to illness are shared by most people in the community. These include stresses that stem from structural problems (e.g., economic adversity, or the absence of men in villages because they have left to find work outside of the country) and common social problems (e.g., domestic violence) or political unrest. However, even in "epidemics" of possession only a minority of people are affected. In any given episode, the first person afflicted (or "index case") may have specific developmental, neurobiological, psychological, and situational reasons to manifest the problem. The phenomenon can spread from one individual who has a compelling experience to others who may be susceptible because of dissociative capacity, trauma exposure, or exposure to particular cultural models. For index cases, conviction about their experience is important because initially they may have little or no external validation of their experience. Convictions may be anchored by compelling experiences that can arise from disordered neurology or from social interaction. These compelling experiences can occur in the moment or after the fact and loop back to alter the way events are remembered (Connors and Halligan 2015).

The example of negative spirit possession or mass psychogenic illness in Nepal illustrates the interaction between neurologically patterned capacity for dissociation and culturally shaped narratives of possession experience mediated by embodied and enacted metaphors (Seligman and Kirmayer 2008). While the example of possession or mass psychogenic illness highlights the sociocultural shaping of symptoms, similar processes may contribute to other psychiatric disorders, including psychosis.

Psychotic symptoms include a family of diverse phenomena that may be related and co-occur in some disorders, including: impaired reality testing; thought disorder; hallucinations; alterations in experience of self and agency; negative symptoms (blunted affect, avolition, poverty of thought); and delusions. Mild forms of these symptoms may be common in the general population, and their pathological significance is uncertain (Larøi et al. 2014; Preti et al. 2014). When these symptoms are especially severe or bizarre or appear in particular patterns or gestalts, they are recognized as part of a psychotic disorder like schizophrenia.

However, the severity and bizarreness of symptoms (whether considered individually or in relation to a syndrome or gestalt) must always be assessed against the backdrop of local norms of experience. Attempts to present prototypical delusional symptoms in psychological tests illustrate the importance of these norms. As examples of bizarre beliefs, the MMPI (Minnesota Multiphasic Personality Inventory) includes statements like: "There are persons who are trying to steal my thoughts"; "Ghosts or spirits can influence other people for good or bad"; "My soul sometimes leaves my body"; and "I dream about things and then they come true." Clearly, the bizarreness (or plausibility) of these statements reflects local beliefs and norms of experience and background ontologies that warrant the existence of mental attacks by others, ghosts or spirits, detachable souls, or precognitive dreams. All of these phenomena are prevalent or expected in some religious contexts, and occasional (nondisabling) experiences would be viewed by many as neither bizarre nor pathological but as within the range of ordinary experience—even if associated with illness, trauma, or major life events. Of course, a person who suffered from frequent experiences of this type that interfered with daily functioning would be viewed as afflicted; but in many cultural contexts this would be understood as a spiritual affliction rather than a medical illness, and the appropriate treatment would involve efforts to restore the spiritual or moral order through ritual acts of propitiation.

Culture can influence psychotic experience through multiple mechanisms, including: processes of embodied experience that sediment collective knowledge in habitus and expectation, which contribute to perceptual salience and conviction; discursive processes of constructing stories that fit within larger cultural systems of meaning, which lend coherence and credibility to accounts of experience; and participating in local worlds with specific ontologies, structures of everydayness, and stressors or demands that influence how the person responds to psychotic experience with a search for meaning (Corin, Thara, and Padmavati 2004). These processes can influence the form and content of hallucinations, delusions, and other psychotic experiences through their effects on basic processes regulating emotion, attention, and social cognition. Hence, a cultural neurophenomenology demands that one adopt a scientific framework to explain psychotic symptomatology in terms of specific neural processes that subserve social cognition *as well as* the dynamics of interpersonal interaction and of the social world itself (Kirmayer 2015b; Schilbach 2016).

Conclusion

Integrating knowledge of culture and context into models of psychopathology requires a better understanding of the ways in which culture is embodied and enacted. Psychopathology presents us with situations where the ongoing production of meaning through action and interpersonal interaction is disrupted and disordered in ways that can reveal underlying structures of experience. The embodied, situated, enacted, and predictive perspectives hold the promise of bridging neurology, phenomenology, and culture in ways that might allow us to see how the neural structures and processes that contribute to pathological experience are also shaped by cultural discourses and practices, which in turn are constitutive of and maintained by social institutions.

Cultural enactments involve both bodily and discursive (or social-symbolic) dimensions. Discursive practices have their own cognitive, sociocultural, and political dynamics. These include discursive processes of constructing stories that fit within larger cultural systems of meaning that lend them credibility and coherence; processes of embodied experience mediated by bodily habitus, metaphors, and narratives that contribute to bodily felt presence, salience, and conviction; living in local worlds with specific ontologies, understood as sets of cultural expectations that shape (and in turn are shaped by) lived experience and material culture; and participation in epistemic communities that warrant specific kinds of knowledge and authority. These processes all shape the social response to experience, and may influence the rigidity and persistence of delusional beliefs through their influence on emotion, attention, and social cognition.

The embodied enactive perspective we have sketched can be applied to understand the cultural shaping of even severe mental disorders that are typically assumed to be patterned by their underlying neurobiology. In addition to a system of neurobiological and psychosocial mechanisms, psychiatry would then rest on a cultural neurophenomenology. Indeed, much the same strategy can be used to make sense of the ordinary social and cultural processes through which we construct and inhabit the cognitive and social niches that constitute our environment.

Acknowledgments

Work on this chapter was supported by grants from the Foundation for Psychocultural Research (*Integrating Ethnography and Neuroscience in Global Mental Health Research*, L. J. Kirmayer, PI) and the Social Sciences and Humanities Research Council of Canada (*Have We Lost Our Minds?*, M. J. D. Ramstead, award holder). We thank Ishan Walpola and Samuel Veissière for helpful discussions and sources.

References

Adams, F., and K. Aizawa. 2001. The bounds of cognition. *Philosophical Psychology* 14 (1): 43–64.

Adams, R. A., K. E. Stephan, H. R. Brown, C. D. Frith, and K. J. Friston. 2013. The computational anatomy of psychosis. *Frontiers in Psychiatry* 4:47.

Adeponle, A. B., D. Groleau, and L. J. Kirmayer. 2015. Clinician reasoning in the use of cultural formulation to resolve uncertainty in the diagnosis of psychosis. *Culture, Medicine, and Psychiatry* 39 (1): 16–42.

Anderson, M. 2014. *After Phrenology: Neural Reuse and the Interactive Brain*. MIT Press.

Bar, M. 2011. *Predictions in the Brain: Using Our Past to Generate a Future*. Oxford University Press.

Barsalou, L. W. 2008. Grounded cognition. *Annual Review of Psychology* 59:617–645.

Bhavsar, V., and D. Bhugra. 2008. Religious delusions: Finding meanings in psychosis. *Psychopathology* 41 (3): 165–172.

Bruineberg, J., and E. Rietveld. 2014. Self-organization, free energy minimization, and optimal grip on a field of affordances. *Frontiers in Human Neuroscience* 8:599. doi:10.3389/fnhum.2014.00599.

Chemero, A. 2009. *Radical Embodied Cognitive Science*. MIT Press.

Clark, A. 1997. *Being There: Putting Brain, Body, and World Together Again*. MIT Press.

Clark, A. 2008. *Supersizing the Mind: Embodiment, Action, and Cognitive Extension*. Oxford University Press.

Clark, A. 2016. *Surfing Uncertainty: Prediction, Action, and the Embodied Mind*. Oxford University Press.

Clark, A., and D. Chalmers. 1998. The extended mind. *Analysis* 58:7–19.

Colombetti, G. 2013. Psychopathology and the enactive mind. In *Oxford Handbook of Philosophy of Psychiatry*, ed. W. Fulford, T. Thornton, and G. Graham, 1083–1102. Oxford University Press.

Coltheart, M., R. Langdon, and R. McKay. 2011. Delusional belief. *Annual Review of Psychology* 62:271–298.

Connors, M. H., and P. W. Halligan. 2015. A cognitive account of belief: A tentative road map. *Frontiers in Psychology* 5:1588.

Corin, E., R. Thara, and R. Padmavati. 2004. Living through a staggering world: The play of signifiers in early psychosis in South India. In *Schizophrenia, Culture, and Subjectivity: The Edge of Experience*, ed. J. J. Jenkins and R. J. Barrett, 110–145. Cambridge University Press.

Csordas, T. J. 1990. Embodiment as a paradigm for anthropology. *Ethos* 18 (1): 5–47.

Csordas, T. J. 1994. *Embodiment and Experience: The Existential Ground of Culture and Self*. Cambridge University Press.

Danesi, M. 2013. On the metaphorical connectivity of cultural sign systems. *Signs and Society* 1 (1): 33–49.

Dennett, D. C. 2013. Expecting ourselves to expect: The Bayesian brain as a projector. *Behavioral and Brain Sciences* 36:209–210.

Desjarlais, R., and J. C. Throop. 2011. Phenomenological approaches in anthropology. *Annual Review of Anthropology* 40:87–102.

Di Paolo, E. A., and E. Thompson. 2014. The enactive approach. In *The Routledge Handbook of Embodied Cognition*, ed. L. Shapiro, 68–78. Routledge.

Ellis, H. D., and M. B. Lewis. 2001. Capgras delusion: A window on face recognition. *Trends in Cognitive Sciences* 5 (4): 149–156.

Festinger, L., H. W. Riecken, and S. Schachter. 1966. *When Prophecy Fails: A Social and Psychological Study of a Modern Group That Predicted the Destruction of the World*. Harper Torchbooks.

Fletcher, P., and C. Frith. 2009. Perceiving is believing: A Bayesian approach to explaining the positive symptoms of schizophrenia. *Nature Reviews: Neuroscience* 10:48–58.

Friston, K. J., K. E. Stephan, R. Montague, and R. J. Dolan. 2014. Computational psychiatry: The brain as a phantastic organ. *Lancet Psychiatry* 1 (2): 148–158.

Frith, C. 2007. *Making Up the Mind: How the Brain Creates Our Mental World*. Blackwell.

Frith, C., and K. J. Friston. 2012. False perceptions and false beliefs: Understanding schizophrenia. In *Neurosciences and the Human Person: New Perspectives on Human Activities*, Pontifical Academy of Sciences, Scripta Varia 121, 1–15.

Froese, T., and E. A. Di Paolo. 2011. The enactive approach: Theoretical sketches from cell to society. *Pragmatics and Cognition* 19:1–36.

Fuchs, T. 2009. Embodied cognitive neuroscience and its consequences for psychiatry. *Poiesis and Praxis: International Journal of Ethics of Science and Technology Assessment* 6:219–233. doi:10.1007/s10202-008-0068-9.

Fuchs, T., and H. De Jaegher. 2009. Enactive intersubjectivity: Participatory sense-making and mutual incorporation. *Phenomenology and the Cognitive Sciences* 8 (4): 465–486.

Fuchs, T., and J. E. Schlimme. 2009. Embodiment and psychopathology: A phenomenological perspective. *Current Opinion in Psychiatry* 22 (6): 570–575.

Gallagher, S. 2005. *How the Body Shapes the Mind*. Oxford University Press.

Geertz, C. 1983. *Local Knowledge*. Basic Books.

Gibbs, R. W., Jr. 2006. *Embodiment and Cognitive Science*. Cambridge University Press.

Glenberg, A. M. 2010. Embodiment as a unifying perspective for psychology. *Wiley Interdisciplinary Reviews: Cognitive Science* 1 (4): 586–596.

Gold, I., and J. Gold. 2014. *Suspicious Minds: How Culture Shapes Madness*. Free Press.

Groleau, D., and L. J. Kirmayer. 2004. Sociosomatic theory in Vietnamese immigrants' narratives of distress. *Anthropology and Medicine* 11 (2): 117–133.

Groleau, D., A. Young, and L. J. Kirmayer. 2006. The McGill Illness Narrative Interview (MINI): An interview schedule to elicit meanings and modes of reasoning related to illness experience. *Transcultural Psychiatry* 43 (4): 671–691.

Hacking, I. 1995. The looping effect of human kinds. In *Causal Cognition: A Multidisciplinary Debate*, ed. D. Sperber, D. Premack, and A. J. Premack, 351–383. Oxford University Press.

Hacking, I. 1998. *Mad Travelers: Reflections on the Reality of Transient Mental Illnesses*. University Press of Virginia.

Hacking, I. 1999. *The Social Construction of What?* Harvard University Press.

Hacking, I. 2002. *Historical Ontology*. Harvard University Press.

Hacking, I. 2004. Between Michel Foucault and Erving Goffman: Between discourse in the abstract and face-to-face interaction. *Economy and Society* 33:277–302.

Halpin, H. 2013. Does the Web extend the mind? In *Proceedings of the 5th Annual ACM Web Science Conference*, 139–147. ACM.

Herman, L., and B. Vervaeck. 2009. Capturing Capgras: *The Echo Maker* by Richard Powers. *Style* 43 (3): 407–428.

Hohwy, J. 2013. *The Predictive Mind*. Oxford University Press.

Husserl, E. 1990. *Ideas Pertaining to a Pure Phenomenology and to a Phenomenological Philosophy: Second Book; Studies in the Phenomenology of Constitution*. Springer Science & Business Media.

Hutto, D. D. 2008. *Folk Psychological Narratives: The Sociocultural Basis of Understanding Reasons*. MIT Press.

Hutto, D. D., and E. Myin. 2013. *Radicalizing Enactivism: Basic Minds without Content*. MIT Press.

Hutto, D. D., and G. Satne. 2015. The natural origins of content. *Philosophia* 43 (3): 521–536.

Kendler, K. S. 2014. The structure of psychiatric science. *American Journal of Psychiatry* 171:931–938.

Kirmayer, L. J. 1992. The body's insistence on meaning: Metaphor as presentation and representation in illness experience. *Medical Anthropology Quarterly* 6 (4): 323–346.

Kirmayer, L. J. 2006. Beyond the "new cross-cultural psychiatry": Cultural biology, discursive psychology, and the ironies of globalization. *Transcultural Psychiatry* 43 (1): 126–144.

Kirmayer, L. J. 2008. Culture and the metaphoric mediation of pain. *Transcultural Psychiatry* 45 (2): 318–338.

Kirmayer, L. J. 2015a. Empathy and alterity in psychiatry. In *Re-visioning Psychiatry: Cultural Phenomenology, Critical Neuroscience, and Global Mental Health*, ed. L. J. Kirmayer, R. Lemelson, and C. Cummings, 141–167. Cambridge University Press.

Kirmayer, L. J. 2015b. Re-visioning psychiatry: Toward an ecology of mind in health and illness. In *Re-visioning Psychiatry: Cultural Phenomenology, Critical Neuroscience, and Global Mental Health*, ed. L. J. Kirmayer, R. Lemelson, and C. Cummings, 622–660. Cambridge University Press.

Kirmayer, L. J., and D. Bhugra. 2009. Culture and mental illness: Social context and explanatory models. In *Psychiatric Diagnosis: Patterns and Prospects*, ed. I. M. Salloum and J. E. Mezzich, 29–37. John Wiley.

Kirmayer, L. J., and D. Crafa. 2014. What kind of science for psychiatry? *Frontiers in Human Neuroscience* 8:435.

Kirmayer, L. J., and I. Gold. 2012. Re-socializing psychiatry: Critical neuroscience and the limits of reductionism. In *Critical Neuroscience: A Handbook of the Social and Cultural Contexts of Neuroscience*, ed. S. Choudhury and J. Slaby, 307–330. Blackwell.

Kirmayer, L. J., and R. Santhanam. 2001. The anthropology of hysteria. In *Contemporary Approaches to the Study of Hysteria: Clinical and Theoretical Perspectives*, ed. P. W. Halligan, C. Bass, and J. C. Marshall, 251–270. Oxford University Press.

Kirmayer, L. J., A. Young, and J. M. Robbins. 1994. Symptom attribution in cultural perspective. *Canadian Journal of Psychiatry* 39 (10): 584–595.

Kleinman, A. 1978. Concepts and a model for the comparison of medical systems as cultural systems. *Social Science and Medicine, Part B: Medical Anthropology* 12:85–93.

Kleinman, A. 1987. Anthropology and psychiatry: The role of culture in cross-cultural research on illness. *British Journal of Psychiatry* 151 (4): 447–454.

Körner, A., S. Topolinski, and F. Strack. 2015. Routes to embodiment. *Frontiers in Psychology* 6:940.

Kövecses, Z. 2015. *Where Metaphors Come From: Reconsidering Context in Metaphor*. Oxford University Press.

Lakoff, G., and M. Johnson. 1980. *Metaphors We Live By*. University of Chicago Press.

Larøi, F., T. M. Luhrmann, V. Bell, W. A. Christian, S. Deshpande, C. Fernyhough, J. Jenkins, et al. 2014. Culture and hallucinations: Overview and future directions. *Schizophrenia Bulletin* 40 (Suppl. 4): S213–S220.

Leung, A. K., L. Qiu, L. Ong, and K. P. Tam. 2011. Embodied cultural cognition: Situating the study of embodied cognition in socio-cultural contexts. *Social and Personality Psychology Compass* 5 (9): 591–608.

Lock, M. 1993. Cultivating the body: Anthropology and epistemologies of bodily practice and knowledge. *Annual Review of Anthropology* 22:133–155.

Mauss, M. [1935] 1979. *Sociology and Psychology: Essays*. Routledge & Kegan Paul.

Meier, B. P., S. Schnall, N. Schwarz, and J. A. Bargh. 2012. Embodiment in social psychology. *Topics in Cognitive Science* 4 (4): 705–716.

Merleau-Ponty, M. [1945] 2012. *Phenomenology of perception*. Trans. D. Landes. Routledge.

Miller, G. A., E. Galanter, and K. H. Pribram. 1960. *Plans and the Structure of Behavior*. Holt, Rinehart & Winston.

Noë, A. 2005. *Action in Perception*. MIT Press.

Odling-Smee, F. J., K. N. Laland, and M. W. Feldman. 2003. *Niche Construction: The Neglected Process in Evolution*. Princeton University Press.

Pezzulo, G. 2014. Why do you fear the bogeyman? An embodied predictive coding model of perceptual inference. *Cognitive, Affective, and Behavioral Neuroscience* 14 (3): 902–911.

Piccinini, G. 2015. *Physical Computation: A Mechanistic Account*. Oxford University Press.

Powers, R. 2006. *The Echo Maker*. Farrar, Straus & Giroux.

Preti, A., D. Sisti, M. B. L. Rocchi, S. Siddi, M. Cella, C. Masala, D. R. Petretto, et al. 2014. Prevalence and dimensionality of hallucination-like experiences in young adults. *Comprehensive Psychiatry* 55 (4): 826–836.

Ramstead, M. J., S. P. Veissière, and L. J. Kirmayer. 2016. Cultural affordances: Scaffolding local worlds through shared intentionality and regimes of attention. *Frontiers in Psychology* 7. doi:10.3389/fpsyg.2016.01090.

Rashed, M. A. 2013. Culture, salience, and psychiatric diagnosis: Exploring the concept of cultural congruence and its practical application. *Philosophy, Ethics, and Humanities in Medicine* 8:5. doi:10.1186/1747-5341-8-5.

Rietveld, E., and J. Kiverstein. 2014. A rich landscape of affordances. *Ecological Psychology* 26 (4): 325–352.

Robbins, P., and M. Aydede. 2009. *The Cambridge Handbook of Situated Cognition.* Cambridge University Press.

Rowlands, M. 2010. *The New Science of the Mind: From Extended Mind to Embodied Phenomenology.* MIT Press.

Rupert, R. 2004. Challenges to the hypothesis of extended cognition. *Journal of Philosophy* 101:389–428.

Ryder, A., and Y. E. Chentsova-Dutton. 2015. Cultural clinical psychology: From cultural scripts to contextualized treatments. In *Re-visioning Psychiatry: Cultural Phenomenology, Critical Neuroscience, and Global Mental Health*, ed. L. J. Kirmayer, R. Lemelson, and C. A. Cummings, 400–433. Cambridge University Press.

Sapkota, R. P., D. Gurung, D. Neupane, S. K. Shah, H. Kienzler, and L. J. Kirmayer. 2014. A village possessed by "witches": A mixed-methods case-control study of possession and common mental disorders in rural Nepal. *Culture, Medicine, and Psychiatry* 38:642–668.

Sass, L., and G. Byrom. 2015. Phenomenological and neurocognitive perspectives on delusions: A critical overview. *World Psychiatry: Official Journal of the World Psychiatric Association (WPA)* 14:164–173.

Schilbach, L. 2016. Towards a second-person neuropsychiatry. *Philosophical Transactions of the Royal Society B: Biological Sciences* 371:20150081.

Searle, J. R. 1995. *The Construction of Social Reality.* Free Press.

Seligman, R., S. Choudhury, and L. J. Kirmayer. 2016. Locating culture in the brain and in the world: From social categories to the ecology of mind. In *Handbook of Cultural Neuroscience*, ed. J. Y. Chiao and R. Seligman, 3–20. Oxford University Press.

Seligman, R., and L. J. Kirmayer. 2008. Dissociative experience and cultural neuroscience: Narrative, metaphor and mechanism. *Culture, Medicine, and Psychiatry* 32:31–64.

Seth, A. K. 2013. Interoceptive inference, emotion, and the embodied self. *Trends in Cognitive Sciences* 17:565–573.

Seth, A. K., K. Suzuki, and H. D. Critchley. 2012. An interoceptive predictive coding model of conscious experience. *Frontiers in Psychology* 2:295. doi:10.3389/fpsyg.2011.00395.

Shapiro, L. 2011. *Embodied Cognition*. Routledge.

Singhal, A., E. Rogers, and M. Mahajan. 1999. The gods are drinking milk! Word-of-mouth diffusion of a major news event in India. *Asian Journal of Communication* 9 (1): 86–107.

Sperber, D. 1996. *Explaining Culture: A Naturalistic Approach*. Blackwell.

Sutton, J., C. B. Harris, P. G. Keil, and A. J. Barnier. 2010. The psychology of memory, extended cognition, and socially distributed remembering. *Phenomenology and the Cognitive Sciences* 9 (4): 521–560.

Thompson, E. 2007. *Mind in Life: Biology, Phenomenology, and the Sciences of Mind*. Harvard University Press.

Thompson, E., and F. J. Varela. 2001. Radical embodiment: Neural dynamics and consciousness. *Trends in Cognitive Sciences* 5 (10): 418–425.

Varela, F. J. 1999. *Ethical Know-How: Action, Wisdom, and Cognition*. Stanford University Press.

Varela, F. J., E. Thompson, and E. Rosch. 1991. *The Embodied Mind: Cognitive Science and Human Experience*. MIT Press.

Veissière, S. 2016. Varieties of Tulpa experiences: The hypnotic nature of human sociality, personhood, and interphenomenality. In *Hypnosis and Meditation: Towards an Integrative Science of Conscious Planes*, ed. A. Raz and M. Lifshitz. Oxford University Press.

Vidal, D. 1998. When the gods drink milk! Empiricism and belief in contemporary Hinduism. *South Asia Research* 18 (2): 149–171.

Williams, L. E., J. Y. Huang, and J. A. Bargh. 2009. The scaffolded mind: Higher mental processes are grounded in early experience of the physical world. *European Journal of Social Psychology* 39:1257–1267.

Wilson, R. A., and A. Clark. 2009. How to situate cognition: Letting nature take its course. In *The Cambridge Handbook of Situated Cognition*, ed. P. Robbins and M. Ayede, 55–77. Cambridge University Press.

Wittgenstein, L. 1969. *On Certainty*. Harper & Row.

Young, A. 1981. When rational men fall sick: An inquiry into some assumptions made by medical anthropologists. *Culture, Medicine, and Psychiatry* 5:317–335.

Young, A. 1982. Rational men and the explanatory model approach. *Culture, Medicine, and Psychiatry* 6 (1): 57–71.

Zatti, A., and C. Zarbo. 2015. Embodied and exbodied mind in clinical psychology: A proposal for a psycho-social interpretation of mental disorders. *Frontiers in Psychology* 6:236.

Contributors

Mark Bickhard, Department of Philosophy and Department of Psychology, Lehigh University

Ingar Brinck, Department of Philosophy and Cognitive Science, Lund University, Sweden

Anna Ciaunica, Institute of Philosophy, Porto; Institute of Philosophy, School of Advanced Study, University of London

Hanne De Jaegher, Department of Logic and Philosophy of Science, IAS-Research Centre for Life, Mind, and Society, University of the Basque Country, San Sebastián, Spain; Department of Informatics, Centre for Computational Neuroscience and Robotics, and Centre for Research in Cognitive Science, University of Sussex, Brighton

Nicolas de Warren, Husserl Archives, KU Leuven, Belgium

Ezequiel Di Paolo, Ikerbasque, Basque Foundation for Science, Bilbao, Spain; Department of Logic and Philosophy of Science, IAS-Research Centre for Life, Mind, and Society, University of the Basque Country, San Sebastián, Spain; Department of Informatics, Centre for Computational Neuroscience and Robotics, and Centre for Research in Cognitive Science, University of Sussex, Brighton

Christoph Durt, Department of Philosophy, University of Vienna; Psychiatric Clinic/Department of Philosophy, University of Heidelberg

John Z. Elias, Department of Philosophy, University of Hertfordshire, UK; Hamilton Holt School, Rollins College, Winter Park, Florida

Joerg Fingerhut, Einstein Group, "Consciousness, Emotions, Values," Berlin School of Mind and Brain, Humboldt-Universität zu Berlin

Aikaterini Fotopoulou, Clinical, Educational, and Health Psychology, Division of Psychology and Language Sciences, University College London

Thomas Fuchs, Psychiatric Clinic/Department of Philosophy, University of Heidelberg

Shaun Gallagher, Philosophy, University of Memphis; School of Humanities and Social Inquiry, University of Wollongong

Vittorio Gallese, Department of Neuroscience, University of Parma; Institute of Philosophy, School of Advanced Study, University of London

Duilio Garofoli, Cognitive Archaeology Unit, Institute for Archaeological Sciences, Eberhard Karls University of Tübingen; Research Center "The Role of Culture in Early Expansions of Humans" of the Heidelberg Academy of Sciences and Humanities, Eberhard Karls University of Tübingen

Katrin Heimann, Interacting Minds Center, Aarhus University

Peter Henningsen, Department of Psychosomatic Medicine and Psychotherapy, Klinikum rechts der Isar, Technical University of Munich

Daniel D. Hutto, School of Humanities and Social Inquiry, Faculty of Law, Humanities, and the Arts, University of Wollongong

Laurence J. Kirmayer, Division of Social and Transcultural Psychiatry, McGill University

Alba Montes Sánchez, Center for Subjectivity Research, University of Copenhagen

Dermot Moran, School of Philosophy, University College Dublin

Maxwell J. D. Ramstead, Division of Social and Transcultural Psychiatry, McGill University

Matthew Ratcliffe, Department of Philosophy, University of Vienna

Vasudevi Reddy, Centre for Situated Action and Communication, Department of Psychology, University of Portsmouth

Zuzanna Rucińska, Institute of Philosophy, Leiden University

Alessandro Salice, Department of Philosophy, University College Cork, Republic of Ireland

Glenda Satne, School of Humanities and Social Inquiry, Faculty of Law, Humanities, and the Arts, University of Wollongong; Department of Philosophy, Alberto Hurtado University, Santiago, Chile

Heribert Sattel, Department of Psychosomatic Medicine and Psychotherapy, Klinikum rechts der Isar, Technical University of Munich

Christian Tewes, Psychiatric Clinic/Department of Philosophy, University of Heidelberg; Department of Philosophy, University of Jena

Dan Zahavi, Center for Subjectivity Research, University of Copenhagen

Name Index

Subject Index

Actions. *See* Praxis
  possible motor (*see* Affordances)
Aesthetic experience, liberated embodied
  simulation and, 323–326
Affective attunement, 181, 186
Affective cognitive science. *See* Embodied,
  embedded, enactive, extended, and affective
  (4EA) cognitive science
Affordance-effectivity pairs, 266, 270–272,
  272n28, 274–275
Affordances, 3, 206n7, 245, 247–253, 257–269,
  259n3, 262n9, 401–404. *See also specific topics*
  canonical, 251, 270–272, 274–275
  dispositions and, 259, 261–264, 266, 267,
    269–270, 273, 274
  enactivism and, 206n7, 258, 261, 261n7
  Gibsonian, 206n5
  joint, 245, 252, 253
  landscape of, 248, 264, 266
Agency, 7, 28, 49, 55–56, 58, 62, 83, 135, 340n8.
  See also Collective agency; Social agency
  body, 175–176
  cooperative, 340n8
  selfhood and the capacity for, 157
  sense of, 83
Agent-environment systems, 254
Alterity, 91
Animals
  cultural history of, 268
  dispositions, 261–264, 266–267, 269

effectivities, 259, 262–263, 267, 269–271, 273,
  275
environment and, 258–260, 269, 272–273
neurobiology in, 311, 313
perception in, 322
training, 267
Anthropogeny, 314
Anthropology, 51, 75, 404
  cultural, 75–77, 323, 334, 340, 341 (*see also*
    Culture)
  medical, 75, 397, 399, 405
  psychological, 397, 404
Anti-Semitism, 57, 58n10
Art, 323. *See also* Aesthetic experience
Attachment, mother-infant, 185
Attunement, 33
Autonomous system, 7–8

Behavioral modernity, 279–280
Behaviorism, 78, 118–119
Bodily formatted (B-formatted) representations,
  320, 321, 325
Body. *See also Körper*
  cinaesthetic, 359
  embodied accomplishments and
    accomplishments of the, 79–80
  entwinement of and the cultural world, mind,
    and, 2–4
Body adornment. *See* Neanderthals
Body agency, 175–176

Printed in the United States
by Baker & Taylor Publisher Services